THE EXTRATERRESTRIAL LIFE DEBATE

THE EXTRATERRESTRIAL LIFE DEBATE

Antiquity to 1915

A SOURCE BOOK

edited with commentary by

MICHAEL J. CROWE

University of Notre Dame Press
Notre Dame, Indiana

Notre Dame, Indiana 46556
www.undpress.nd.edu

Title page art: detail from drawing by J. J. Grandville of a Bridge between Worlds, from *Un Autre Monde* (1844).

Library of Congress Cataloging-in-Publication Data

Crowe, Michael J.
The extraterrestrial life debate, antiquity to 1915 : a source book / edited by Michael J. Crowe.
p. cm.
Includes bibliographical references and index.
ISBN-13: 978-0-268-02368-3 (pbk. : alk. paper)
ISBN-10: 0-268-02368-9 (pbk. : alk. paper)
1. Life on other planets. 2. Plurality of worlds. I. Title.
QB54.C764 2008
576.8'39—dc22

2008026865

∞ *The paper in this book meets the guidelines for permanence and durability of the Committee on Production Guidelines for Book Longevity of the Council on Library Resources.*

To my brother,

Patrick H. Crowe

CONTENTS

FIGURES

TABLES

PREFACE

Do extraterrestrials exist? This book does not set forth an answer to that interesting question, nor was this the goal for which these materials were assembled. Nonetheless, the book may shed a little light on the question and on some who have sought to answer it. One claim that is suggested in these materials is that, whether or not extraterrestrials exist, they have long since invaded the Earth.

A similar claim is made in another book, which is related in a number of ways to this book. In 1986, I published *The Extraterrestrial Life Debate 1750–1900*,[1] which I based on research in over forty libraries around the world and designed to withstand scholarly scrutiny and to provide references for future scholarship. Despite its seven hundred pages, this earlier volume included only brief quotations from the hundreds of writings published before 1900 on issues regarding extraterrestrials. The present volume, as a source book, has different functions. It is designed to offer readers direct contact with some of the very engaging original sources in the extraterrestrial life debate. Some of these sources appear in English translation for the first time in this volume. Another goal is to provide commentaries that will help

1. M. J. Crowe, *The Extraterrestrial Life Debate, 1750–1900: The Idea of a Plurality of Worlds from Kant to Lowell* (Cambridge: Cambridge University Press, 1986; new ed., Mineola, N.Y.: Dover Publications, 1999). Far fuller discussions and references on many of the figures presented in this work will be found in that longer volume.

readers see the full significance of these writings. And it is also aimed at bringing the most important results of this earlier research to a broader audience, including college classrooms. I have written it with the conviction that nothing can more effectively convey a sense of the ideas, controversies, and characters of the past than reading their original writings.

This book complements the earlier book in other ways as well. It has been possible in some cases to include in this volume discoveries made after 1986 by others or myself and thereby to enhance parts of the history. In some relatively rare cases, I have reused phrasings or presentations from the earlier volume when I found that no changes seemed necessary. Although these two volumes have been designed to be complementary, each can stand alone; one need not read the one to understand the other.

To return to the claim that extraterrestrials have invaded: this book can be contrasted with two other types of books that have appeared in recent years. One type raises the question: what would be the effects of the discovery of extraterrestrial life; how, for example, would the detection of evidence of extraterrestrial life influence our religion, literature, systems of morals, and so forth? Another type of book, usually written in a sensationalist manner, is characterized by claims that extraterrestrials have already invaded, that they, for example, directed the building of such objects as the pyramids or provided humanity with various scientific discoveries or religious revelations or located stores of precious metals.

The approach taken here is quite different: although making no claims about whether extraterrestrials exist, I shall cite evidence to show that they have long since invaded and that their effects can be uncovered by historical research. Extraterrestrials have, for example, deeply influenced a number of religious doctrines and denominations. One instance of this concerns the writings of Emanuel Swedenborg and his New Jerusalem Church. Not only are the Swedenborgian scriptures filled with extraterrestrials, so also are the founding documents of the Seventh-day Adventist Church and of the Church of Jesus Christ of Latter-day Saints (the Mormon Church). Moreover, preachers of traditional denominations frequently employed extraterrestrials to expand their congregations, or devoted their efforts to fending off extraterrestrials marshaled in opposition to orthodoxy. Although extraterrestrials may not have built the pyramids, it can be shown they helped provide motivation and funding for some of the major astronomical observatories established in modern times. These materials reveal that for centuries, literary figures have embraced extraterrestrials, hoping thereby to enliven their prose, to enrich their imagery, and/or to give grandeur to their conceptions. And as a result, extraterrestrials have put gold and other precious metals into their pockets. Belief in extraterrestrial beings, in short, has been widespread in Western thought and has already influenced it in very significant ways.

It bears noting that to be interested in these materials, the reader need not be passionate about planets or entranced by extraterrestrials. In fact, readers who may be unable to distinguish R2-D2 from Darth Vader or to translate "Nanu Nanu," or who have not yet laid down dollars to bring *ET* into their living rooms need not abandon these pages. The readings, although focused on extraterrestrials, should shed light on aspects of the thoughts of such terrestrials as Aristotle, Lucretius, Augustine, Oresme, Cusanus, Bruno, Galileo, Kepler, Descartes, Fontenelle, Huygens, Newton, Paine, Whewell, and Lowell. And these materials should illuminate how various religious writers viewed the doctrines of Christ's incarnation and redemption. In fact, students who have read these materials report many surprises, including how frequently religious issues have entered the debate.

These readings serve to refute the widespread misconception that although the life sciences have strongly interacted with religion, the same cannot be said about astronomy. In fact, throughout nearly every period of Western thought, astronomy has influenced and been influenced by religious notions. Sometimes these interactions have been friendly, at other times hostile.

A substantial portion of these interactions centered on the question of extraterrestrial intelligent life. It is sometimes assumed that belief in extraterrestrial life began in the twentieth century. We shall see that this is far from the truth. Evidence is available to show that nearly half the leading intellectuals of the eighteenth and nineteenth centuries discussed extraterrestrial life issues in their writings. Consider, for example, the Enlightenment period. To test the claim that extraterrestrials danced in the heads of many Enlightenment intellectuals, I examined the eight anthologies of Enlightenment thought available in my personal library, taking the list of authors included in each as an indication of the editor's judgment as to who were the leading thinkers of the Enlightenment. I then checked off on each editor's list those authors whom my researches have shown treated the topic of other inhabited worlds. As the following table shows, at least 41 percent of the authors included by each of the eight editors met this criterion; moreover, in five of the eight cases, the percentage exceeds 53 percent. This test seems adequate to justify the claim that not only many, but, in fact, close to a majority of Enlightenment intellectuals actively engaged in a debate that many persons currently assume first arose in their own lifetimes.

It is noteworthy that the percentages given in this table are all lower bounds, which can only increase as future research finds extraterrestrials in the pages of other authors in these anthologies. Moreover, it is significant that relatively few scientists

Table 1. Table of information on extraterrestrials in authors in Enlightenment anthologies

Editor of the Anthology	*Title of the Anthology*	*No. of Authors Included*	*No. of These in EIL Debate*	*Percentage*
Berlin, Isaiah	*The Age of Enlightenment*	9	6	67%
Brinton, Crane	*The Portable Age of Reason Reader*	76	32	42%
Crocker, Lester	*The Age of Enlightenment*	29	12	41%
Gay, Peter	*The Enlightenment*	41	20	49%
Manuel, Frank	*The Enlightenment*	17	9	53%
Schneider, Isidor	*The Enlightenment*	45	24	53%
Snyder, Louis	*The Age of Reason*	17	14	82%
Torrey, Norman	*Les Philosophes*	11	6	54%

and no astronomers appear in these anthologies; had at least the latter been included, the percentages would have risen significantly.

The materials that follow not only demonstrate how successfully extraterrestrials have propagated, but also their extraordinary adaptability. Hundreds of authors have found ways to use them in their thought and writings. So pliable is the doctrine of a plurality of worlds that almost any author can create extraterrestrials suited to his or her system. Fiction writers beset by such limitations as characters with a single head or civilizations with only two sexes, or frustrated by the small scale of merely terrestrial catastrophes, have long since learned that extraterrestrials can rescue their writings from such restrictions. Writers from almost every era, and nearly all the authors appearing in these pages, have, as we shall see, succeeded in creating extraterrestrials suited to their needs. An interesting question to ask of the authors of many of the selections provided is: what work did this author wish to have the extraterrestrials do?

What image best characterizes the extraterrestrial life debate over its long history? My personal candidate is the image of a "night fight." Imagine two armies clashing in the depths of night. Think, for example, of Thucydides' account of the nocturnal siege of the Athenian camp in the Syracusan campaign of the Peloponnesian War. Here two armed forces clashed in a terrible battle made all the worse by the fact that darkness made it impossible to distinguish ally from enemy until close conflict commenced. Combatants could not be certain until the last moment whether a figure they were about to face was friend or foe. This image points to a distinctive feature of the extraterrestrial life debate. Authors who had been allies on an array of issues frequently found themselves at odds over extraterrestrials, or at least over the messages they extracted from them. One example from the mid-Victorian period is that Adam Sedgwick, John Herschel, Richard Owen, and Alfred Lord Tennyson, who had been compatriots of William Whewell in countless campaigns, found themselves at odds with him over extraterrestrials. Moreover, Whewell, whose fundamental inclinations separated him on nearly all issues from Thomas Paine shared with him a number of convictions about the significance of extraterrestrial life. An especially striking case concerns Richard A. Proctor, who by 1875 found himself opposed to extraterrestrials whom he had created less than a decade earlier. The ambiguity present in the debate stemmed in part from the fact that some authors saw extraterrestrials as supportive of theism, but in tension with Christianity. In particular, extraterrestrials appeared before the eyes of some as witnesses for the generosity and beneficence of the Creator, but to others extraterrestrials suggested the unlikelihood that such a God would become incarnated and act personally to redeem thousands of earths in the universe.

Some of the issues about extraterrestrial life discussed in this collection may not yet be moribund. Whether or not this is the case, the materials in this collection combine to form an exciting story and to reveal important features of Western thought.

This book is part of what can be called the "history of extraterrestrial life ideas research program." An invitation to join this program appears in an appendix to this volume. Scholars and teachers may wish to peruse it before reading the volume; students may wish to postpone it until afterwards.

A NOTE ON THE SOURCES

A few words about the reproduction of the primary sources: though I have not tried to produce facsimile copies of the original sources, I have tried faithfully to reproduce the contents of the sources. This has led to some apparent errors in the text, whether they be grammatically or stylistically awkward phrases, or outright misspellings. For the most part, I have not corrected these; for example, in the Huygens selection, I have left the seventeenth-century spelling of "knowlege." I have on occasion inserted letters within square brackets or noted misspellings with "[*sic*],” especially where it seemed misunderstandings could easily occur. A reader who wishes to check whether apparent mistakes exist in the original sources is invited to consult those sources.

ACKNOWLEDGMENTS

The research and writing of this book were facilitated by many institutions and persons.

This volume is based in part on my earlier *The Extraterrestrial Life Debate, 1750–1900,* which in turn was quarried from research carried out in over forty major libraries in six countries, made use of information, insights, and criticism from numerous colleagues, and received funding from both the National Science Foundation and the University of Notre Dame. Consequently, I continue to be aware of the role these institutions and persons have played, albeit indirectly, in the composition of the present volume.[2]

In recent years, many additional persons have contributed to this volume. I wish especially to thank Dr. Matthew F. Dowd, Keith R. Lafortune, and Christina Turner, who at various times while graduate students at the University of Notre Dame worked with me in preparing sections of this book. Their energy, generosity, and encouragement were such that my fondest wish for them is that if they become professors, they may be similarly blessed. The suggestions and comments of the undergraduate and graduate students in my courses at Notre Dame in which I've used this text have substantially improved its contents. Most recently, two undergraduates, Benjamin Gavel and Joanna Thurnes, have contributed significantly to locating typographical errors in the text. Moreover, some funding from the Office of the Dean of Notre Dame's College of Arts and Letters made it possible for Ben Gavel to help in the final stages of the preparation of this book. In particular, he collaborated fully and effectively at the proofs stage and in preparing the index. Professor Peter Ramberg of Truman State University has made a number of extremely valuable suggestions regarding this volume. Not only have his insights enriched its contents, so also have written comments made by his students in the courses in which he used it as a text. Very warm thanks are due to Professor Marie George of St. John's University, whose helpful and detailed reading of the entire text has improved it in many ways, and to Steven J. Dick, now Chief Historian at the National Aeronautics and Space Administration, who for three decades has shared with me his interest in and insights about the history of the extraterrestrial life debate. Special thanks are also

2. For a listing, see Crowe, *Extraterrestrial Life Debate*, xix.

due to Margaret Jasiewicz and Cheryl Reed, whose typing skills produced many of the selections included in this volume. I am especially conscious of my debt to Matthew Dowd, now an editor with Notre Dame Press, whose painstaking efforts and perceptive suggestions have turned my manuscript into the volume that is now before you. I am also indebted to the other staff of the Press who have worked professionally and pleasantly toward the completion of this volume. And last but not least, thanks to my wife, Professor Marian Crowe, who supported this project in many ways.

Michael J. Crowe
Cavanaugh Professor Emeritus in the Humanities
Program of Liberal Studies and Graduate Program
in History and Philosophy of Science
University of Notre Dame
January 2008

PART ONE

ANTIQUITY TO NEWTON

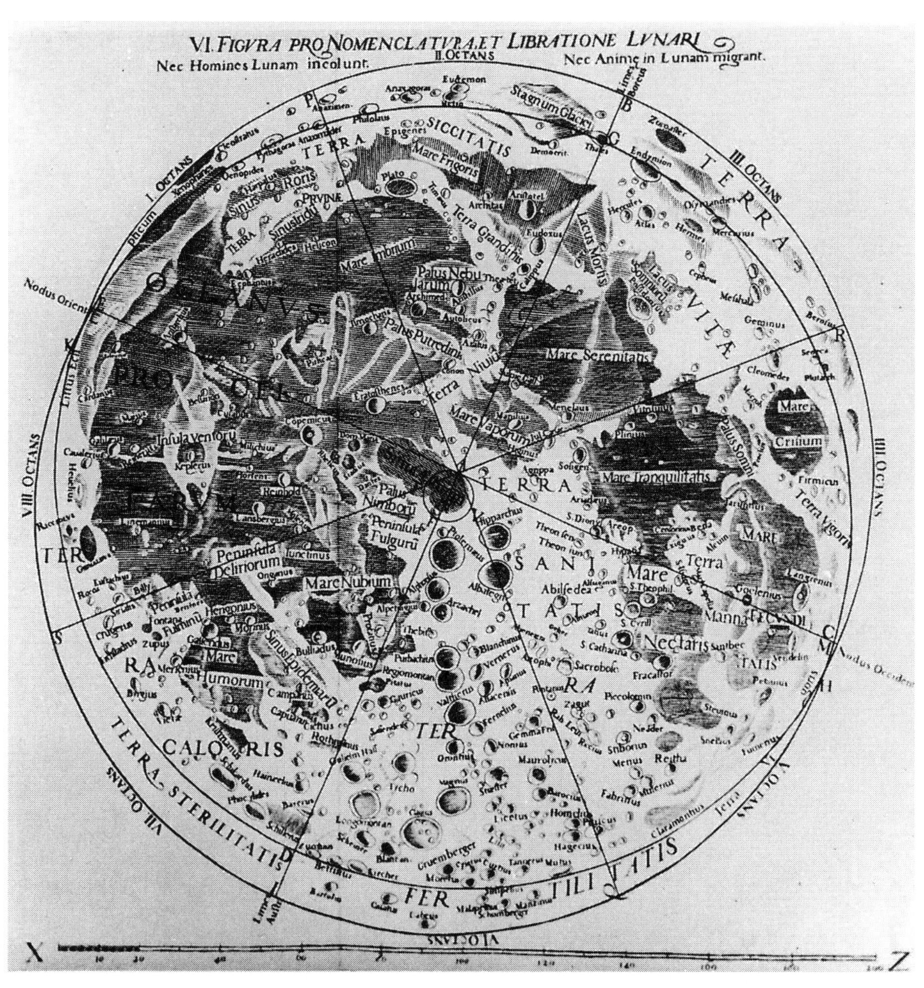

Fig. 1. Moon map prepared by F. J. Grimaldi, S.J., for the *Almagestum novum* (1651) of G. B. Riccioli, S.J. Notes at top say "No Men Dwell on the Moon" and "No Souls Migrate to the Moon."

ONE

THE DEBATE IN ANTIQUITY

THE ATOMISTS

When in Western thought did the debate over the existence of extraterrestrial intelligent life commence? Classical scholars date that development from Greek antiquity, in particular, from the fifth century BCE, when Leucippus and Democritus advocated the existence of such beings.[1] Although only fragments from their writings have been preserved, ample testimony from antiquity indicates that they advocated the existence of other worlds. For example, Hippolytus, a third century CE theologian, recorded that

> Democritus holds the same view as Leucippus about the elements, full and void . . . he spoke as if the things that are were in constant motion in the void; and there are innumerable worlds, which differ in size. In some worlds there is no sun and moon, in others they are larger than in our world, and in others more numerous. (3) The intervals between the worlds are unequal; in some

1. G. S. Kirk and J. E. Raven, *The Presocratic Philosophers* (Cambridge: Cambridge University Press, 1964), 121–26 and 409–14.

> parts there are more worlds, in others fewer; some are increasing, some at their height, some decreasing; in some parts they are arising, in other failing. They are destroyed by collision one with another. There are some worlds devoid of living creatures or plants or any moisture.[2]

Leucippus and Democritus were atomists, that is, they believed that matter is ultimately composed of tiny particles in constant motion. In fact, their belief in "innumerable worlds" derived from their atomistic doctrine. Given the random motions of an infinite number of atoms, chance would produce the formation of other worlds.

In understanding this passage it is important to recognize that by the term *world* they (as well as essentially all other ancient authors) did not mean another solar system comparable to our own and at least in principle visible to us. Rather they were referring to a cosmos comparable to our own with an earth at the center and with planets and a stellar vault surrounding it.

The atomist (or as it is sometimes called, the materialist) tradition in philosophy remained alive for many centuries in the ancient world. Two prominent later atomists were Epicurus and Lucretius, both of whom also supported the doctrine of innumerable inhabited worlds.

Epicurus, who died around 270 BCE, championed the doctrine of a plurality of worlds in a number of his writings, for example, in a letter he wrote to his disciple Herodotus. Early in that letter, Epicurus presents his atomic theory. Soon thereafter, he expresses his views on other worlds:

> Furthermore, there are infinite worlds both like and unlike this world of ours. For the atoms being infinite in number . . . are borne far out into space. For those atoms, which are of such nature that a world could be created out of them or made by them, have not been used up either on one world or on a limited number of worlds, nor again on all the worlds which are alike, or on those which are different from these. So that there nowhere exists an obstacle to the infinite number of worlds.[3]

That extraterrestrial beings inhabit these worlds is clear from a passage later in the letter:

---

2. As translated in Kirk and Raven, *Presocratic Philosophers*, 411.

3. Epicurus, "Epicurus to Herodotus," trans. C. Bailey, in *The Stoic and Epicurean Philosophers*, ed. Whitney J. Oates (New York: Modern Library, 1957), 5.

> Furthermore, we must believe that in all worlds there are living creatures and plants and other things we see in this world; for indeed no one could prove that in a world of one kind there might or might not have been included the kinds of seeds from which living things and plants and all the rest of the things we see are composed, and that in a world of another kind they could not have been.[4]

Given Epicurus's claims that atoms and the void are the sole constituents of the universe, that all things come to be through the chance collision of these atoms, and that (and thereby) a plurality of worlds is formed, it is not surprising that Epicurus goes on to reject traditional theological views. Such views had been used to explain many phenomena that Epicurus believed were explicable by his atomic theory. Toward the conclusion of his letter, he expressed this doctrine by stating:

> we must grasp this point, that the principal disturbance in the minds of men arises because they think that these celestial bodies are blessed and immortal, and yet have wills and actions and motives which are inconsistent with these attributes; and because they are always expecting or imagining some everlasting misery.[5]

The leading disciple of Epicurus among his contemporaries was Metrodorus of Chios, who also espoused the doctrine of a plurality of worlds, asserting:

> It would be strange if a single ear of corn grew in a large plain or were there only one world in the infinite. And that worlds are infinite in number follows from the causes [i.e., atoms] being infinite.[6]

The doctrines espoused by Epicurus attracted a number of disciples; in fact, Epicureanism spread to the Roman Empire, where the most prominent proponent of this position was the poet and philosopher Titus Lucretius Carus (ca. 99–55 BCE), who employed the richness of the Latin language and the elegance of his style to present those doctrines in his famous didactic poem *De rerum natura (On the Nature of the Universe)*. The following selection from that treatise contains a substantial exposition of the Epicurean theory of a plurality of worlds.

---

4. Epicurus, "Epicurus to Herodotus," 13.
5. Epicurus, "Epicurus to Herodotus," 14.
6. As quoted from Simplicius in F. M. Cornford, "Innumerable Worlds in the Pre-Socratic Philosophy," *Classical Quarterly* 28 (1964): 13.

Titus Lucretius Carus, *On the Nature of the Universe,* trans. and introduced by R. E. Latham (Middlesex: Penguin Books, 1975), 90–93.

Give your mind now to the true reasoning I have to unfold. A new fact is battling strenuously for access to your ears. A new aspect of the universe is striving to reveal itself. But no fact is so simple that it is not harder to believe than to doubt at the first presentation. Equally, there is nothing so mighty or so marvellous that the wonder it evokes does not tend to diminish in time. Take first the pure and undimmed lustre of the sky and all that it enshrines: the stars that roam across its surface, the moon and the surpassing splendour of the sunlight. If all these sights were now displayed to mortal view for the first time by a swift unforeseen revelation, what miracle could be recounted greater than this? What would men before the revelation have been less prone to conceive as possible? Nothing, surely. So marvellous would have been that sight—a sight which no one now, you will admit, thinks worthy of an upward glance into the luminous regions of the sky. So has satiety blunted the appetite of our eyes. Desist, therefore, from thrusting out reasoning from your mind because of its disconcerting novelty. Weigh it, rather, with discerning judgement. Then, if it seems to you true, give in. If it is false, gird yourself to oppose it. For the mind wants to discover by reasoning what exists in the infinity of space—that lies out there, beyond the ramparts of this world—that region into which the intellect longs to peer and into which the free projection of the mind does actually extend its flight.

Here, then, is my first point. In all dimensions alike, on this side or that, upward or downward through the universe, there is no end. This I have shown, and indeed the fact proclaims itself aloud and the nature of space makes it crystal clear. Granted, then, that empty space extends without limit in every direction and that seeds innumerable in number are rushing on countless courses through an unfathomable universe under the impulse of perpetual motion, *it is in the high-*

est degree unlikely that this earth and sky is the only one to have been created and that all those particles of matter outside are accomplishing nothing. This follows from the fact that our world has been made by nature through the spontaneous and casual collision and the multifarious, accidental, random and purposeless congregation and coalescence of atoms whose suddenly formed combinations could serve on each occasion as the starting point of substantial fabrics—earth and sea and sky and the races of living creatures. On every ground, therefore, you must admit that there exists elsewhere other congeries of matter similar to this one which the ether clasps in ardent embrace.

When there is plenty of matter in readiness, when space is available and no cause or circumstance impedes, then surely things must be wrought and effected. You have a store of atoms that could not be reckoned in full by the whole succession of living creatures. You have the same natural force to congregate them in any place precisely as they have been congregated here. You are bound therefore to acknowledge that in other regions there are other earths and various tribes of men and breeds of beasts.

Add to this the fact that nothing in the universe is the only one of its kind, unique and solitary in its birth and growth; everything is a member of a species comprising many individuals. Turn your mind first to the animals. You will find the rule apply to the brutes that prowl the mountains, to the children of men, the voiceless scaly fish and all the forms of flying things. So you must admit that sky, earth, sun, moon, sea and the rest are not solitary, but rather numberless. For a firmly established limit is set to their lives also and their bodies also are a product of birth, no less than that of any creature that flourishes here according to its kind.

Bear this well in mind, and you will immediately perceive that *nature is free and uncontrolled by proud masters* and runs the universe by herself without the aid of gods. For who—by the sacred hearts of the gods who pass their unruffled lives, their placid aeon, in calm and peace!—who can rule the sum total of the measureless? Who can hold in coercive hand the strong reins of the unfathomable? Who can spin all the firmaments alike and foment with the fires of ether all the fruitful earths? Who can be in all places at all times, ready to darken the clear sky with clouds and rock it with a thunderclap—to launch bolts that may often wreck his own temples, or retire and spend his fury letting fly at deserts with that missile which often passes by the guilty and slays the innocent and blameless?

ARISTOTLE (384–322 BCE)

Both Plato (428–348 BCE) and Aristotle opposed the idea of a plurality of worlds. The main discussion of this issue in Plato's writings occurs in his *Timaeus,* which contains the fullest exposition of any of the Platonic dialogues of his views on the physical world. At one point in the dialogue, the question arises: "Have we, then, been right to call it one Heaven, or would it have been true rather to speak of many and indeed of an indefinite number?"[7] The answer provided by the spokesperson for Plato's view is: "One we must call it, if we are to hold that it was made according to its pattern." His argument is basically that the Demiurge, who created this world, which resembles the Demiurge in being a kind of organism, must have made the world resemble himself in being unique. Moreover, Plato urges that were there to be more than one world, creation would be composite and thus subject to dissolution and decay.[8]

Aristotle gave the question of other worlds much fuller treatment than his teacher Plato had done. Moreover, Aristotle's position was strongly influential for many centuries on discussions of this issue. His main analyses on the topic appear in his *On the Heavens* and in his *Metaphysics.* In the first two books of the former work, Aristotle, after briefly setting out his doctrine of the four elements (earth, air, fire, and water), suggests that there exists a fifth element composed of the matter of the celestial realm. Whereas the motions of the four terrestrial elements are by nature upwards or downward (fire and air moving upward, earth and water downward), the natural motion of the celestial materials is circular. He then considers whether

7. F. M. Cornford, *Plato's Cosmology* (New York: Liberal Arts Press, 1957), 41. See *Timaeus* 31A.

8. Cornford, *Plato's Cosmology,* 52. See *Timaeus* 33.

the universe can be spatially infinite, concluding that it cannot be. At the end of chapter 6 of book 1, Aristotle comments: "This will lead us to a further question. Even if the total mass is not infinite, it may yet be great enough to admit a plurality of universes."[9] This question becomes the central concern of chapters 8 and 9.

Aristotle, *On the Heavens*, trans. J. L. Stocks, in *The Basic Works of Aristotle*, ed. Richard McKeon (New York: Random House, 1941), 413–15. See *On the Heavens*, bk. 8. Reproduced with permission of Oxford University Press.

8

We must now proceed to explain why there cannot be more than one heaven—the further question mentioned above. For it may be thought that we have not proved universally of bodies that none whatever can exist outside our universe, and that our argument applied only to those of indeterminate extent.

Now all things rest and move naturally and by constraint. A thing moves naturally to a place in which it rests without constraint, and rests naturally in a place to which it moves without constraint. On the other hand, a thing moves by constraint to a place in which it rests by constraint, and rests by constraint in a place to which it moves by constraint. Further, if a given movement is due to constraint, its contrary is natural. If, then, it is by constraint that earth moves from a certain place to the centre here, its movement from here to there will be natural, and if earth from there rests here without constraint, its movement hither will be natural. And the natural movement in each case is one. Further, these worlds, being similar in nature to ours, must all be composed of the same bodies as it. Moreover each of the bodies, fire, I mean, and earth and their intermediates, must have the same power as in our world. For if these names are used equivocally, if the

9. Aristotle, *On the Heavens*, trans. J. L. Stocks, in *The Basic Works of Aristotle*, ed. Richard McKeon (New York: Random House, 1941), 409. See *On the Heavens* 274a25.

identity of name does not rest upon an identity of form in those elements and ours, then the whole to which they belong can only be called a world by equivocation. Clearly, then, one of the bodies will move naturally away from the centre and another towards the centre, since fire must be identical with fire, earth with earth, and so on, as the fragments of each are identical in this world. That this must be the case is evident from the principles laid down in our discussion of the movements; for these are limited in number, and the distinction of the elements depends upon the distinction of the movements. Therefore, since the movements are the same, the elements must also be the same everywhere. The particles of earth, then, in another world move naturally also to our centre and its fire to our circumference. This, however, is impossible, since, if it were true, earth must, in its own world, move upwards, and fire to the centre; in the same way the earth of our world must move naturally away from the centre when it moves towards the centre of another universe. This follows from the supposed juxtaposition of the worlds. For either we must refuse to admit the identical nature of the simple bodies in the various universes, or, admitting this, we must make the centre and the extremity one as suggested. This being so, it follows that there cannot be more worlds than one.

To postulate a difference of nature in the simple bodies according as they are more or less distant from their proper places is unreasonable. For what difference can it make whether we say that a thing is this distance away or that? One would have to suppose a difference proportionate to the distance and increasing with it, but the form is in fact the same. Moreover, the bodies must have some movement since the fact that they move is quite evident. Are we to say then that all their movements, even those which are mutually contrary, are due to constraint? No, for a body which has no natural movement at all cannot be moved by constraint. If then the bodies have a natural movement, the movement of the particular instances of each form must necessarily have for goal a place numerically one, i.e. a particular centre or a particular extremity. If it be suggested that the goal in each case is one in form but numerically more than one, on the analogy of particulars which are many though each undifferentiated in form, we reply that the variety of goal cannot be limited to this portion or that but must extend to all alike. For all are equally undifferentiated in form, but any one is different numerically from any other. What I mean is this: if the portions in this world behave similarly both to one another and to those in another world, then the portion which is taken hence will not behave differently either from the portions in another world or from those in the same world, but similarly to them, since in form no portion differs from another. The result is that we must either abandon our present assumptions or assert that the centre and the extremity are each numerically one. But this being so, the heaven, by the same evidence and the same necessary inferences, must be one only and no more.

A consideration of the other kinds of movement also makes it plain that there is some point to which earth and fire move naturally. For in general that which is moved changes from something into something, the starting-point and the goal being different in form, and always it is a finite change. For instance, to recover health is to change from disease to health, to increase is to change from smallness to greatness. Locomotion must be similar: for it also has its goal and starting-point—and therefore the starting-point and the goal of the natural movement must differ in form—just as the movement of coming to health does not take any direction which chance or the wishes of the mover may select. Thus, too, fire and earth move not to infinity but to opposite points; and since the opposition in place is between above and below, these will be the limits of their movement. (Even in circular movement there is a sort of opposition between the ends of the diameter, though the movement as a whole has no contrary: so that here too the movement has in a sense an opposed and finite goal.) There must therefore be some end to locomotion: it cannot continue to infinity.

This conclusion that local movement is not continued to infinity is corroborated by the fact that earth moves more quickly the nearer it is to the centre, and fire the nearer it is to the upper place. But if movement were infinite speed would be infinite also; and if speed then weight and lightness. For

> as superior speed in downward movement implies superior weight, so infinite increase of weight necessitates infinite increase of speed.

In Chapter 9 of his "On the Heavens," Aristotle presents another argument, introducing it by noting that

> it might seem impossible that the heaven should be one and unique, since in all formations and products whether of nature or of art we can distinguish the shape in itself and the shape in combination with matter. For instance the form of the sphere is one thing and the gold or bronze sphere another; the shape of the circle again is one thing, the bronze or wooden circle another. For when we state the essential nature of the sphere or circle we do not include in the formula gold or bronze, because they do not belong to the essence, but if we are speaking of the copper or gold sphere we do include them.[10]

This generalization, however, appears to lead to a problem for Aristotle, when it is applied not to circles or spheres, but to the physical entity of the universe itself. In particular, the generalization suggests that just as there are typically many physical circles or spheres, so also the universe, being a physical entity, must have a multiplicity of embodiments; in short, "there either are, or may be, more heavens [or universes] than one." Aristotle attempts to combat this difficulty by asserting that "this world contains the entirety of matter." He elaborates and qualifies this point by stating that

> Now the universe is certainly a particular and a material thing: if however it is composed not of a part but of the whole of matter, then though the being of 'universe' and of 'this universe' are still distinct, yet there is no other universe, and no possibility of others being made, because all the matter is already included in this. It remains, then, only to prove that it is composed of all natural perceptible body.[11]

To prove his last point, Aristotle asks the reader to imagine a body outside our universe. Such a body would have to be composed of one of the four elements or of the

10. Aristotle, *On the Heavens*, 416. See *On the Heavens* 277b30–278a6.
11. Aristotle, *On the Heavens*, 417. See *On the Heavens* 278b2–9.

special celestial matter that he believes exists in the heavens and that moves with a circular motion. In opposition to this idea he reverts to the argument presented in his chapter 8, which he states shows that no such matter can exist because it would have to have the motions appropriate to each of those forms of matter, which are by their nature defined in relation to the position of the Earth, which is at the center of the universe.

Aristotle presents another argument against the idea of other worlds in chapter 8 of book 12 of his *Metaphysics*. This proof derives from Aristotle's claim that the celestial motions have their origins in a Prime Mover who is the moving principle of the universe. Aristotle suggests that were there more than one universe, there would in consequence have to be more than one Prime Mover, a notion that he took to be impossible. As he states: "So the unmovable first mover is one both in definition and in number: so too, therefore, is that which is moved always and continuously; therefore there is one heaven alone."[12] From this it is clear that Aristotle's position concerning the question of other worlds was linked not only to his physics and cosmology, but also to his theology. This should not, however, be seen as surprising because the same was true for the Epicurean philosophers.

As the above information suggests, major differences separated the Epicureans from the Aristotelians, as is evident in the following table:

Table 2. Comparing and contrasting views of Atomists and Aristotelians

Universe of Atomists	*Universe of Aristotelians*
homogeneous	hierarchical
infinite	finite
random	governed by design
purposeless	teleological
atheistic	theistic
heavens are changing and terrestrial	heavens are unchanging and attain perfection
vacua exist	vacua do not exist
plurality of worlds	denial of the idea of a plurality of worlds.

As will be clear in what follows, the polarities present in this table persisted even after Epicureanism (i.e., atomistic or materialist traditions) and Aristotelianism fell from favor.

12. Aristotle, *Metaphysics*, trans. W. D. Ross, in *The Basic Works of Aristotle*, ed. Richard McKeon (New York: Random House, 1941), 884. See *Metaphysics* 1074a37–39.

TWO

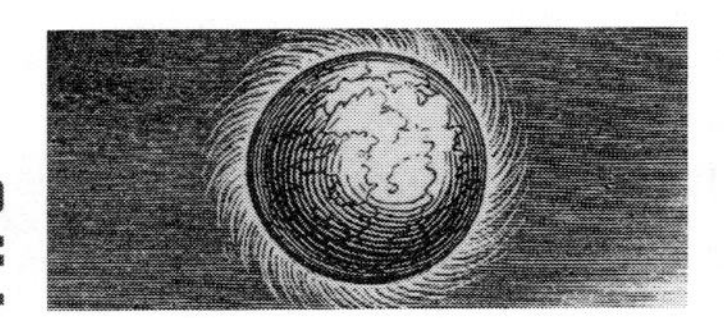

FROM AUGUSTINE TO THE FIFTEENTH CENTURY

THE CHURCH FATHERS, ESPECIALLY ORIGEN AND AUGUSTINE

In its first half dozen centuries CE, the dynamic new religion known as Christianity came into contact with the impressive and sophisticated systems of thought worked out by the Greeks and Romans. One of these points of contact was the question of a plurality of worlds. The reaction of the early Christian thinkers to the idea of other worlds is suggested by the title of the work quoted in the previous chapter by the third century CE theologian Hippolytus. He titled the work in which he discussed the ideas of Democritus and Leucippus *Refutation of All Heresies.* Another church father who expressed similar sentiments was Philastrius, Bishop of Brescia (fourth century, died before 397), who lists pluralism in his book on heresies, commenting:

> There is another heresy that says that there are infinite and innumerable worlds, according to the empty opinion of certain philosophers — since Scripture has said that there is one world and teaches us about one world — taking this view from the apocrypha of the prophets, that is from the secrets, as the pagans themselves called them; there was also the Democritus who asserted

> there to be many worlds; he agitated the souls of many people and stirred up doubtful opinions with his diverse errors, since he proclaimed this [i.e., the plurality of worlds] as proceeding from his own wisdom.[1]

In fact, one study concludes that "the early Church Fathers (from the first through the eighth century) . . . with the sole exception of Origen, all of them who explicitly raise the question of whether there is one world or many, or at least take an explicit stand on the question, reject [the idea of a plurality of worlds]."[2]

The one exception is the Alexandrian theologian Origen (c. 185–c. 254), who has been described as the "author of the first Christian systematic theology."[3] The position adopted by Origen supports the idea not of a plurality of spatially separated worlds, but rather the idea (associated with the Stoics) of a succession of worlds in time. Origen's statement of his position is:

> But this is the objection which they generally raise: they say, "If the world had its beginning in time, what was God doing before the world began? For it is at once impious and absurd to say that the nature of God is inactive and immovable. . . ." [W]e can give an answer in accordance with the standard of religion, when we say that not then for the first time did God begin to work when He made this visible world; but as, after its destruction, there will be another world, so also we believe that others existed before the present world came into being. And both of these positions will be confirmed by the authority of holy Scripture. For that there will be another world after this, is taught by Isaiah, who says, "There will be new heavens, and a new earth, which I shall make to abide in my sight, saith the Lord;" and that before this world others also existed is shown by Ecclesiastes [1:9, 10], in the words: "What is that which hath been / Even that which shall be / And what is that which has been created / Even this which is to be created: and there is nothing altogether new under the sun. Who shall speak and declare, Lo, this is new? It hath already been in the ages which have been before us." By these testimonies it is established both that there were

1. As quoted and translated by Marie I. George in *Christianity and Extraterrestrials? A Catholic Perspective* (New York: iUniverse, 2006), 66, citing Philastrius, *Sancti Filastrii episcopi Brixiensis Diversarvm hereseon liber* (Vindobonae, Austria: F. Tempsky, 1898), no. 86, 79. See also Marie I. George, "The Early Church Fathers on the Plurality of Worlds," which will be published in *Fides et ratio*.

2. George, "Early Church Fathers," page 2 of author's manuscript.

3. Thomas O'Meara, "Christian Theology and Extraterrestrial Intelligent Life," *Theological Studies* 60 (1999): 3–30, at 8.

ages before our own, and that there will be others after it. It is not, however, to be supposed that several worlds existed at once, but that, after the end of this present world, others will take their beginning.[4]

[I]t seems to me impossible for a world to be restored for the second time, with the same order and with the same amount of births, and deaths, and actions; but that a diversity of worlds may exist with changes of no unimportant kind.[5]

The most influential church father was Augustine (354–430), Bishop of Hippo and author of *The City of God,* which many scholars regard as the most outstanding treatise composed by any of the church fathers. That work contains a number of discussions of the question of other worlds. Early in its eleventh book, Augustine presents his objections based on his Christian faith to the traditional Greek and Roman doctrine that the world has always existed. From this he proceeds to a complex argument designed to respond to those thinkers (not identified) who, although accepting that God created the world, "have difficulties about the time of its creation." His response begins by raising the parallel problem as to why God chose to create the world at this place, rather than at some other. Augustine argues that the latter question leads to an absurdity, namely, "Epicurus' dream of innumerable worlds."

For if they imagine infinite spaces of time before the world, during which God could not have been idle, in like manner they may conceive outside the world infinite realms of space, in which, if any one says the Omnipotent cannot hold His hand from working, will it not follow that they must adopt Epicurus' dream of innumerable worlds? with this difference only, that he assert that they are formed and destroyed by the fortuitous movements of atoms while they will hold that they are made by God's hand, if they maintain that, throughout the boundless immensity of space, stretching interminably in every direction round the world, God cannot rest, and that the worlds which they suppose him to make cannot be destroyed.[6]

---

4. As quoted and translated in George, *Christianity and Extraterrestrials?* 64–65, from Origen, *De Principiis,* in *Fathers of the Third Century,* vol. 4, ed. A. Cleveland Coxe (Grand Rapids, Mich.: Wm. B. Eerdmans Publishing Company, 1985), bk. 3, chap. 5, #3, 341–42.

5. As quoted and translated in George, "Early Church Fathers," 5, from Origen, *De Principiis,* bk. 2, chap. 4, 272.

6. Augustine, *The City of God,* trans. Marcus Dods (New York: Modern Library, 1950), 349. See book 11, chap. 5.

Two points in this paragraph deserve special notice because they adumbrate ideas influential in the later history of the extraterrestrial life debate. The first is the juxtapositioning of the "infinitude" of space with that of time. The second is Augustine's discussion of the idea of God's immense powers and the question whether those powers would not have been fully used had God *not* created a plurality of worlds.

Augustine maintains that consideration of the problem as to where in creating the world God placed us in an infinity of space should lead those concerned about our placement within an infinity of time to see that their problem was ultimately misconceived. Augustine also, though briefly, attempts to deal with the problem in a second way by citing his notion that time is nothing else than the motion of the material. Without matter, particularly the celestial bodies, there could be no time. Thus there was no problem as to why God created the world at a certain time because time itself began only with that creation.[7]

Elsewhere in his *City of God*, especially in book 12, Augustine mentions the doctrine of a plurality of worlds, but a careful reading of these passages indicates that the view that he is most concerned to refute is not the doctrine of a plurality of worlds in space but rather the idea, espoused by various Stoic philosophers and by Origen, of a plurality of worlds in time. Among the many arguments he presents against this view, the one to which he attaches the greatest significance for Christians is, as he puts it, "the fact of the eternal life of the saints." In other words, he is rejecting in this way the doctrine of metempsychosis or transmigration of souls, according to which human souls return to and are reincarnated on the Earth.[8]

THOMAS AQUINAS (1224–1274)

A remarkable rebirth of learning occurred in Western Europe in the twelfth and thirteenth centuries. In the first of these centuries, many classic scientific works of antiquity came to be translated into Latin by various scholars, the most prominent of whom was Gerard of Cremona. Among the authors thus made accessible to Western thinkers was Aristotle, only a fraction of whose writings (and none of his writings in natural philosophy) had been known in the earlier middle ages. Of particular relevance for present purposes was Gerard's translation, made around 1170, of Aristotle's *On the Heavens*. The availability of such ancient works in Latin translation presented Christian authors with the momentous, difficult, and exciting task

7. Augustine, *City of God*, book 11, chap. 5.
8. Augustine, *City of God*, book 12, chap. 19, 402.

of finding whether and how these brilliant systems of classical thought could be reconciled with Christian belief.

In the thirteenth century, various authors faced this challenge in an energetic and productive manner. Two of the most important were Albertus Magnus (1193–1280) and his student Thomas Aquinas. The enthusiasm with which Albert took up the question of the plurality of worlds is evident in the following comment: "Since one of the most wondrous and noble questions about Nature is whether there is one world or many, a question that the human mind desires to understand *per se*, it seems desirable for us to inquire about it."[9] Despite his enthusiasm for the question, the answer that Albert provided to it was negative, his arguments deriving largely from the discussion in Aristotle.

Thomas Aquinas took up the question in his famous *Summa theologica* and like his teacher came down in opposition to the doctrine. Aquinas's main arguments are given in the next selection. As in other articles of his *Summa*, Thomas in this article specifies an issue, then presents a section containing various objections to the position he favors, which in turn leads to Thomas's statement of his position and his main reasons for advocating it. Finally, he offers his replies to the earlier objections.

Thomas Aquinas, Part I, Question 47, Article 3 of his *Summa Theologica*, from Anton C. Pegis, ed., *Basic Writings of Saint Thomas Aquinas* (New York: Random House, 1945), 1:462–63.

Third Article

WHETHER THERE IS ONLY ONE WORLD?

We Proceed thus to the Third Article:—

Objection 1. It would seem that there is not only one world, but many. Because, as Augustine says, it is unfitting to say that God has created things without a reason.[10] But for

---

9. As translated and quoted in Steven J. Dick, *Plurality of Worlds: The Origin of the Extraterrestrial Life Debate from Democritus to Kant* (Cambridge: Cambridge University Press, 1982), 23.

10. *Lib. 83 Quaest.*, q. 46 (PL [Patrologia Latina] 40, 30).

the same reason that He created One, He could create many, since His power is not limited to the creation of one world, but is infinite, as was shown above.[11] Therefore God has produced many worlds.

Obj. 2. Further, nature does what is best, and much more does God. But it is better that there be many worlds than that there be one because many good things are better than a few. Therefore many worlds have been made by God.

Obj. 3. Further, everything which has a form in matter can be multiplied in number, while the species remains the same, because multiplication in number comes from matter. But the world has form in matter. Thus, when I say man, I mean the form, and when I say this man, I mean the form in matter; so when we say world, the form is signified, and when we say this world, the form in matter is signified. Therefore there is nothing to prevent the existence of many worlds.

On the contrary, It is said (*Jo.* i. 10): *The world was made by Him,* where the world is named as one, as if only one existed.

I answer that, The very order of things created by God shows the unity of the world. For this world is called one by the unity of order, whereby some things are ordered to others. But whatever things come from God have relation of order to each other and to God Himself, as was shown above.[12] Hence it is necessary that all things should belong to one world. Therefore only those were able to assert the existence of many worlds who did not acknowledge any ordaining wisdom, but rather believed in chance; as did Democritus, who said that this world, besides an infinite number of other worlds, was made from a coming together of atoms.[13]

Reply Obj. 1. This argument proves that the world is one because all things must be arranged in one order, and to one end. Therefore from the unity of order in things Aristotle

11. Q. 25, a. 2.

12. Q. II, a. 3; q. 21, a. I, ad 3.

13. Cf. Aristotle, *De Caelo,* III, 4 (303a 4); Cicero, *De Nat. Deor.,* I, 26 (p. 28); St. Ambrose, *In Hexaem.,* I, I (PL 14, 135).

infers the unity of God governing all.[14] In the same way, Plato proves the unity of the world, as a sort of copy, from the unity of the exemplar.[15]

Reply Obj. 2. No agent intends material plurality as the end; for material multitude has no certain limit, but of itself tends to infinity, and the infinite is opposed to the notion of end. Now when it is said that many worlds are better than one, this has reference to material multitude. But the best, in this sense, is not the intention of the divine agent; since for the same reason it might be said that if He had made two worlds, it would be better if He had made three; and so on to infinity.

Reply Obj. 3. The world is composed of the whole of its matter. For it is not possible for there to be another earth than this one, since every earth would naturally be carried to this central one, wherever it was. The same applies to the other bodies which are parts of the world.

It is clear from this article that Thomas was well aware not only of the atomist position favoring the idea of a plurality of worlds, but also of various versions of the principle of plenitude. Thus the second objection contains the argument that an infinite God should create an infinity of worlds. The first objection also makes this point: an all powerful God should create an infinity of worlds. In countering these objections, Thomas draws on Plato and on Aristotle. Following the former, Thomas suggests that the oneness of God makes it appropriate for God to create only one world, which would mirror his perfection. Similarly, following Aristotle, Thomas argues that it would be more appropriate for God to create one world because perfection is associated with order and with unity rather than diversity. In his reply to the third objection, Thomas draws on Aristotle's argument in terms of natural place. As his statement of his position makes clear, Thomas was convinced that an "ordaining wisdom" governs the world and that this was irreconcilable with those who "believed in chance; as did Democritus, who said that this world, besides an infinite number of other worlds, was made from a coming together of atoms."

---

14. *Metaph.*, XI, 10 (1076a 4).

15. Cf. St. Thomas, *In De Caelo*, I, lect. 19.—Cf. also Plato, *Timaeus* (p. 31a).

THE CONDEMNATION OF 1277 AND NICOLE ORESME (1325–1382)

The last quarter of the thirteenth century saw a surprising turn of events. In 1277, Étienne Tempier, Bishop of Paris, under pressure from various theologians who were troubled by the increasing dominance of Aristotelian thought, especially as interpreted by Thomas Aquinas, issued what is known as the Condemnation of 1277. In it, Tempier condemned 219 propositions and threatened those who held them with excommunication. An issue that concerned Tempier was that theologians in attempting to delineate God's relationships with the world had produced rational schemes that could be seen as denying that God *could* act in certain manners. Among the propositions condemned was number 34, which stated "[t]hat the first cause [God] could not make several worlds."[16] In other words, it seemed a short step from arguments that the Christian idea of God could be reconciled with what was known of the universe to a denial that despite God's omnipotence, God could not create more than one world.[17]

Tempier's proclamation created in the late-thirteenth and fourteenth centuries a situation in which the question of other worlds, as well as other questions about the physical world, could be discussed more openly. In fact, before the end of the thirteenth century, Henry of Ghent, Godfrey of Fontaine, Richard of Middleton, William of Ware, Jean of Bassols, and Thomas of Strasbourg had all urged that a plurality of worlds was not a theological impossibility. Moreover, in the fourteenth century, a number of the most gifted scholars discussed the question of other worlds, revealing that they were at least open to the possibility that God could have created a plurality of worlds. Among these authors were William of Ockham (ca. 1290–1349), John Buridan (d. 1358), and Nicole Oresme. What is most striking in the discussions by these three authors is that although they show themselves to be open to the possibility that God could have created a plurality of worlds and although they marshal an array of powerful rebuttals to Aristotle's arguments against this doctrine, they all ultimately conclude against the pluralist position. The most highly regarded fourteenth-century discussion of this issue is that by Oresme, who served as tutor to

---

16. As quoted in *Source Book in Medieval Science*, ed. Edward Grant (Cambridge, Mass.: Harvard University Press, 1974), 48.

17. It is relevant to note that some church fathers, for example, Athanasius (297–373), John Chrysostom (345–407), and Basil (329–379) were aware of the distinction between what God could and would do. For further information, see George, "Early Church Fathers," 9–12.

the future King Charles V of France and eventually as bishop of Lisieux. Oresme's discussion appears in his *Le livre du ciel et du monde* (*The Book of the Heavens and the World*), which consists of a French translation of and commentary on Aristotle's *On the Heavens*. As will be evident in reading this discussion, Oresme's analyses have significance for far more than the question of a plurality of worlds.

Nicole Oresme, *Le livre du ciel et du monde*, ed. Albert D. Menut and Alexander J. Denomy, C. S. B., trans. Albert Menut (Madison: University of Wisconsin Press, 1968), 167, 169, 171, 173, 175, 177, and 179.

Now we have finished the chapters in which Aristotle undertook to prove that a plurality of worlds is impossible, and it is good to consider the truth of this matter without considering the authority of any human but only that of pure reason. I say that, for the present, it seems to me that one can imagine the existence of several worlds in three ways.

[THE IDEA OF A SUCCESSION OF WORLDS IN TIME]

One way is that one world would follow another in succession of time, as certain ancient thinkers held that this world had a beginning because previous to this all was a confused mass without order, form, or shape. Thereafter, by love or concord, this mass was disentangled, formed, and ordered, and thus was the world created. And finally after a long time this world will be destroyed by discord and will return to the same confused mass, and again, through concord, another world will then be made. Such a process will take place in the future an infinite number of times, and it has been thus in the past. But this opinion is not touched upon here and was reproved by Aristotle in several places in his philosophical works. It cannot happen in this way naturally, although God could do it and could have done it in the past by His own omnipotence, or He could annihilate this world and create another thereafter.

And, according to St. Jerome, Origen used to say that God will do this innumerable times.

[THE IDEA OF A NESTED SERIES OF WORLDS]

Another speculation can be offered which I should like to toy with as a mental exercise. This is the assumption that at one and the same time one world is inside another so that inside and beneath the circumference of this world there was another world similar but smaller. Although this is not in fact the case, nor is it at all likely, nevertheless, it seems to me that it would not be possible to establish the contrary by logical argument; for the strongest arguments against it would, it seems to me, be the following or similar ones.

In a section omitted, Oresme lets his imagination roam over seven versions of the idea of a nested series of worlds. Among these are the ideas that there may exist a world under the surface of the Earth with a god governing it different from the god who oversees the Earth, or a world within the Moon, or worlds beyond both the Earth and Moon. To give plausibility to such notions as a world beneath the surface of the Earth, Oresme provides an interesting discussion of the total relativity of size.

To show that these and similar speculations do not preclude the possibility of such a thing, I will posit, first of all, that every body is divisible into parts themselves endlessly divisible, as appears in Chapter One; and I point out that *large* and *small* are relative, and not absolute, terms used in comparisons. For each body, however small, is large with respect to the thousandth part of itself, and any body whatsoever, however large, would be small with respect to a larger body. Nor does the larger body have more parts than the smaller, for the parts of each are infinite in number. Also from this it follows that, were the world to be made between now and tomorrow 100 or 1,000 times larger or smaller than it is at present, all its parts being enlarged or diminished proportionally, everything would appear tomorrow exactly as now, just as though nothing had been changed.

In a section omitted, Oresme, after further discussion of these versions of the idea of nested worlds, concludes with the following comment.

I say that the contrary cannot be proved by reason nor by evidence from experience, but also I submit that there is no proof from reason or experience or otherwise that such worlds do exist. Therefore, we should not guess nor make a statement that something is thus and so for no reason or cause whatsoever against all appearances; nor should we support an opinion whose contrary is probable; however, it is good to have considered whether such opinion is impossible.

[THE IDEA OF A SERIES OF WORLDS IN SPACE]

The third manner of speculating about the possibility of several worlds is that one world should be [conceived] entirely outside the other in space imagined to exist, as Anaxagoras held. This solitary type of other world is refuted here by Aristotle as impossible. But it seems to me that his arguments are not clearly conclusive, for his first and principal argument states that, if several worlds existed, it would follow that the earth in the other world would tend to be moved to the center of our world and conversely, etc., as he has loosely explained in Chapters Sixteen and Seventeen. To show that this consequence is not necessary, I say in the first place that, although *up* and *down* are said with several meanings, as will be stated in Book II, with respect to the present subject, however, they are used with regard to us, as when we say that one-half or part of the heavens is up above us and the other half is down beneath us. But up and down are used otherwise with respect to heavy and light objects, as when we say the heavy bodies tend downward and the light tend upward. Therefore, I say that up and down in this second usage indicate nothing more than the natural law concerning heavy and light bodies, which is that all the heavy bodies so far as possible are located in the middle of the light bodies without setting up for them any other motionless or natural place. This can be understood from a later statement and from an explanation in the fourth chapter, where it was shown how a portion of air could rise up naturally from the center of the earth to the heavens and could descend naturally from the heavens to the center of the earth. Therefore, I say that a heavy body to which no light body is attached would not move of itself; for in such a place as that in which this heavy body is resting, there would be neither up nor down because, in this case, the natural law stated above would not operate and, consequently, there would not be any up or down in that place. This can be clarified by what Aristotle says in Book Four of the *Physics*, namely, that in a void there is no difference of place with respect to up or down. Therefore, Aristotle says that a body in a vacuum would not move of itself. In the eleventh chapter of this first book it appears, according to Aristotle, that, since nothing is lower than the center of the earth, nothing is or can be higher than the circumference or the concavity of the lunar sphere, the place proper to fire, as we have often said. Thus, taking up in the second sense above, beyond or outside of this circumference or heaven there is no up nor down. From this it follows clearly that, if God in His infinite power created a portion of earth and set it in the heavens where the stars are or beyond the heavens, this earth would have no tendency whatsoever to be moved toward the center of our world. So it appears that the consequence stated above by Aristotle is not necessary. I say, rather, that, if God created another world like our own, the earth and the other elements of this other world would be present there just as they are in our own world. But Aristotle confirms his conclusion by another argument in Chapter Seventeen and it is briefly this: all parts of the earth tend toward a single natural place, one in number; therefore, the earth of the other world would tend toward the center of this world. I answer that this argument has little appearance of truth, considering what is now said and what was said in Chapter Seventeen. For the truth is that in this world a part of the earth does not tend toward one center and another part toward another center, but all heavy bodies in this world tend to be united in one mass such that the center of the weight of this mass is at the center of this world, and all the parts constitute one body, numerically speaking. Therefore, they have one single place. And if some part of the earth in the other world were in this world, it would tend toward the center of this world and become united with the mass, and conversely. But it does not have to follow that the portions of earth or of the heavy bodies of the other world, if it existed, would tend to the center of this world because in their world they would form a single mass possessed of a single place and would be arranged in up and down order, as we have indicated, just like the mass of heavy bodies in this world. And these two bodies or masses would be of one kind, their natural places would be formally identical, and likewise the two worlds. In Chapter Twenty Aristotle mentions another argument from what was said in the *Metaphysics*—namely, that there cannot be more than one God and, therefore, it seems there can be only one world. I reply that God is infinite in His immensity, and, if several worlds existed, no one of them would be outside Him nor outside His power; but surely other intelligences would exist in one world and others in the other world, as already stated.

In a section omitted, Oresme continues his insightful critique of the hierarchical arrangement of the Aristotelian cosmos. The section ends with a statement that can only be seen as remarkable.

Therefore, I conclude that God can and could in His omnipotence make another world besides this one or several like or unlike it. Nor will Aristotle or anyone else be able to prove completely the contrary. But, of course, there has never been nor will there be more than one corporeal world, as was stated above.

WILLIAM VORILONG (d. 1464) AND NICHOLAS OF CUSA (1401–1464)

The Franciscan philosopher and theologian William Vorilong seems to have been the first author to have raised the question of whether the idea of a plurality of worlds can be reconciled with the Christian doctrines of the divine incarnation and redemption. Vorilong wrote:

If it be inquired whether a whole world is able to be made more perfect than this universe, I answer that not one world alone, but that infinite worlds, more perfect than this one, lie hid in the mind of God. If Democritus, who posits actual infinite worlds, rightly understood this fact, he would have understood rightly. If it then is asked how the second world cleaves to this one, I answer that it would be possible for the species of this world to be distinguished from that of the other world. If it be further inquired where it would exist, I answer that it would be able to be placed above any part of the heaven, south, or north, east or west.

Now doubt arises. By what means are we able to have knowledge of that world. I answer by angelic revelation or by divine means. If it be inquired whether men exist on that world, and whether they have sinned as Adam sinned, I answer no, for they would not exist in sin and did not spring from Adam. But it is shown that they would exist from the virtue of God, transported into that world, as Enoch and Elias in the earthly paradise. As to the question whether Christ by dying on this earth could redeem the inhabitants of another world, I answer that he is able to do this even if the worlds were infinite, but it would not be fitting for Him to go into another world that he must die again.[18]

Remarkable as is Oresme's discussion of the issue of a plurality of worlds, that written by Nicholas of Cusa, a prominent German philosopher, theologian, and administrator, surpasses it in the radicalness of its claims. It was in 1440 that Cusa published his most famous work, his brilliant if enigmatic *De docta ignorantia* (*Of Learned Ignorance*), a chapter of which appears next.

Nicholas of Cusa, *Of Learned Ignorance*, trans. Germain Heron (London: Routledge and Kegan Paul, 1954), 111–18.

CHAPTER XII. CONDITIONS OF THE EARTH

The ancient philosophers did not reach these truths we have just stated, because they lacked learned ignorance. It is now evident that this earth really moves though to us it seems

18. Grant McColley and H. W. Miller, "Saint Bonaventure, Francis Mayron, William Vorilong and the Doctrine of a Plurality of Worlds," *Speculum* 12 (1937): 386–89, at 388.

stationary. In fact, it is only by reference to something fixed that we detect the movement of anything. How would a person know that a ship was in movement, if, from the ship in the middle of the river, the banks were invisible to him and he was ignorant of the fact that water flows? Therein we have the reason why every man, whether he be on earth, in the sun or on another planet, always has the impression that all other things are in movement whilst he himself is in a sort of immovable centre; he will certainly always choose poles which will vary accordingly as his place of existence is the sun, the earth, the moon, Mars, etc. In consequence, there will be a machina mundi whose centre, so to speak, is everywhere, whose circumference is nowhere, for God is its circumference and centre and He is everywhere and nowhere.

Even the earth is not a sphere as some have maintained, though it is inclined to be spherical. The figure of the world, like its movement, is limited in its parts; and when an infinite line is thought of as limited in such a way that, as limited, it is incapable of greater perfection or extension, then it is circular, for their beginning and end are coincident. Circular movement, therefore, is the more perfect, from which it follows that the more perfect solid figure is the sphere. Therefore the movement of any part is for the perfection of the whole; e.g. heavy bodies move towards the earth, light things upwards, earth towards earth, water to water, air to air, fire to fire; and in its movement the whole tends as much as possible to become circular and every figure inclines to the spherical, as we perceive in the members of animals, in trees and in the heavens. One movement, in consequence, is more circular and more perfect than another; and figures differ in like manner.

The earth, then, is a stately and spherical figure whose movement is circular; but it could be more perfect. From what has just been said it is clear that in the world there is neither a maximum nor a minimum in perfections, movements and figures, so that it is untrue to say that this earth is the basest and lowest planet; if, in fact, it seems more central than the other planets of the world, it is also for that very reason nearer the pole, as was already said. The earth is not a comparative part of the world, nor is it an aliquot part of the world; for the world has neither a maximum nor a minimum and, in consequence, has neither a middle nor aliquot parts. With man or animal we have the same thing, for, though the hand of man seems to bear a relation to his body through its weight it is not an aliquot part of man; and the same assertion has to be made of magnitude and figure. Nor is the darkness of colour a proof of the earth's baseness; for the brightness of the sun, which is visible to us, would not be perceived by anyone who might be in the sun. Like the earth, the sun has its peculiar constitution. On examination the body of the sun is found to be disposed like this: nearer the centre there is a sort of earth, at the circumference a sort of fiery brightness and midway between them a kind of watery cloud and clearer air. Consequently if one were outside the region of fire, our earth at the circumference of the region would appear through the midst of the fire as a bright star, in much the same way as the sun appears brightest to us who are around the circumference of the solar region. Possibly the explanation of the moon's not appearing so bright lies in the fact that we are on the near side of its circumference towards the more central parts—in its watery region, as it were. It is for that reason that its light is not apparent, though it has its own light, which is visible to those existing on the extremities of its circumference; to us only the reflected light of the sun is visible. On that account, too, the moon's heat—the indubitable effect of its movement and greater in the circumference where the movement is greater—is not communicated to us like the sun's heat. Our earth, therefore, seems to lie between the region of the sun and the moon and through these it is influenced by other stars that are invisible to us, since we are outside their region. The regions of the stars that sparkle are the only ones we see.

The earth, then, is a brilliant star having a light, heat and influence distinctively its own and different from those of all other stars just as each star differs from every other in light, nature and influence. Each star communicates to another light and influence; but this communication is not the purpose of the stars, for all stars move and sparkle for one sole purpose: the realization of the best possible existence for them; and from that communication follows as a consequence. Likewise, light gives light not that I may see but it is of its very nature to give light; the communication of light takes place as a consequence when I use it for the purpose of seeing. God, ever to be blessed, has so created all things that whilst each thing strives to conserve its own being as a gift of God, it does so in participation with other things; e.g. the foot has only one purpose, viz. walking, but in carrying that out it is of service not only to the foot but to the eye, hands, body and the whole man. Similar examples of this are the eye and the other members; and the parts of the world equally exemplify it. Plato said the world was an animal. If, without his being immersed in it, you conceive God as its soul, much of what has been said will be clear to you.

Because the earth is smaller than the sun and is influenced by it is not a reason for calling it baser, for the entire region of the earth, which stretches to the circumference fire, is great. From shadow and eclipses we know that the earth is smaller than the sun; yet despite that we do not know to what extent the region of the sun is greater or smaller than the region of the earth. It cannot be exactly equal to it, for no star can be equal to another. And the earth is not the smallest star, for eclipses have shown us that it is larger than the moon; and some say that it is larger even than Mercury, larger, perhaps, than all other stars. From size, therefore, no proof can be alleged of its baseness.

Even the influence exerted on it is not a proof of its imperfection; perhaps it, as a star, has a similar influence on the sun and its region, as already stated. We have no knowledge from experience of that influence, since we have no experience beyond that of our existence in the centre where the influences merge. Even if, in fact, we consider the earth as potency, the sun as its formal act or soul, and the moon as their connecting medium, the result would be the mutual relation of the influences of those stars situated within the one region (others like Mercury, Venus, etc., are above this region, according to the ancients and even some moderns); then the correlation of influence is clearly such that one could not exist without another. Likewise, in different degrees this influence, one and threefold, will be found in all. As regards these points, it is evidently impossible for man to discover whether the region of the earth is in degree more perfect or less perfect by comparison with the regions of the other stars such as the sun, moon and the others.

Nor can place furnish an argument for the earth's baseness. Life, as it exists here on earth in the form of men, animals and plants, is to be found, let us suppose, in a higher form in the solar and stellar regions. Rather than think that so many stars and parts of the heavens are uninhabited and that this earth of ours alone is peopled—and that with beings, perhaps, of an inferior type—we will suppose that in every region there are inhabitants, differing in nature by rank and all owing their origin to God, who is the centre and circumference of all stellar regions. Now, even if inhabitants of another kind should exist in the other stars, it seems inconceivable that, in the line of nature, anything more noble and perfect could be found than the intellectual nature that exists here on this earth and its region. The fact is that man has no longing for any other nature but desires only to be perfect in his own.

Were we to suppose that, for the realization of the plan of the universe, the whole region of the other inhabited stars stands in some relation of comparison, unknown to us, to the whole region of this earth; and that, in consequence, through the intermediary of the universal region a certain relationship springs up from both sides between the inhabitants of this earth or region and the inhabitants of the other stars—in the same way as through the intermediary of the hand there exists a relation of comparison between the particular joints of the fingers and the foot, and through the intermediary of the foot between the particular joints of the foot and the hand, so that all be suitably adapted to the whole animal; not even then with this supposition could we find a relation of comparison between those inhabitants of the other stars, of whatever nature they be, and the natives of this world.

For since that whole region is unknown to us, its inhabitants remain wholly unknown. To go no further than this earth:—animals of a given species unite to form a common home of the species and share the common characteristics of their habitat, knowing nothing of or caring nothing for strangers. Their idea of strangers, even if it reaches some kind of vocal expression, is wholly exterior and conjectural and, such as it is, conceivable only after lengthy experience. Of the inhabitants then of worlds other than our own we can know still less, having no standards by which to appraise them. It may be conjectured that in the area of the sun there exist solar beings, bright and enlightened intellectual denizens, and by nature more spiritual than such as may inhabit the moon—who are possibly lunatics—whilst those on earth are more gross and material. It may be supposed that those solar intelligences are highly actualized and little in potency, while the earth-denizens are much in potency and little in act, and the moon-dwellers betwixt and between.

We make these conjectures from a consideration of the fiery nature of the sun, the water and air elements in the moon and the weighty bulk of the earth. And we may make parallel surmise of other stellar areas that none of them lack inhabitants, as being each, like the world we live in, a particular area of one universe which contains as many such areas as there are uncountable stars. In these local areas (we may guess), so countless that only He who has created all things in number can enumerate them, the whole cosmos suffers a triple contraction in its downward fourfold progress.

Nor is the physical decay which we see upon the earth a convincing proof that our earth is vile. Given a single cosmos with the action and reaction of star upon star, we can never pronounce any one thing to be pure, irredeemable corruption. Stellar influences when focused upon an individual may sometimes fall away into its constituent elements, so that one or other mode of existence disappears; and therefore, it were better to regard corruption as different modes of being, and to pronounce with Virgil that there is no room for death anywhere. For death would appear to be no more than the resolution of a composite into its elements. And who shall say that such resolution occurs only upon this earth?

It has been asserted that there is a separate species on the earth to correspond with each one of the stars. Now if the earth provides in each species a focus for the action of each star, why may not a similar provision be made among other heavenly bodies that are subject to the action of their fellows? And who shall say that such stellar activity, now contracted in a composition, upon the resolution of that composition into its elements does not return whence it came; that the individual animal of a particular species upon the earth does not, when withdrawn from all stellar action, dissolve into its aboriginal matter, its form alone returning to that particular star from which that species derived actual being upon mother earth? Or that the form alone does not revert to the exemplar, the soul of the world, as say the Platonists, and the matter to the state of possibility? The energy which united them, the spirit of union that they receive from the movement of the stars, might, upon the maladjustment of organs or other cause of corruption, give place to a separatist movement and withdraw to the stars, while the form climbed above astral influence and the matter sank below it. Or who shall say that the forms of a particular area of creation do not come to rest in a higher form, say an intellectual one, and by its means attain the ultimate purpose of the universe? And as this purpose is attained in God by inferior forms by means of this intellectual one, so it itself is to rise to the circumference, which is God, while the body sinks to the centre where God also is, so that all movement may be Godward. Centre and circumference are one in God, and in Him at the right moment the centre-seeking body and circumference-tending soul shall be united. Meanwhile movement being stilled, not indeed all movement but such as makes for generation, these essential elements of the world, without which the world could not be, and, with the end of generation in time, possible matter shall be re-united with its form through the concurrence of the reviving energy of union.

These things, however, no man can know unless he be specially instructed by God. Doubtless, the God of infinite excellence has created all things for Himself and wills not that anything should perish of those things that He has made. Doubtless, also, He is the most bountiful rewarder of all who worship Him. But the manner of His present operation and of

> His bountifulness to come only God knows: for He is His own activity. Later I shall add a little more, as far as God gives me to understand, to what it must for the moment suffice to have touched upon in ignorance.

In reflecting on this discussion and its significance, it may be useful to keep in mind remarks on it made by two historians of science well qualified to assess its significance. The first remark, that made by Thomas Kuhn in his *The Copernican Revolution,* suggests that Cusa's adoption of the doctrine of a plurality of worlds may have had a major significance for the history of astronomy. Kuhn's statement is: "Nicholas of Cusa . . . derived the motion of the earth from the plurality of worlds in an unbounded Neoplatonic universe."[19] The second remark was made by Pierre Duhem and relates to the question that might easily arise in the mind of a reader of Cusa's chapter, especially a reader aware of the fate that a century and a half later befell Giordano Bruno, who was significantly influenced by Cusa. As is well known, Bruno not only adopted the idea of extraterrestrial life, but also was burned at the stake by the Roman Catholic Church for his heterodox views. Did Cusa suffer similarly? Duhem's answer to that query is:

> When for the first time in Latin Christianity, one hears a person speak of the plurality of inhabited worlds, it is proposed by a theologian who a few years earlier had spoken at an ecumenical council; a person who in a very celebrated book sought to divine the characteristics of the sun and of the moon, went on to be honored by the confidence of popes [and by] the most elevated ecclesiastical honors.[20]

In particular, Cusa's radical views in his 1440 *Of Learned Ignorance* did not prevent the church from elevating him to the status of cardinal in 1448.

19. Thomas Kuhn, *The Copernican Revolution: Planetary Astronomy in the Development of Western Thought* (Cambridge, Mass.: Harvard University Press, 1971), 143.

20. Pierre Duhem, *Le système du monde* (Paris: A. Hermann et Fils, 1958), 10:324.

THREE

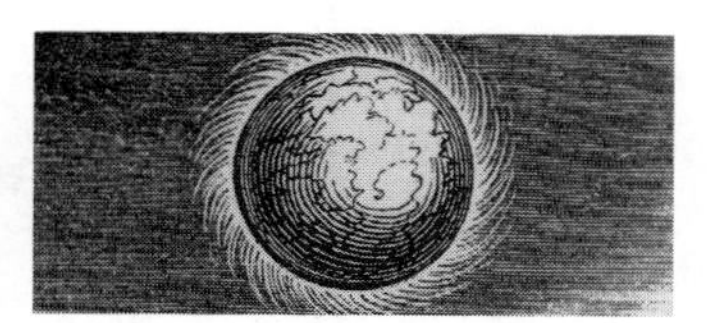

FROM COPERNICUS TO BRUNO

What factors between 1500 and about 1750 transformed the idea of extraterrestrial life from being a radical notion accepted by few to being a doctrine taught in college classrooms, championed by preachers, and celebrated by poets? Scholars have offered two contrasting explanations. Some suggest that astronomical developments, especially the heliocentric theory of Copernicus, Kepler, and Galileo, along with the observations made possible by the invention early in the seventeenth century of the telescope, were the key factors in bringing about this change.[1] Others have, however, suggested that a supportive shift in the religious and philosophical mentality of the period was crucial. Arthur Lovejoy championed the latter view, urging that the features that distinguish the modern from the medieval cosmos

> owed their introduction, and for the most part, their eventual general acceptance, not to the actual discoveries or to the technical reasonings of astronomers,

1. For an effective presentation of this point of view see Steven J. Dick, *Plurality of Worlds: The Origins of the Extraterrestrial Life Debate from Democritus to Kant* (Cambridge: Cambridge University Press, 1982) and also his "The Origins of the Extraterrestrial Life Debate and Its Relation to the Scientific Revolution," *Journal of the History of Ideas* 41 (1980): 3–27.

> but to those originally Platonic metaphysical preconceptions which . . . had been always repressed and abortive in medieval thought. . . . These features [were chiefly] derivative from philosophical and theological premises. They were, in short, manifest corollaries of the principle of plenitude.[2]

Lovejoy's formulation of this principle is that "no genuine potentiality of being can remain unfulfilled, that the extent and the abundance of the creation must be as great as the possibility of existence and commensurate with the productive capacity of a 'perfect' and inexhaustible Source, and that the world is better, the more things it contains."[3] Theists applying the principle of plenitude to the celestial realm concluded that God must have placed living beings wherever conditions comparable to those on Earth occurred. Allied to this principle was the belief that planets or stellar systems devoid of life would entail the unacceptable conclusion that God's creative energies had in those instances been wasted. Readings in this and in later chapters will offer evidence bearing on this debate.

NICHOLAS COPERNICUS (1473–1543)

Who is most responsible for having opened the door through which extraterrestrials came into the modern world? The surprising answer offered by many scholars is that this resulted from the ideas of a canon of the Catholic Church working in a remote corner of Europe on a book that was so mathematical that few could comprehend it and so radical in its central claim that fewer cared to try. Moreover, neither his famous book nor any of his writings, published or unpublished, give any clue as to whether he had thought at all about the question of extraterrestrial life. This was Nicholas Copernicus, the Polish astronomer who at Frauenburg Cathedral composed his *De revolutionibus orbium coelestium* (*On the Revolutions of the Celestial Orbs*), published in 1543 and according to a credible account was first seen by its author on the day of his death.

The revolution launched by Copernicus resulted from his removing the Earth from the center of the universe, setting it spinning on its axis, and revolving around the Sun. This action transformed the Earth into a planet, which in turn suggested that the planets may be earths, in other words, that they may be inhabited. In the

---

2. Arthur Lovejoy, *The Great Chain of Being: A Study of the History of an Idea* (New York: Harper & Brothers, 1960 reprinting of the 1936 original), 99, 111.

3. Lovejoy, *Great Chain*, 52.

longer run, it transformed stars into suns, which some suggested may themselves be surrounded by inhabited planets. This was no rapid revolution; in fact, scholars have succeeded in identifying only ten astronomers from the entire sixteenth century who adopted the Copernican theory.[4] Such information indicates why some historians have suggested that the so-called "Copernican revolution" might more appropriately be described as the Copernican disturbance followed by the Keplerian or Galilean revolution.[5]

Our present concern is, however, chiefly with the process by which ideas of extraterrestrial life came first to recognition, then to respectability, and eventually to widespread acceptance. At the time when Copernicus published his famous book, well-informed authors could bring many arguments—scientific, philosophical, and religious—against its daring thesis. Among the earliest objections was that raised in 1550, only seven years after the appearance of Copernicus's book, by the prominent Lutheran reformer Philipp Melanchthon, who challenged Copernican cosmology by warning that it might foster the idea that Christ's incarnation and redemption could have occurred on other planets:

> [T]he Son of God is One; our master Jesus Christ was born, died, and resurrected in this world. Nor does He manifest Himself elsewhere, nor elsewhere has He died or resurrected. Therefore it must not be imagined that Christ died and was resurrected more often, nor must it be thought that in any other world without the knowledge of the Son of God, that men would be restored to eternal life.[6]

It would, however, be a serious mistake to believe that it was only on religious grounds that sixteenth-century authors opposed the heliocentric hypothesis. Many significant physical and astronomical objections could also be marshaled against it.

GIORDANO BRUNO (1548–1600)

Perhaps the most vocal and certainly the most prolific of the early Copernicans was Giordano Bruno, who simultaneously championed the idea of a plurality of worlds. This he did in such books as his *La cena de la ceneri* (*The Ash Wednesday Supper*)

---

4. Robert Westman, "The Astronomers' Role in the Sixteenth Century: A Preliminary Study," *History of Science* 17 (1980): 105–47, at 136n6.

5. N. R. Hanson, "The Copernican Disturbance and the Keplerian Revolution," *Journal of the History of Ideas 22* (1961): 169–184.

6. As translated and quoted in Dick, *Plurality of Worlds*, 89.

and his *De l'infinito universo et mondi* (*On the Infinite Universe and Worlds*), both of which were published in 1584, and in his *De immenso et innumerabilibus* (*Of the Immense and Innumerable*) of 1591. What seems to have made Bruno best known in later times is the fact that in 1600 he was burned at the stake by the Inquisition of the Roman Catholic Church.

The selection that follows consists of an opening summary Bruno provided for his *De l'infinito universo et mondi*. While reading it, a good question to ask comes from the introduction to this chapter, in particular, whether scientific or metaphysical factors played a larger role in Bruno's advocacy of extraterrestrial life. In fact, Bruno's enthusiasm for extraterrestrials was such that he attributed souls not only to the planets, but also to stars, meteors, and the universe as a whole.

Giordano Bruno, *On the Infinite Universe and Worlds*, in Dorothea Waley Singer, *Giordano Bruno: His Life and Thought with Annotated Translation of His Work "On the Infinite Universe and Worlds"* (New York: Henry Schuman, 1950), 229–31, 233, 237–40, 246.

INTRODUCTORY EPISTLE addressed to the most illustrious Monsieur Michel de Castelnau, Seigneur de Mauvissière, de Concressault and de Joinville, Chevalier of the Order of the Most Christian King, Privy Councillor, Captain of 50 men at arms, and Ambassador to Her most Serene Majesty the Queen of England.

IF, O most illustrious Knight, I had driven a plough, pastured a herd, tended a garden, tailored a garment: none would regard me, few observe me, seldom a one reprove me; and I could easily satisfy all men. But since I would survey the field of Nature, care for the nourishment of the soul, foster the cultivation of talent, become expert as Daedalus concerning the ways of the intellect; lo, one doth threaten upon beholding me, another doth assail me at sight, another doth bite upon reaching me, yet another who hath caught me would devour me; not one, nor few, they are many, indeed almost all. If you would know why, it is because I hate the mob, I loathe the vulgar herd and in the multitude I find no joy. It is Unity that doth enchant me. By her power I am free though thrall, happy in sorrow, rich in poverty, and quick even in death.

In a comparably flowery paragraph and a half omitted at this point, Bruno continues his invocations and discusses his interests and hopes, but says nothing of his cosmological views until the final two sentences of his epistle.

This I shall make manifest by conclusive arguments, dependent on lively reasonings derived from regulated sensation, instructed by true phenomena; for these as trustworthy ambassadors emerge from objects of Nature, rendering themselves present to those who seek them, obvious to those who gaze attentively on them, clear to those who apprehend, certain and sure to those who understand. Thus I present to you my contemplation concerning the infinite universe and innumerable worlds.[7]

ARGUMENT OF THE FIRST DIALOGUE

YOU learn from the first Dialogue FIRSTLY, that the inconstancy of sense-perception doth demonstrate that sense is no source of certainty, but can attain thereto only through comparison and reference from one sensible percept to another, from one sense to another, so that truth may be inferred from diverse sources.

SECONDLY, the demonstration is begun of the infinity of the universe;[8] and the first argument is derived from the failure to limit the world by those whose fantasy would erect around it boundary walls.

THIRDLY, it will be shown that it is unfitting to name the world finite, and contained within itself, since this condition

7. Bruno's most appealing dedication to Mauvissiere is reserved for the conclusion of the Arguments. [This and other footnotes to Bruno's text are by Dorothea Waley Singer, its editor and translator. MJC]

8. Bruno uses "universo" for the infinite universe. His word "mondo" is throughout translated "world." Bruno uses "mondo" not only for our terrestrial globe, but for the universe as apprehended by our senses, and as conceived by the Aristotelians. Thus he speaks of our world (*questo mondo*)—including the stars that we see occupying our space, bounded by the vault of heaven. This, together with innumerable other worlds—i.e., other systems of heavenly bodies, each system occupying its own space—forms for Bruno the one infinite universe (*universo*).

belongeth only to immensity, as shown by the second argument. Moreover, the third argument is based on the inconvenience and indeed impossibility of imagining the world to occupy no position. For inevitably it would follow that it was without being, since everything whether corporeal or incorporeal doth occupy corporeally or incorporeally some position.

The FOURTH argument is based on a demonstration or urgent question put by the Epicureans:

> Moreover, suppose now that all space were created finite; if one were to run on to the end, to its furthest coasts, and throw a flying dart, would you have it that the dart, hurled with might and main, goeth on whither it is sped, flying afar, or think you that something can check and bar its way? . . . For whether there be something to check it and bring about that it arriveth not whither it was sped, and planteth not itself in the goal, or whether it fareth forward, yet it set not forth from the end.[9]

After discussing whether it is appropriate to assert that voids exist and after urging that "sense doth present to us an infinite universe" (232), Bruno invokes the principle of plenitude.

TENTHLY, since it is well that this world doth exist, no less good is the existence of each one of the infinity of other worlds.

Among the other arguments summarized by Bruno in this introductory section is that "if omnipotence maketh not the world infinite, it is impotent to do so; and if it hath not power to create it infinite, then it must lack vigour to preserve it to eternity" (235).

---

9. Lucretius, *De rerum natura,* I, 968–73, 977–79. Bruno quotes the Latin text. The English translations of this and all other passages from Lucretius are based on Dr. C. Bailey, *Lucretius on the Nature of Things* (Clarendon Press, 1950). It will be noticed that Bruno does not here give the name of Lucretius. He has a few textual deviations from the received Latin text. These are noted by Gentile, *Op. ital.,* I, 271. Throughout his edition, Gentile cites parallel passages in the other works of Bruno.

ARGUMENT OF THE THIRD DIALOGUE

IN THE third Dialogue there is first denied that base illusion of the shape of the heavens, of their spheres and diversity. For the heaven is declared to be a single general space, embracing the infinity of worlds, though we do not deny that there are other infinite 'heavens' using that word in another sense. For just as this earth hath her own heaven (which is her own region), through which she moveth and hath her course, so the same may be said of each of the innumerable other worlds. The origin is shown of the illusion of so many moving bodies subordinated to each other[10] and so shaped as to have two external surfaces and one internal cavity,[11] and of other *nostrums* and medicines, which bring nausea and horror even to those who concoct and dispense them, not less than to the wretches who swallow them.

SECONDLY, we expound how both general motion and that of the above-mentioned eccentrics, and as many as may be referred to the aforesaid firmament are all pure illusion, deriving from the motion of the centre of the earth along the ecliptic and from the four varieties of motion which the earth taketh around her own centre. Thus it is seen that the proper motion of each star resulteth from the difference in position, which may be verified subjectively within the star as a body moving alone spontaneously through the field of space. This consideration maketh it understood that all their arguments concerning the *[primum] mobile* and infinite motion are vain and based on ignorance of the motion of this our own globe.

THIRDLY, it will be propounded that every star hath motion even as hath our own and those others which are so near to us that we can sensibly perceive the differences in their orbits and in their motions: but those suns, bodies in which fire doth predominate, move differently to the earths

10. *Mobili deferenti.*

11. i.e., whose orbit depends on the circular path both of deferent and of epicycle and is itself circular.

in which water predominateth; thus may be understood whence is derived the light diffused by stars, of which some glow of themselves and others by reflection.

FOURTHLY, it is shewn how stars at vast distances from the sun can, no less than those near to it, participate in the sun's heat, and fresh proof is given of the opinion attributed to Epicurus, that one sun may suffice for an infinite universe.[12] Moreover, this explaineth the true difference between stars that do and stars that do not scintillate.

FIFTHLY, the opinion of the Cusan is examined concerning the material and the habitability of other worlds and concerning the cause of light.

SIXTHLY, it is shewn that although some bodies are luminous and hot of their own nature, yet it doth not follow that the sun illumineth the sun and the earth illumineth herself, or that water doth illumine itself. But light proceedeth always from the opposed star; just as, when looking down from lofty eminences such as mountains, we sensibly perceive the whole sea illuminated; but were we on the sea, and occupying the same plane thereof, we should see no illumination save over a small region where the light of the sun and the light of the moon were opposed to us.

SEVENTHLY, we discourse concerning the vain notion of quintessences; and we declare that all sensible bodies are no other, and composed of no different proximate or primal principles than those of our earth, nor have they other motion, either in straight lines or circles. All this is set forth with reasons attuned to the senses, while Fracastoro doth accommodate himself to the intelligence of Burchio. And it is shewn clearly that there is no accident here which may not be expected also on those other worlds; just as if we consider well we must recognize that naught there can be seen from here which cannot also be seen here from there. Consequently, that

12. Cf. *Letter to Pythokles* in Diogenes Laertius, X. The first Latin translation of Diogenes Laertius appeared in Paris, probably in 1510. But Bruno's inspiration was probably Lucretius, V.

beautiful order and ladder of nature[13] is but a charming dream, an old wives' tale.

EIGHTHLY, though the distinction between the elements be just, yet their order as commonly accepted is by no means perceptible to the senses or intelligible. According to Aristotle, the four elements are equally parts or members of this globe—unless we would say that water is in excess, wherefore with good cause the stars are named now water, now fire, both by true natural philosophers, and by prophets, divines and poets, who in this respect are spinning no tales nor forging metaphors, but allow other wiseacres to spin their tales and to babble. These worlds must be understood as heterogeneous bodies, animals, great globes in which earth is no heavier than the other elements. In them all particles move, changing their position and respective arrangement, just as the blood and other humours, spirits and smallest parts which ebb and flow are absorbed and again exhaled by us and other minor animals. In this connection a comparison is adduced shewing that the earth is no heavier by virtue of the attraction of her mass toward her own centre than is any other simple body of similar composition; that moreover the earth in herself is neither heavy, nor doth she ascend or descend; and that it is water which unifieth, and maketh density, consistency and weight.

NINTHLY, since the famous order of the elements is seen to be vain, the nature is deduced of these sensible compound bodies which as so many animals and worlds are in that spacious field which is the air or the heaven, or the void, in which are all those worlds which contain animals and inhabitants no less than can our own earth, since those worlds have no less virtue nor a nature different from that of our earth.

TENTHLY, after it hath been seen how the obstinate and the ignorant of evil disposition are accustomed to dispute, it will further be shewn how disputes are wont to conclude; although others are so wary that without losing their composure, but with a sneer, a smile, a certain discreet malice, that which

13. Of Aristotle.

they have not succeeded in proving by argument—nor indeed can it be understood by themselves—nevertheless by these tricks of courteous disdain they [pretend to have proven], endeavouring not only to conceal their own patently obvious ignorance but to cast it on to the back of their adversary. For they dispute not in order to find or even to seek Truth, but for victory, and to appear the more learned and strenuous upholders of a contrary opinion. Such persons should be avoided by all who have not a good breastplate of patience.

Late in Bruno's introductory summary of his fifth and final dialogue, he suggests: "Thus is the excellence of God magnified and the greatness of his kingdom made manifest; he is glorified not in one, but in countless suns; not in a single earth, a single world, but in a thousand thousand, I say in an infinity of worlds" (246).

At the beginning of his Third Dialogue, Bruno presents a discussion that is especially important for the issue of a plurality of worlds. The dialogue takes the form of a discussion among various interlocutors including Philotheo (sometimes called Theophilo), who speaks for Bruno and expounds the doctrine, and Elphinor, who is an inquirer. The most historically significant part of this dialogue is the section in which Bruno makes the claim that stars are suns and that they are surrounded by planets. This seems to be the first published statement of these two claims. Moreover, Bruno also maintains that the planets in the extrasolar planetary systems are inhabited.

Giordano Bruno, *On the Infinite Universe and Worlds,* in Dorothea Waley Singer, *Giordano Bruno: His Life and Thought with Annotated Translation of His Work "On the Infinite Universe and Worlds"* (New York: Henry Schuman, 1950), 302–6.

THIRD DIALOGUE

PHILOTHEO. [The whole universe] then is one, the heaven, the immensity of embosoming space, the universal envelope, the ethereal region through which the whole hath course and motion. Innumerable celestial bodies, stars, globes, suns and earths may be sensibly perceived therein by us and an infinite number of them may be inferred by our own reason. The universe, immense and infinite, is the complex of this [vast] space and of all the bodies contained therein.

ELPHINOR. So that there are no spheres with concave and convex surfaces nor deferent orbs; but all is one field, one universal envelope.

PHIL. So it is.

ELP. The opinion of diverse heavens hath then been caused by diverse motions of the stars and by the appearance of a sky filled with stars revolving around the earth; nor can these luminaries by any means be seen to recede one from another; but, maintaining always the same distance and relation one to another, and a certain course, they [appear to] revolve around the earth, even as a wheel on which are nailed innumerable mirrors revolveth around his own axis. Thus it is considered obvious from the evidence of our eyes that these luminaries have no motion of their own; nor can they wander as birds through the air; but they move only by the revolution of the orbs to which they are fixed, whose motion is effected by the divine pulse of some [supreme] intelligence.

THEOPHILO. Such is the common opinion. But once the motion is understood of our own mundane star which is fixed to no orb, but impelled by her own intrinsic principle, soul and nature, taketh her course around the sun through the vastness of universal space, and spinneth around her own centre, then this opinion will be dispelled. Then will be opened the gate of understanding of the true principles of nature, and we shall be enabled to advance with great strides along the path of truth which hath been hidden by the veil of sordid and bestial illusions and hath remained secret until to-day, through the injury of time and the vicissitudes of things, ever since there succeeded to the daylight of the ancient sages the murky night of the foolhardy sophists.

• • •

ELP. Indubitable that the whole fantasy of spheres bearing stars and fires, of the axes, the deferents, the functions of the epicycles, and other such chimeras, is based solely on the belief that this world occupieth as she seemeth to do the very centre of the universe, so that she alone being immobile and fixed, the whole universe revolveth around her.

PHIL. This is precisely what those see who dwell on the moon and on the other stars in this same space, whether they be earths or suns.

ELP. Suppose then for the moment that the motion of our earth causeth the appearance of daily world motion, and that by her own diverse motions the earth causeth all those motions which seem to appertain to the innumerable stars, we should still say that the moon, which is another earth, moveth by her own force through the air around the sun. Similarly, Venus, Mercury and the others which are all earths, pursue their courses around the same father of life.

PHIL. It is so.

ELP. The proper motions of each of these are those of their apparent motions which are not due to our so-called world motion; and the proper motions of the bodies known as fixed stars (though both their apparent fixity and the world motion should be referred to our earth) are more diverse and more numerous than the celestial bodies themselves. For if we could observe the motion of each one of them, we should find that no two stars ever hold the same course at the same speed; it is but their great distance from us which preventeth us from detecting the variations. However much these stars circulate around the solar flame or spin round their own centres in order to participate in the vital heat [of a sun], it is impossible for us to detect their diverse approach toward and retreat from us.

PHIL. That is so.

ELP. There are then innumerable suns, and an infinite number of earths revolve around those suns, just as the seven we can observe revolve around this sun which is close to us.

PHIL. So it is.

ELP. Why then do we not see the other bright bodies which are earths circling around the bright bodies which are suns? For beyond these we can detect no motion whatever; and why do all other mundane bodies (except those known as comets) appear always in the same order and at the same distance?

PHIL. The reason is that we discern only the largest suns, immense bodies. But we do not discern the earths because, being much smaller, they are invisible to us. Similarly it is not impossible that other earths revolve around our sun and are invisible to us on account either of greater distance or of smaller size, or because they have but little watery surface, or because such watery surface is not turned toward us and opposed to the sun, whereby it would be made visible as a crystal mirror which receiveth luminous rays; whence we perceive that it is not marvellous or contrary to nature that often we hear that the sun hath been partially eclipsed though the moon hath not been interpolated between him and our sight. There may be innumerable watery luminous bodies—that is, earths consisting in part of water—circulating around the sun, besides those visible to us; but the difference in their orbits is indiscernible by us on account of their great distance, wherefore we perceive no difference in the very slow motion discernible of those visible above or beyond Saturn; still less doth there appear any order in the motion of all around the centre, whether we place our earth or our sun as that centre.

ELP. How then wouldst thou maintain that all of these bodies, however far from their centre, that is from the sun, can nevertheless participate in the vital heat thereof?

PHIL. Because the further they are from the sun, the larger is the circle of their orbit around it; and the greater their orbit, the more slowly they accomplish their journey round the sun; the more slowly they move, the more they resist the hot flaming rays of the sun.

ELP. You maintain then that though so distant from the sun, these bodies can derive therefrom all the heat that they need. Because, spinning at a greater rate around their own centre and revolving more slowly around the sun, they can derive not only as much heat but more still if it were needed; since by the more rapid spin around her own centre, such part of the convexity of the earth as hath not been sufficiently heated is the more quickly turned to a position to receive heat; while from the slower progress around the fiery central body, she stayeth to receive more firmly the impression therefrom, and thus she will receive fiercer flaming rays.

PHIL. That is so.

ELP. Therefore you consider that if the stars beyond Saturn are really motionless as they appear, then they are those innumerable suns or fires more or less visible to us around which travel their own neighbouring earths which are not discernible by us.

THEO. Yes, we should have to argue thus, since all earths merit the same amount of heat, and all suns merit the same amount.

ELP. Then you believe that all those are suns?

PHIL. Not so, for I do not know whether all or whether the majority are without motion, or whether some circle around others, since none hath observed them. Moreover they are not easy to observe, for it is not easy to detect the motion and progress of a remote object, since at a great distance change of position cannot easily be detected, as happeneth when we would observe ships in a high sea. But however that may be, the universe being infinite, there must ultimately be other suns. For it is impossible that heat and light from one single body should be diffused throughout immensity, as was supposed by Epicurus if we may credit what others relate of him.[14] Therefore it followeth that there must be innumerable suns, of which many appear to us as small bodies; but that star will appear smaller which is in fact much larger than that which appeareth much greater.

ELP. All this must be deemed at least possible and expedient.

PHIL. Around these bodies there may revolve earths both larger and smaller than our own.

ELP. How shall I know the difference? How, I say, shall I distinguish fiery bodies from earths?

---

14. Perhaps suggested by Lucretius, *De rerum natura*, V, 597–601.

PHIL. Because fiery bodies are fixed and earths are in motion; because fiery bodies scintillate and earths do not; of which indications, the second is more easily perceptible than the first.

ELP. They say that the appearance of scintillation is caused by the great distance from us.

PHIL. If that were so, the sun would not scintillate more than all the others; and the small stars which are more remote would scintillate more than the larger which are nearer to us.

ELP. Do you believe that fiery worlds are inhabited even as are watery bodies?

PHIL. Neither more nor less.

ELP. But what animals could live in fire?

PHIL. You must not regard these worlds as compounded of identical parts, for then they would be not worlds but empty masses, vain and sterile. Therefore it is convenient and natural to assume that their parts are diverse just as our own and other earths comprise diverse parts, though some celestial bodies have the appearance of illumined water as others of shining flames.

ELP. You believe then that the prime matter of the sun differeth not in consistency and solidity from that of the earth? (For I know that you do not doubt that a single prime matter is the basis of all things.)

PHIL. This indeed is certain. . . .

A final question: Based on these readings, how would one expect Copernicus and Bruno to view each other? It is striking that never in the selections from Bruno does this famous early Copernican ever mention Copernicus. A hint as to why this might be comes from a comment on Copernicus made by Bruno in his *Ash Wednesday Supper*. Regarding the founding father of modern astronomy, Bruno states that he

> was possessed of a grave, elaborate, careful, and mature mind; a man who was not inferior, except by succession of place and time, to any astronomer who had been before him; a man who in regard to natural judgment was far superior to Ptolemy, Hipparchus, Eudoxus, and all the others who walked in the footsteps of these. . . . But for all that he did not move too much beyond them;

> being more intent on the study of mathematics than of nature, he was not able to go deep enough and penetrate beyond the point of removing from the way the stumps of inconvenient and vain principles, so as to resolve completely the difficult objections, and to free both himself and others from so many vain investigations, and to set attention on things constant and certain.[15]

It is impossible to determine exactly what Copernicus would have thought of Bruno, who was born five years after Copernicus's death. One can, however, get some idea from a remark made by a prominent Bruno scholar, Dame Frances Yates, who commented that, had he seen it, "Copernicus might have well bought up and destroyed all copies of [Bruno's *Ash Wednesday Supper*] had he been alive."[16]

---

15. Giordano Bruno, *Ash Wednesday Supper*, trans. and introduced by Stanley L. Jaki (The Hague: Mouton, 1975), 56–57.

16. Francis A. Yates, *Giordano Bruno and the Hermetic Tradition* (New York: Random House, 1975), 297.

FOUR

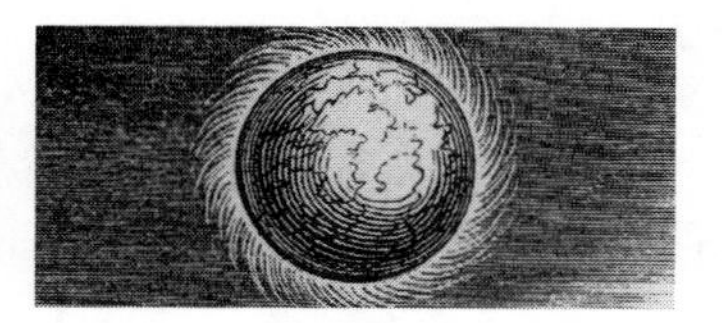

GALILEO, KEPLER, DESCARTES, AND PASCAL

GALILEO GALILEI (1564–1642)

When around 1608 Galileo Galilei turned the newly invented telescope to the heavens, he made a number of sensational discoveries, which he published in 1610 in his *Sidereus nuncius* (*Starry Messenger*). Among the discoveries reported in that book are: (1) that the Moon has mountains and seas, (2) that Jupiter is orbited by four moons, and (3) that the number of stars far exceeds what had previously been seen. Striking as these results were, their impact becomes all the more evident when they are interpreted, as eventually some authors did, in terms of the idea of a plurality of worlds: (1) that the Moon has a number of earthlike features, (2) that at least Jupiter also has definite similarities to the Earth, and (3) that if the stars are suns and if suns are orbited by inhabited planets, then our universe must be extremely richly stocked with life.

It is interesting to ask how Galileo himself interpreted these and other observations in regard to the question of extraterrestrial life. Did he see them, for example, as a vindication of or support for the pluralist universe championed by Giordano Bruno? The first fact to be noted is that never in his published writings or letters did Galileo ever mention Bruno.[1] Another indication comes from 1613, when Galileo published his *Letter on Sunspots*, directed to the Jesuit astronomer Christopher Scheiner. Galileo states:

> I agree with Apelles [Scheiner] in regarding as false and damnable the view of those who would put inhabitants on Jupiter, Venus, Saturn, and the moon, meaning by "inhabitants" animals like ours, and men in particular. Moreover, I think I can prove this. If we could believe with any probability that there were living beings and vegetables on the moon or any planet, different not only from terrestrial ones but remote from our wildest imaginings, I should for my part neither affirm it nor deny it, but should leave the decision to wiser men than I.[2]

That Galileo's reservations about, possibly even opposition to, claims for extraterrestrials had not diminished by 1616 is indicated in a letter he wrote on 28 February of that year to Giacomo Muti:

> A few days ago, when paying my respects to the illustrious Cardinal Muti, a discussion arose on the inequalities of the moon's surface. Signor Alessandro Capoano, in order to disprove the fact, argued that if the lunar superfecies be unequal and mountainous, one may say as a consequence that, since Nature has made our earth mountainous for the benefit of plants and animals beneficial to man, so on the moon there must be other plants and other animals beneficial to other intellectual creatures. Such a consequence, he said, being most false, therefore the fact from which it is drawn must also be false, therefore lunar mountains do not exist! To this I replied: As to the inequalities of the moon's surface we have only to look through a telescope to be convinced of their existence; as to the 'consequences,' I said, they are not only not necessary but absolutely false and impossible, for I was in a position to prove that neither men nor animals, nor plants as on this earth, nor anything else at all like them can exist on the moon. I said then, and I say now, that I do not believe that the body of the moon is composed of earth and water, and wanting these two elements we must necessarily conclude that it wants all the other things which without these other things cannot exist or subsist. I add further: even allowing that the

---

1. Alexandre Koyré, *From the Closed World to the Infinite Universe* (New York: Harper and Brothers, 1957), 99.

2. Galileo Galilei, "Letter on Sunspots," in *Discoveries and Opinions of Galileo*, trans. Stillman Drake (Garden City, New York: Doubleday Anchor, 1957), 137.

matter of the moon may be like that of the earth (a most improbable supposition), still not one of those things which the earth produces can exist on the moon, since to their production other things besides earth and water are necessary—namely, the sun—the greatest agent in nature—and the resulting vicissitudes of heat and cold, and of day and night. Now, such vicissitudes are on the moon very different from those on the earth. In the latter case, to produce a diversity of seasons, the sun rises and falls more than 47° (in passing from one tropic to the other); in the former case the variation is only 5° on each side of the ecliptic. While, therefore, on the earth the sun in every 24 hours illuminates all parts of its surface, each half of the moon is alternately in sunshine and darkness for 15 continuous days of 24 hours. Now, if our plants and animals were exposed to ardent sunshine every month for 360 consecutive hours, and then for a similar time were plunged in cold and darkness, they could not possibly preserve themselves, much less produce and multiply. We must, therefore, conclude that, what would be impossible on our earth under the circumstances we have *supposed* to exist, must be impossible on the moon where those conditions *do* exist.[3]

JOHANNES KEPLER (1571–1630)

The most mathematically gifted astronomer of the early seventeenth century was Johannes Kepler, who was also an early advocate of the Copernican theory. He endorsed Copernicanism in 1596 in his *Mysterium cosmographicum* (*Cosmographical Mystery*), in which he reported his "discovery" of correlations among the planetary radii and the five regular or Platonic solids. These are the only geometrical solids that are composed of identical faces, edges, and angles. The cube and tetrahedron are the simplest such figures. Kepler believed that this correlation explained both why only six planets exist and why the planets orbit at specific distances from the Sun.

In his *Mysterium cosmographicum*, Kepler provided a representation of this structure of the solar system and included a statement that he had composed at the time of his discovery:

The earth is the circle which is the measure of all. Construct a dodecahedron round it. The circle surrounding that will be Mars, round Mars construct a tetrahedron. The circle surrounding that will be Jupiter. Round Jupiter construct a cube. The circle surrounding that will be Saturn. Now construct an icosahedron

3. As given in J. J. Fahie, *Galileo: His Life and Work* (London: J. Murray, 1903), 135–36.

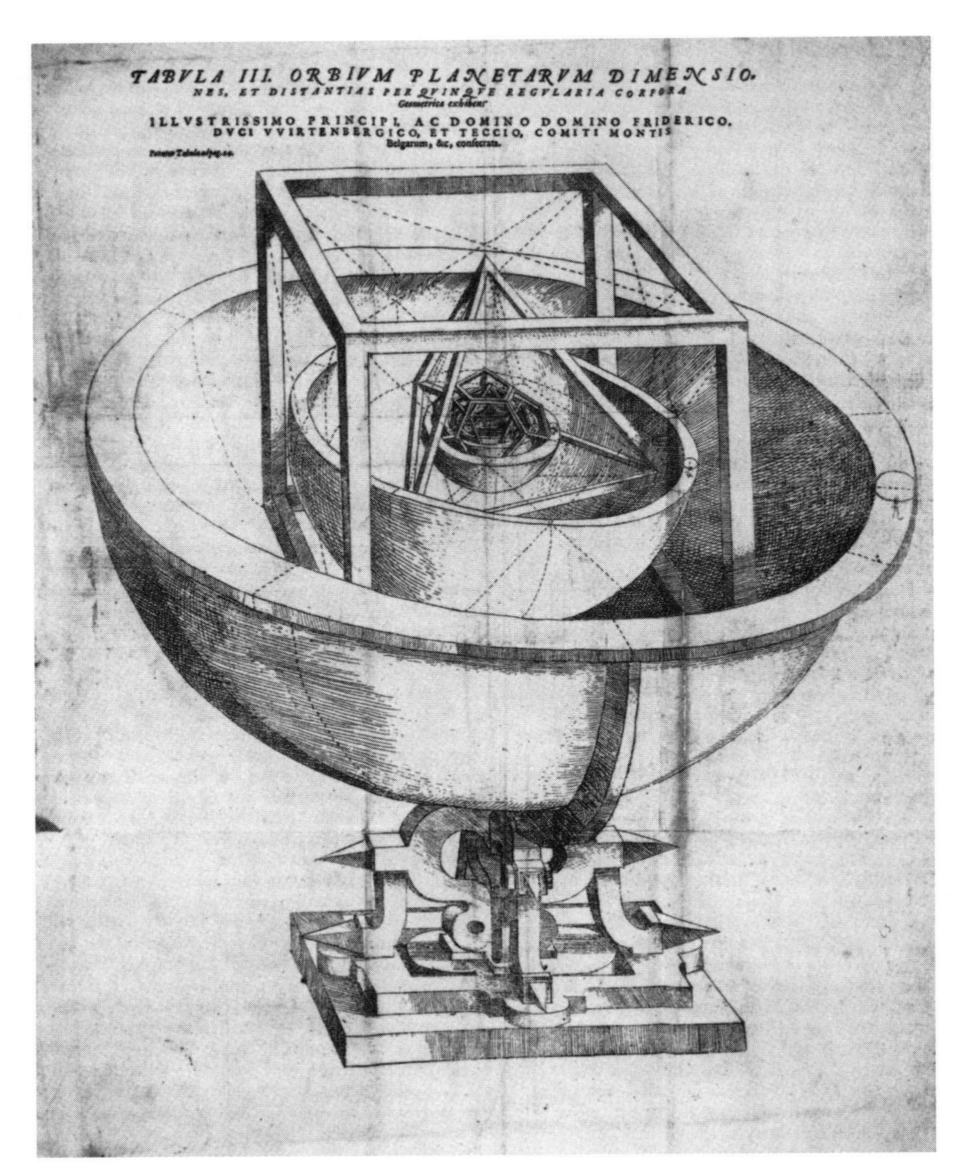

Fig. 2. Kepler's diagram of the solar system

inside the earth. The circle inscribed within that will be Venus. Inside Venus inscribe an octahedron. The circle inscribed within that will be Mercury.[4]

Kepler was able to marshal moderately impressive empirical support for his theory that God had designed the solar system on the basis of the regular solids, a claim that Kepler maintained for the remainder of his life. Although this view is no longer taken seriously in astronomy, other results attained by Kepler were and are of major significance for astronomy. Among the most important such results are the three Keplerian laws of planetary motion. Kepler brought out the first two of these laws in 1609 in his *Astronomia nova* (*New Astronomy*), and revealed the third in 1619 in his *Harmonices mundi* (*Harmony of the World*).

The consistently cautious approach to the question of extraterrestrial life evident in Galileo's writings and letters is strikingly different from what one encounters in Kepler's. Perhaps the best known expression of Kepler's views is his *Somnium*, a work of science fiction about life on the Moon that Kepler drafted in 1609 but that was first published only in 1634, four years after Kepler's death. This work contains ingenious speculations by Kepler about life on the Moon. It is not, however, the only statement we have of Kepler's views on extraterrestrials. Fascinating claims and ideas and a quite different perspective appear in a letter to Galileo that Kepler published in 1610 in response to Galileo's report in his *Starry Messenger* of lunar mountains, Jovian satellites, and much else.

Johannes Kepler, *Kepler's Conversation with Galileo's Sidereal Messenger*, trans. and annotated by Edward Rosen (New York: Johnson Reprint Company, 1965), 33–46.

VI

Kepler begins with a short introductory section discussing how stars and planets differ in appearance, noting in particular Galileo's comment that when viewed with a telescope, stars appear smaller than when viewed with the naked eye.

4. Johannes Kepler, *Mysterium Cosmographicum: The Secret of the Universe*, trans. A. M. Duncan (New York: Abaris Books, 1981), 69.

Your second highly welcome observation concerns the sparkling appearance of the fixed stars, in contrast with the circular appearance of the planets. What other conclusion shall we draw from this difference, Galileo, than that the fixed stars generate their light from within, whereas the planets, being opaque, are illuminated from without; that is, to use Bruno's terms, the former are suns, the latter, moons or earths?

Nevertheless, let him not lead us on to his belief in infinite worlds, as numerous as the fixed stars and all similar to our own. Your third observation comes to our support: the countless host of fixed stars exceeds what was known in antiquity. You do not hesitate to declare that there are visible over 10,000 stars. The more there are, and the more crowded they are, the stronger becomes my argument against the infinity of the universe, as set forth in my book on the "New Star," Chapter 21, page 104. This argument proves that where we mortals dwell, in the company of the sun and the planets, is the primary bosom of the universe; from none of the fixed stars can such a view of the universe be obtained as is possible from our earth or even from the sun. For the sake of brevity, I forbear to summarize the passage. Whoever reads it in its entirety will be inclined to assent.

Let me add this consideration to buttress my case. To my weak eyes, any of the larger stars, such as Sirius, if I take its flashing rays into account, seems to be only a little smaller than the diameter of the moon. But persons with unimpaired vision, using astronomical instruments that are not deceived by these wavy crowns, as is the naked eye, ascertain the dimensions of the stars' diameters in terms of minutes and fractions of minutes. Suppose that we took only 1000 fixed stars, none of them larger than 1' (yet the majority in the catalogues are larger). If these were all merged in a single round surface, they would equal (and even surpass) the diameter of the sun. If the little disks of 10,000 stars are fused into one, how much more will their visible size exceed the apparent disk of the sun? If this is true, and if they are suns having the same nature as our sun, why do not these suns collectively outdistance our sun in brilliance? Why do they all together transmit so dim a light to the most accessible places? When sunlight bursts into a sealed room through a hole made with a tiny pin point, it outshines the fixed stars at once. The difference is practically infinite; if the whole room were removed, how great would it become? Will my opponent tell me that the stars are very far away from us? This does not help his cause at all. For the greater their distance, the more does every single one of them outstrip the sun in diameter. But maybe the intervening aether obscures them? Not in the least. For we see them with their sparkling, with their various shapes and colors. This could not happen if the density of the aether offered any obstacle.

Hence it is quite clear that the body of our sun is brighter beyond measure than all the fixed stars together, and therefore this world of ours does not belong to an undifferentiated swarm of countless others. I shall have more to say about this subject later on.

You have a large number of eyewitnesses to the innumerability of the stars. The rabbis are said to enumerate more than 12,000. A clergyman of my acquaintance one moonless night counted over 40 in Orion's Shield. The larger stars in the Pleiades are arranged in order by Maestlin to the number of fourteen, if I am not mistaken, none of them below the limits of the traditional magnitudes.

VII

You have conferred a blessing on astronomers and physicists by revealing the true character of the Milky Way, the nebulae, and the nebulous spirals. You have upheld those writers who long ago reached the same conclusion as you: they are nothing but a mass of stars, whose luminosities blend on account of the dullness of our eyes.

Accordingly, scientists will henceforth cease to create comets and new stars out of the Milky Way, after the manner of Brahe, lest they irrationally assert the passing away of perfect and eternal celestial bodies.

VIII

Finally I move on with you to the new planets, the most wonderful topic in your little book. On this subject I shall say only a few words to you in addition to what I wrote at the outset.

In the first place, I rejoice that I am to some extent restored to life by your work. If you had discovered any planets revolving around one of the fixed stars, there would now be waiting for me chains and a prison amid Bruno's innumerabilities, I should rather say, exile to his infinite space. Therefore, by reporting that these four planets revolve, not around one of the fixed stars, but around the planet Jupiter, you have for the present freed me from the great fear which gripped me as soon as I had heard about your book from my opponent's triumphal shout.

In a section omitted at this point, Kepler criticizes Bruno and praises Copernicus, noting that the latter had corrected the ancient astronomical authors. Kepler also mentions his claim made in his *Mysterium cosmographicum* that the distances of the planets from the Sun can be explained by the idea that God designed the solar system on the basis of the regular solids. Kepler not only suggests that this idea fits the data very well, but that it also reveals the rationality, indeed the beauty, of God's design for the solar system.

Therefore, Galileo, you will not envy our predecessors their due praise. What you report as having been quite recently observed by your own eyes, they predicted, long before you, as necessarily so. Nevertheless, you will have your own fame. Copernicus and I, as a Copernican, pointed out to the ancients the mistaken way in which they considered the five solids to be expressed in the world, and we substituted the authentic and true way. Similarly, you correct and, in part, unsettle Bruce's doctrine, borrowed from Bruno. These men thought that other celestial bodies have their own moons revolving around them, like our earth with its moon. But you prove that they were talking in generalities. Moreover, they supposed it was the fixed stars that are so accompanied. Bruno even expounded the reason why this must be so. The fixed stars, forsooth, have the quality of sun and fire, but the planets, of water. By an indefeasible law of nature these opposites combine. The sun cannot be deprived of the planets; the fire, of its water; nor in turn the water, of the fire. Now the weakness of his reasoning is exposed by your observations. In the first place, suppose that each and every fixed star is a sun. No moons have yet been seen revolving around them. Hence this will remain an open question until this phenomenon too is detected by someone equipped for marvelously refined observations. At any rate, this is what your success threatens us with, in the judgment of certain persons. On the other hand, Jupiter is one of the planets, which Bruno describes as earths. And behold, there are four other planets around Jupiter. Yet Bruno's argument made this claim not for the earths, but for the suns.

Meanwhile I cannot refrain from contributing this additional feature to the unorthodox aspects of your findings. It is not improbable, I must point out, that there are inhabitants not only on the moon but on Jupiter too or (as was delightfully remarked at a recent gathering of certain philosophers) that those areas are now being unveiled for the first time. But as soon as somebody demonstrates the art of flying, settlers from our species of man will not be lacking. Who would once have thought that the crossing of the wide ocean was calmer and safer than of the narrow Adriatic Sea, Baltic Sea, or English Channel? Given ships or sails adapted to the breezes of heaven, there will be those who will not shrink from even that vast expanse. Therefore, for the sake of those who, as it were, will presently be on hand to attempt this voyage, let us establish the astronomy, Galileo, you of Jupiter, and me of the moon.

Let the foregoing pleasantries be inserted on account of the miracle of human courage, which is evident in the men of the present age especially. For the revered mysteries of sacred history are not a laughing matter for me.

I have also thought it worth while, in passing, to tweak the ear of the higher philosophy. Let it ponder the questions whether the almighty and provident Guardian of the human race permits anything useless and why, like an experienced steward, He opens the inner chambers of his building to us at this particular time. Such was the opinion put forward by my good friend Thomas Seget, a man of wide learning. Or does God the creator, as I replied, lead mankind, like some growing youngster gradually approaching maturity, step by step from one stage of knowledge to another? (For example, there was a period when the distinction between the planets and the fixed stars was unknown; it was quite some time before Pythagoras or Parmenides perceived that the evening star and the morning star are the same body, the planets are not mentioned in Moses, Job, or the Psalms). Let the higher philosophy reflect, I repeat, and glance backward to some extent. How far has the knowledge of nature progressed, how much is left, and what may the men of the future expect?

But let us return to humbler thoughts, and finish what we began. There are in fact four planets revolving around Jupiter at different distances with unequal periods. For whose sake, the question arises, if there are no people on Jupiter to behold this wonderfully varied display with their own eyes? For, as far as we on the earth are concerned, I do not know by what arguments I may be persuaded to believe that these planets minister chiefly to us, who never see them. We should not anticipate that all of us, equipped with your telescopes, Galileo, will observe them hereafter as a matter of course.

In a section omitted at this point, Kepler raises the question of whether Galileo's discovery of the moons of Jupiter should have an effect on astrology. His concluding comment on this issue is: "In this way astrology maintains its standing. At the same time it becomes evident that these four new planets were ordained not primarily for us who live on the earth, but undoubtedly for the Jovian beings who dwell around Jupiter." He then presents evidence that both our Moon and the moons of Jupiter are relatively small compared to their primaries (i.e., the bodies they orbit) and that these satellites are very near their primaries.

The conclusion is quite clear. Our moon exists for us on the earth, not for the other globes. Those four little moons exist for Jupiter, not for us. Each planet in turn, together with its occupants, is served by its own satellites. From this line of reasoning we deduce with the highest degree of probability that Jupiter is inhabited. Tycho Brahe likewise drew the same inference, based exclusively on a consideration of the hugeness of those globes.[5]

A further consequence was very brilliantly pointed out by Wackher. Like our earth, Jupiter also rotates about its axis. That rotation is accompanied by the revolution of the four moons, just as the rotation of our earth is accompanied by the revolution of our moon in the same direction. Therefore Wackher now finally accepts the magnetic principles by which, in my recent "Commentary on Celestial Physics," I explained that the motions of the planets are caused by the rotation of the sun about its axis and poles.

While Jupiter moves along its 12-year orbit, four satellites encompass it before and behind. What was absurd, then, in Copernicus' statement (as you neatly remark, Galileo) that while the earth performs its annual revolution, a single moon clings to it in the same way?

Well, then, someone may say, if there are globes in the heaven similar to our earth, do we vie with them over who occupies the better portion of the universe? For if their globes are nobler, we are not the noblest of rational creatures. Then how can all things be for man's sake? How can we be the masters of God's handiwork?

It is difficult to unravel this knot, because we have not yet acquired all the relevant information. We shall hardly escape being labeled foolish if we expatiate at length on this subject.

Yet I shall not pass over in silence those philosophical arguments which, it seems to me, can be brought to bear. They will establish not merely in general, as was done above, that

---

5. [Kepler has seriously misunderstood Brahe's argument on this point. Brahe had made the following reduction to absurdity argument: If Copernican heliocentrism is correct, then stars must be at incredibly great distances. This in turn entails that God's creative powers would have been wasted unless there were stellar inhabitants. It being absurd that such exist, it follows that the Copernican theory is false. Kepler, in his enthusiasm for extraterrestrials, seems to have entirely missed the reduction to absurdity character of Brahe's argument. See Steven J. Dick, *Plurality of Worlds: The Origins of the Extraterrestrial Life Debate from Democritus to Kant* (Cambridge: Cambridge University Press, 1982), 73–74, 77, 204–5. MJC]

this system of planets, on one of which we humans dwell, is located in the very bosom of the world around the heart of the universe, that is, the sun. These arguments will also establish in particular that we humans live on the globe which by right belongs to the primary rational creature, the noblest of the (corporeal) creatures.

In support of the former proposition concerning the inmost bosom of the world, see the evidence cited above. It was based, first, on the fixed stars, which by their vast numbers truly enclose this area like a wall and, secondly, on our sun, which is more splendid than the fixed stars. To the foregoing may be added a third consideration, which Wackher elicited from me during the past few days, and to which he seemed by his silence to assent.

Geometry is unique and eternal, and it shines in the mind of God. The share of it which has been granted to man is one of the reasons why he is the image of God. Now in geometry the most perfect class of figures, after the sphere, consists of the five Euclidean solids. They constitute the very pattern and model according to which this planetary world of ours was apportioned. Suppose then that there is an unlimited number of other worlds. They will be either unlike ours or like it. You would not say, "like it." For what is the use of an unlimited number of worlds, if every single one of them contains all of perfection within itself? Surely the situation is different with regard to the creatures which perpetuate themselves by a succession of generations. Even Bruno, the defender of infinity, holds that each world must differ from the rest in the kinds of motion, although these are of like number. If the worlds differ in their motions, then they must differ also in their distances, which determine the periods of the motions. If they differ in their distances, then they must differ also in the arrangement, type, and perfection of their solids, from which the distances are derived. Indeed, if you establish universes similar to one another in all respects, you will also produce similar creatures, and as many Galileos, observing new stars in new worlds, as there are worlds. But of what use is this? Briefly, it is better to avoid the march to the infinite permitted by the philosophers. Since it is agreed that there is a limit to the regress in the direction of the smaller, why not also in the direction of the larger? For example, take the sphere of the fixed stars. One three-thousandth part of it, perhaps, is the sphere of Saturn. Of this, in turn, $^{1}/_{10}$th part is the sphere of the earth. One three-hundred-thousandth part of the earth's diameter, again, is man. A tiny part of man is the little pore beneath his skin. Here we stop. Nature goes no lower. Now let us tackle the other horn of the dilemma. Suppose those infinite worlds are unlike ours. Then they will be supplied with something different from the five perfect solids. Hence they will be less noble than our world. Therefore it follows that this world of ours is the most excellent of them all, if there should be a plurality of worlds.

Let us now also indicate why the earth surpasses Jupiter and better deserves to be the abode of the predominant creature.

In the center of the world is the sun, heart of the universe, fountain of light, source of heat, origin of life and cosmic motion. But it seems that man ought quietly to shun that royal throne. Heaven was assigned to the lord of heaven, the sun of righteousness; but earth, to the children of man. God has no body, of course, and requires no dwelling place. Yet more of the force which rules the world is revealed in the sun (in the heaven, as various passages of Scripture put it) than in all the other globes. Because man's house is otherwise, therefore, let him recognize his own wretchedness and the opulence of God. Let him acknowledge that he is not the source and origin of the world's splendor, but that he is dependent on the true source and origin thereof. Moreover, as I said in the "Optics," in the interests of that contemplation for which man was created, and adorned and equipped with eyes, he could not remain at rest in the center. On the contrary, he must make an annual journey on this boat, which is our earth, to perform his observations. So surveyors, in measuring inaccessible objects, move from place to place for the purpose of obtaining from the distance between their positions an accurate base line for the triangulation.

After the sun, however, there is no globe nobler or more suitable for man than the earth. For, in the first place, it is exactly in the middle of the principal globes (if we exclude, as we should, Jupiter's satellites and the moon revolving around the earth). Above it are Mars, Jupiter, and Saturn. Within the embrace of its orbit run Venus and Mercury, while at the center the sun rotates, instigator of all the motions, truly an Apollo, the term frequently used by Bruno.

Secondly, the five solids divide into two groups: the three major bodies, cube, tetrahedron, and dodecahedron and the two minor bodies, icosahedron and octahedron. The earth's orbit separates the two groups like a partition, by touching the centers of the 12 faces of the dodecahedron above it, and the 12 vertices of the corresponding icosahedron below it. Merely by its position amidst the solids, the sphere of the earth is more distinguished than the other spheres.

Thirdly, we on the earth have difficulty in seeing Mercury, the last of the principal planets, on account of the nearby, overpowering brilliance of the sun. From Jupiter or Saturn, how much less distinct will Mercury be? Hence this globe seems assigned to man with the express intent of enabling him to view all the planets. Will anyone then deny that, to make up for the planets concealed from the Jovians but visible to us earth-dwellers, four others are allocated to Jupiter, to match the four inferior planets, Mars, Earth, Venus, and Mercury, which revolve around the sun within Jupiter's orbit?

Let the Jovian creatures, therefore, have something with which to console themselves. Let them even have, if it seems right, their own four planets arranged in conformity with a group of three rhombic solids. Of these, one is the cube (a quasi-rhombic); the second is cuboctahedral; the third is icosidodecahedral; with 6, 12, and 30 quadrilateral faces, respectively. Let the Jovians, I repeat, have their own planets. We humans who inhabit the earth can with good reason (in my view) feel proud of the pre-eminent lodging place of our bodies, and we should be grateful to God the creator.

RENÉ DESCARTES (1596–1650)

An analysis of the writings, published and unpublished, of René Descartes reveals an author who hesitated to endorse the doctrine of a plurality of worlds, but did not shrink from formulating a system of the cosmos that in the hands of others could be seen as supporting that doctrine. In this context, Arthur Lovejoy commented: "It was primarily to the vogue of Cartesianism rather than to any direct influence of the writings of Bruno that the rapidly growing acceptance of theories of the plurality of worlds in the second half of the seventeenth century was chiefly due."[6]

The most important of Descartes's writings for the extraterrestrial life debate is his *Principles of Philosophy* of 1644. In principle 21 of part 2 of that work, Descartes argues that the world is indefinite in extension, suggesting that, beyond any limit we may conceive, we can in fact imagine matter still further extended. In principle 22, he states: "*Thus the matter of the heavens and of the earth is one and the same, and there cannot be a plurality of worlds.*"[7] This statement, however, can easily be misunderstood. What Descartes is denying is only a plurality of worlds in the atomist sense. He is maintaining that there is only one matter and "even were there an infinitude of worlds, they would all be formed of this matter."[8] Thus at this point he is denying only that there can be an infinity of worlds composed of different sorts of matter.

Early in Part Three of his *Principles*, Descartes urges a cosmic humility on his readers, warning that we must not think that all things were created for terrestrials. He suggests that such may be a pious thought, but it is also an incorrect one for "it is yet not at all probable that all things were created for us in such a manner that God has had no other end in creating them . . . for we cannot doubt that an infinitude of things exist, or did exist, . . . which have never been beheld or comprehended by man and which have never been of any use to him."[9] Later in part 3, Descartes sets out his cosmology. According to Descartes, all space is filled with matter. Thus, unlike the atomists but like Aristotle, he denies the possibility of a vacuum. Each star is a sun comparable to our own and each star-sun is at the center of a vortex of whirling material. Our Earth and the other planets are carried along in the Sun's vortex like a pebble in a whirlpool. (See diagram from Descartes's

---

6. As quoted in Dick, *Plurality of Worlds*, 165.

7. René Descartes, *The Philosophical Works of Descartes*, trans. by Elizabeth S. Haldane and G. R. T. Ross (Cambridge: Cambridge University. Press, 1973), 1:265.

8. Descartes, *Works*, 1:265.

9. Descartes, *Works*, 1:271.

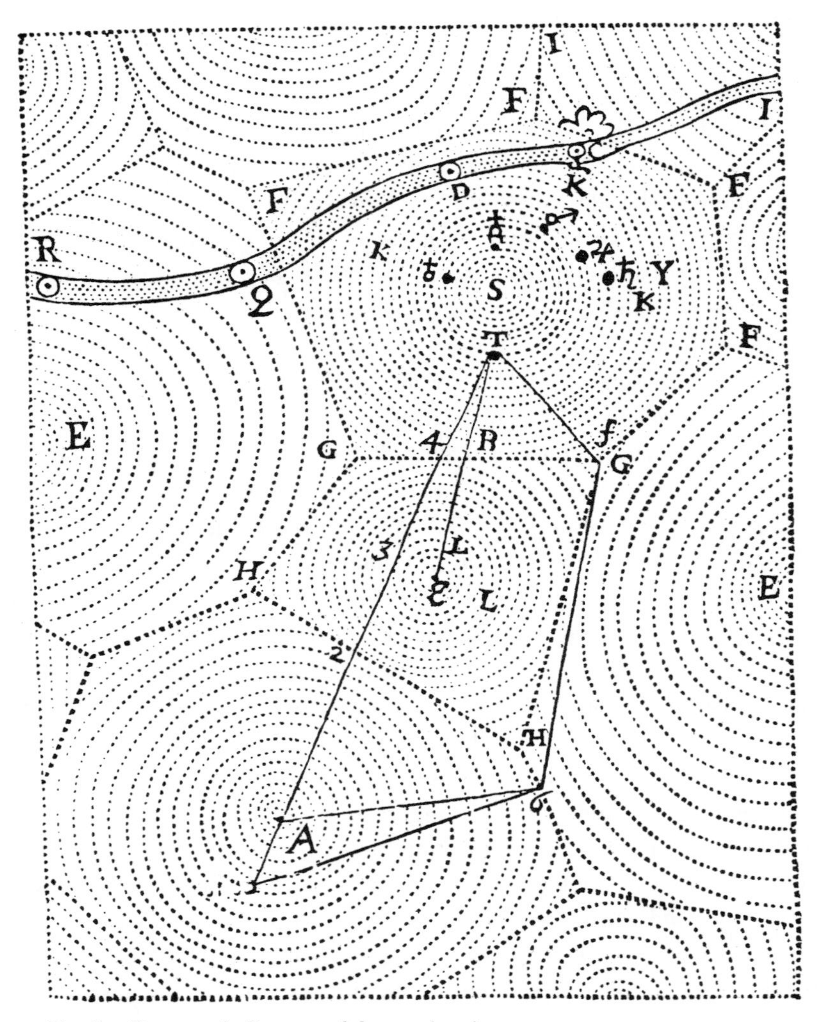

Fig. 3. Descartes's diagram of the vortices in space

Le Monde of 1664.) The various vortices are immediately adjacent to each other. Descartes never maintains that planets, but only matter, moves in the vortices around distant stars. This, however, is a natural inference, and many drew it. In short, in Descartes's system, there exist a plurality of suns, which made it natural for persons to assume a plurality of solar systems. But Descartes himself never postulated planets circling these suns. His followers were not so cautious.

Among Descartes's unpublished writings, perhaps his most interesting comment on the question of a plurality of worlds occurs in an exchange of letters with

Pierre Chanut. In 1647, three years after Descartes published his *Principles of Philosophy*, Chanut wrote its author a letter in which Chanut comments:

> If we conceive the world in the vast extension that you attribute to it, it is impossible that man retains an honorable rank there. [Man] will most probably hold that all these stars have inhabitants or, still better, that they have earths around them, full of creatures more intelligent and better than he.[10]

Descartes responded by arguing that our belief that all was created for humans must be given up, adding, however, that

> I do not see at all that the mystery of the Incarnation, and all the other advantages that God has brought forth for man obstruct him from having brought forth an infinity of other very great advantages for an infinity of other creatures. And although I do not at all infer from this that there would be intelligent creatures in the stars or elsewhere, I also do not see that there would be any reason by which to prove that there were not; but I always leave undecided questions of this kind rather than denying or affirming anything.[11]

Thus Descartes's bold imagination created a cosmos into which a plurality of worlds could easily fit, but his cautiousness and dislike of controversy led him to leave this speculation to others.

BLAISE PASCAL (1623–1662)

The chief objection to the Copernican theory, it is sometimes maintained, was that it "infinitized" the universe. The source of this "infinitization" is the issue of stellar parallax. Astronomers recognized that if the Earth were to orbit the Sun, then stars should appear to shift in position as the Earth approaches or moves farther from them. Because detection of this shift would constitute compelling evidence for the heliocentric system, astronomers from ancient through early modern times

---

10. As quoted in Paolo Rossi, "Nobility of Man and Plurality of Worlds," in Allen Debus, ed., *Science, Medicine and Society in the Renaissance* (New York: Science History Publications, 1972), 2:131–62, at 153.

11. As translated in M. J. Crowe, *The Extraterrestrial Life Debate, 1750–1900: The Idea of a Plurality of Worlds from Kant to Lowell* (Cambridge: Cambridge University Press, 1986; new ed., Mineola, N.Y.: Dover Publications, 1999), 16.

sought it, but without success. The magnitude of the effect depends on a number of factors, chiefly the radius of the Earth's orbit and the distance of the stars. Copernicans explained the failure to find stellar parallax by claiming that the stars are so distant that, although parallax occurs, it is so small that it is undetectable. When the effect was finally detected in 1838, it was recognized that the parallactic shift is less than one second of arc, which explains why it was so difficult to measure. During the three centuries before 1838, astronomers were able to estimate how far the stars would have to be from the Earth for this effect to escape their efforts at detection. They realized that the stars, rather than being located just beyond the orbit of Saturn, must be thousands of times more distant. For example, the English astronomer James Bradley in 1728 concluded from his failure to find a stellar parallax that the nearest stars must be 400,000 times more distant from the Sun than is the Earth. This not only increased the diameter of the universe, it increased its volume by the cube of that amount.

Various astronomers, for example, the famous observational astronomer Tycho Brahe (1546–1601), were very influenced by the problem of the "infinitization" of the universe. He worried that this would entail that the vast space between our solar system and the stars would be wasted space, which he viewed as irreconcilable with his notion of how the Creator designed the universe.

Blaise Pascal has long been recognized as one of the most brilliant mathematicians and scientists of the seventeenth century. Not only did he do pioneering work in atmospheric physics, projective geometry, and probability theory, he also designed and built an early calculating machine. Moreover, his *Pensées* (*Thoughts*), a defense of Christianity, is regarded as a religious and literary classic. Among the most striking features of the *Pensées* are the passages in which Pascal reacts to the "infinitization" of the universe. These passages, some of which appear below, provide a clue to an issue that has sometimes been raised about this brilliant scientist: Was Pascal a Copernican? In my judgment, the information provided above about Copernicanism entailing the "infinitization" of the universe points strongly to the conclusion that Pascal, although he never stated it explicitly, must have been a convinced Copernican. Although the Ptolemaic universe was not small by most standards, it was vastly smaller than that entailed for Copernicans by the difficulty in finding parallax. It seems impossible to believe that the author who by general consensus wrote more vividly than any of his contemporaries about the vastness of our universe was *not* a Copernican.

It was during the last years of his short life that Pascal began to compose his *Pensées*. Published posthumously in 1669, this work consists entirely of such relatively short "thoughts" as those provided in the selection that follows. Pascal's writings do not reveal what position he adopted regarding the question of extraterrestrial life, but such passages as those provided below very effectively indicate his sensitivity to the new notions of the universe in which the extraterrestrial life debate was intensifying.

Blaise Pascal, *Thoughts*, trans. W. F. Trotter, Harvard Classics edition (New York: P. F. Collier and Son, 1938). Note: The Trotter translation cites each item by the number in the ordering of Léon Brunschvicg. The numberings of the *pensées* assigned in the more authoritative ordering of Louis Lafuma are also provided.

From Pascal's *Pensées*, #72 (Brunschvicg numeration) = #199 (Lafuma numeration)

Let man then contemplate the whole of nature in her full and grand majesty, and turn his vision from the low objects which surround him. Let him gaze on that brilliant light, set like an eternal lamp to illumine the universe; let the earth appear to him a point in comparison with the vast circle described by the sun; and let him wonder at the fact that this vast circle is itself but a very fine point in comparison with that described by the stars in their revolution round the firmament. But if our view be arrested there, let our imagination pass beyond; it will sooner exhaust the power of conception than nature that of supplying material for conception. The whole visible world is only an imperceptible atom in the ample bosom of nature. No idea approaches it. We may enlarge our conceptions beyond all imaginable space; we only produce atoms in comparison with the reality of things. It is an infinite sphere, the centre of which is everywhere, the circumference nowhere. In short it is the greatest sensible mark of the almighty power of God, that imagination loses itself in that thought.

Returning to himself, let man consider what he is in comparison with all existence; let him regard himself as lost in this remote corner of nature; and from the little cell in which he finds himself lodged, I mean the universe, let him estimate at their true value the earth, kingdoms, cities, and himself. What is a man in the Infinite?

From Pascal's *Pensées*, #205 (Brunschvicg) = #68 (Lafuma)

When I consider the short duration of my life, swallowed up in the eternity before and after, the little space which I fill, and even can see, engulfed in the infinite immensity of spaces of which I am ignorant, and which know me not, I am frightened, and am astonished at being here rather than there; for there is no reason why here rather than there, why now rather than then. Who has put me here? By whose order and direction have this place and time been allotted to me?

From Pascal's *Pensées*, #206 (Brunschvicg) = #201 (Lafuma)

The eternal silence of these infinite spaces frightens me.[12]

From Pascal's *Pensées*, # 242 (Brunschvicg) = #781 (Lafuma)

I admire the boldness with which these persons undertake to speak of God. In addressing their argument to infidels, their first chapter is to prove Divinity from the works of nature. I should not be astonished at their enterprise, if they were addressing their argument to the faithful; for it is certain that those who have the living faith in their heart see at once that all existence is none other than the work of the God whom they adore. But for those in whom this light is extinguished, and in whom we purpose to rekindle it, persons destitute of faith and grace, who, seeking with all their light whatever they see in nature that can bring them to this knowledge, find only obscurity and darkness; to tell them that they have only to look at the smallest things which surround them, and they will see God openly, to give them, as a complete proof of this great

12. [Pascal remarked (Brunschvicg #23 = Lafuma #784) that "meanings differently arranged have a different effect." This remark sheds light on this *pensée*. Brunschvicg placed it immediately after Brunschvicg #205, which appears immediately above it in this text. This positioning suggests that #205 should be interpreted in a cosmic sense. In the more authoritative (and less interpretive) ordering done by Lafuma, this *pensée* follows a purely terrestrial *pensée*, which positioning removes it from a cosmic context and correspondingly alters its meaning. MJC]

and important matter, the course of the moon and planets, and to claim to have concluded the proof with such an argument, is to give them ground for believing that the proofs of our religion are very weak. And I see by reason and experience that nothing is more calculated to arouse their contempt.

From Pascal's *Pensées*, #346 (Brunschvicg) = #759 (Lafuma)

Thought constitutes the greatness of man.

From Pascal's *Pensées*, #347 (Brunschvicg) = #200 (Lafuma)

Man is but a reed, the most feeble thing in nature; but he is a thinking reed. The entire universe need not arm itself to crush him. A vapour, a drop of water suffices to kill him. But, if the universe were to crush him, man would still be more noble than that which killed him, because he knows that he dies and the advantage which the universe has over him; the universe knows nothing of this.

All our dignity consists, then, in thought. By it we must elevate ourselves, and not by space and time which we cannot fill. Let us endeavour then to think well; this is the principle of morality.

From Pascal's *Pensées*, #348 (Brunschvicg) = #113 (Lafuma)

A thinking reed.—It is not from space that I must seek my dignity, but from the government of my thought. I shall have no more if I possess worlds. By space the universe encompasses and swallows me up like an atom; by thought I comprehend the world.

FIVE

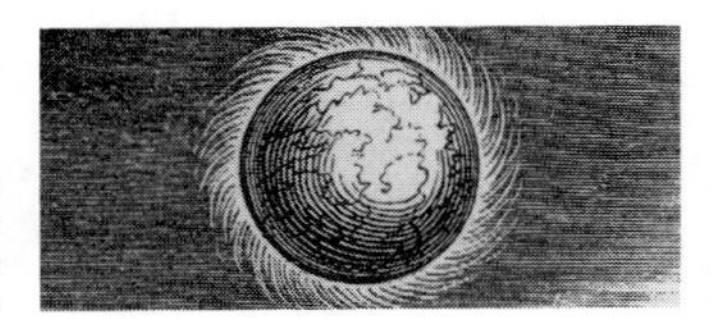

FONTENELLE AND HUYGENS

BERNARD LE BOVIER DE FONTENELLE (1657–1757)

Two books published toward the end of the seventeenth century greatly stimulated interest in and probably acceptance of extraterrestrials. The authors of these books, which are the focus of this chapter, were Fontenelle and Huygens. The earlier book appeared in 1686, when Bernard le Bovier de Fontenelle published his *Entretiens sur la pluralité des mondes* (*Conversations on the Plurality of Worlds*). The popularity of the book was such that translations soon appeared in Danish, Dutch, German, Greek, Italian, Polish, Russian, Spanish, Swedish, and English. In fact, by 1688, three different English translations of it were available, with five more following. Overall, this extraordinarily popular volume has gone through approximately one hundred editions.[1] Although the public proclaimed Fontenelle's book delightful, the Roman Catholic Church deemed it dangerous, placing it on the Index of Prohibited Books in 1687, removing it in 1825, but reinstating it in 1900!

1. Nina Rattner Gelbart, "Introduction" to Bernard le Bovier de Fontenelle, *Conversations on the Plurality of Worlds*, trans. H. A. Hargreaves (Berkeley: University of California Press, 1990), vii.

Written in the form of a dialogue between a philosopher and a charming marquise, it is one of the first works dealing with a scientific subject that was fully accessible to women. Espousing Copernican astronomy and the Cartesian vortex cosmology, Fontenelle incorporated a number of devices developed by seventeenth-century pioneers of science fiction.

Fontenelle's sixth dialogue, which he added in 1687, includes a helpful summary of five main arguments employed in his book:

> [1] the similarities of the planets to the earth which is inhabited; [2] the impossibility of imagining any other use for which they were made; [3] the fecundity and magnificence of nature, [4] the consideration she seems to show for the needs of their inhabitants as having given moons to planets distant from the sun, and more moons to those more remote; and [5] that which is very important—all that which can be said on one side and nothing on the other.[2]

Bernard le Bovier de Fontenelle, *A Plurality of Worlds,* trans. John Glanvill ([London]: Nonesuch Press, 1929), 12–13, 37–39, 55–57, 64–67, 79–81, 88–91, 95–100, 104–6, 113–15, 120–21, and 137–38.

THE FIRST EVENING

. . . I confess, Madam, *said I,* the night hath somewhat a more melancholy Air, than the day; we fancy the Stars march more silently than the Sun, and our thoughts wander with the more liberty, whilst we think all the world at rest but our selves: Besides the day is more uniform, we see nothing but the Sun, and light in the Firmament; whilst the night gives us variety of Objects, and shews us ten thousand Stars, which inspire us with as many pleasant Ideas. What you say is true, *said she,* I love the Stars, there is somewhat charming in them, and I could almost be angry with the Sun for effacing 'em.

2. Bernard le Bovier de Fontenelle, *Entretiens sur la pluralité des mondes,* ed. by Alexandre Calame (Paris: M. Didier, 1966), 161.

I can never pardon him, *I cry'd*, for keeping all those Worlds from my sight: What Worlds, *said she*, looking earnestly upon me, what Worlds do you mean?

I beg your Pardon, Madam, *said I*, you have put me upon my folly, and I begin to rave: what Folly, *said she*, I discover none? Alas, *said I*, I am asham'd, I must own it, I have had a strong Fancy every Star is a World. I will not swear it is true, but must think so, because it is so pleasant to believe it; 'Tis a fancy come into my head, and it is diverting. If your folly be so diverting, *said the Countess*, Pray make me sensible of it; provided the pleasure be so great, I will believe of the Stars all you would have me.

The philosopher proceeds to explain, among other matters, the Copernican system, in which the Earth rotates on its own axis while it orbits the immobile Sun.

THE SECOND EVENING

IN the Morning, I sent to the Countes's Apartment, to know how she had rested, and whether the Motion of the Earth had not disturb'd her? she answer'd, she began to be accustom'd to it, and that she had slept as well as *Copernicus* himself. Soon after there came some Neighbours to dine with her, but they went away in the Evening; so that after Supper we walk'd again into the Park, and immediately fell upon our Systemes. She so well conceiv'd what I told her the night before, that [if] she desir'd I would proceed without any repetition. Well, Madam, *said I*, Since the Sun, which is now immoveable, hath left off being a Planet, and the Earth, which turns round him, is now become one, you will not be surprized when you hear that the Moon is an Earth too, and that she is inhabited as ours is. I confess, *said she*, I have often heard talk of the World in the Moon, but I always look'd upon it as Visionary and meer Fancy. And it may be so still, *said I*. I am in this case as People in a Civil War, where the uncertainty of what may happen makes 'em hold intelligence with the opposite Party; for tho' I verily believe the Moon is inhabited, I live civilly with those who do not believe it; and I am (as some honest Gentlemen in point of Religion) still ready to embrace the prevailing Opinion, but till the Unbelievers have a more considerable Advantage, I am for the People in the Moon.

Suppose there had never been any Communication between *London* and *Greenwich,* and a Cockney, who was never beyond the Walls of *London,* saw *Greenwich* from the top of the Pyramid; you ask him if he believes *Greenwich* is inhabited as *London* is? He presently answers, No; for, *saith he,* I see People at *London,* but none at *Greenwich;* nor did I ever hear of any there. 'Tis true, you tell him, that from the Pyramid he cannot perceive any Inhabitants at *Greenwich,* because of the distance; but all that he doth discover of *Greenwich* very much resembleth what he sees at *London,* the Steeples, Houses, Walls; so that it may very well be inhabited as *London* is; all this signifies nothing, my Cockney still persists *Greenwich* is not inhabited, because he sees no body there. The Moon is our *Greenwich,* and every one of us as meer Cockneys as he that never was out of the sound of Bow-Bell. You are too severe, *said she,* upon your Fellow-Citizens; we are not all sure so silly as your Cockney; since *Greenwich* is just as *London* is, he is a Fool if he doth not think it inhabited: But the Moon is not at all like the Earth. Have a care what you say, *I reply'd,* for if the Moon resembleth the Earth, you are under a necessity to believe it inhabited.

• • •

What sort of People will they be then, *said the Countess?* Troth, Madam, *said I,* I know not; for put the case that we our selves inhabited the Moon, and were not Men, but rational Creatures; could we imagin, do you think, such fantastical People upon the Earth, as Mankind is? Is it possible we should have an Idea of so strange a Composition, a Creature of such foolish Passions, and such wise Reflections? So Learned in things of no use, and so stupidly ignorant of what most concerns him? So much concern for Liberty, and yet such great inclinations to Servitude? So desirous of Happiness, and yet so very incapable of being so? the People in the Moon must be wise indeed to suppose all this of us. But do we not see ourselves continually, and cannot so much as guess how we were made? So that we are forc'd to say, the Gods when they created us were drunk with Nectar, and when they were sober again, could not chuse but laugh at their own handy-work. Well, well, *said the Countess*, we are safe enough then, they in the Moon know nothing of us; but I could wish we were a little better acquainted with them, for it troubles me that we should see the Moon above us, & yet not know what is done there. Why, *said I*, are you not as much concern'd for that part of the Earth which is not yet discover'd? What Creatures inhabit it, and what they do there? for we and they are carry'd in the same Vessel; they possess the Prow, and we the Poop, and yet there is no manner of Communication between us; they do not know at one end of the Ship who lives or what is done at the other end; and you would know what passeth in the Moon, which is another great Vessel, sailing in the Heavens at a vast distance from us.

• • •

THE THIRD EVENING

. . . Well, Madam, *said I*, I have great News for you; that which I told you last Night, of the Moon's being inhabited, may not be so now: There is a new Fancy got into my Head, which puts those People in great Danger. I cannot suffer it, *said she;* yesterday you were preparing me to receive a Visit from 'em, and now there are no such People in Nature: Once you would have me believe the Moon was inhabited; I surmounted the Difficulty I had, and will now believe it. You are a little too nimble, *I reply'd;* did I not advise you never to be entirely convinc'd in things of this nature, but to reserve half of your understanding free and disengag'd, that you may admit of the contrary opinion, if there be any occasion. I care not for your Sentences, *said she*, let us come to matter of Fact. Are we not to consider the Moon as *Greenwich?* No, *said I*, the Moon doth not so much resemble the Earth, as *Greenwich* doth *London:* The Sun draws from the Earth and Water, Exhalations and Vapours, which mounting to a certain height in the Air, do there assemble, and form the Clouds; these uncertain Clouds are driven irregularly round the Globe, sometimes shadowing one Countrey, and sometimes another; he then who beholds the Earth from a-far off; will see frequent alterations upon its surface, because a great Country overcast with Clouds, will appear dark or light, as the Clouds stay, or pass over it; he will see the Spots on the Earth often change their Place, & appear, or disappear as the Clouds remove; but we see none of these changes wrought upon the Moon, which would certainly be the same, were there but Clouds about her; but on the contrary, all her Spots are fix'd and certain, and her light parts continue where they were at first, which truly is a great misfortune; for by this reason, the Sun draws no Exhalations or Vapours above the Moon; so that it appears she is a Body infinitely more hard, and solid than the Earth, whose subtile parts are easily separated from the rest, and mount upwards as soon as Heat puts them in Motion: But it must be a heap of Rock and Marble, where there is no Evaporation; besides, Exhalations are so natural and necessary where there is Water, that there can be no Water at all, where there is no Exhalation; and what sort of Inhabitants must those be, whose Countrey affords no Water, is all Rock, and produceth nothing? Very fine, *said she*, you have forgot since you assur'd me, we might from hence distinguish Seas in the Moon; nay, You or your Friends were Godfathers to some of 'em. Pray, what is become of your *Caspian* Sea, and your Black Lake? All Conjecture, Madam, *I reply'd*, tho' for your Ladyships sake, I am very sorry for it; for those dark places we took to be Seas, may perhaps be nothing but large Cavities; 'tis hard to guess aright at so great a distance. But will this suffice then, *said she*, to extirpate the People in the Moon? Not altogether, *I reply'd*, we will neither determine for, nor against them.

• • •

Coming out of the Moon, towards the Sun, we see *Venus*, which puts me again in mind of *Greenwich*. *Venus* turns upon her self, and round the Sun, as well as the Moon; they likewise discover by their Tellescopes, that *Venus* like the Moon, (if I may speak after the same manner) is sometimes new, sometimes full, and sometimes in the wain, according to the diverse situations she is in, in respect of the Earth. The Moon, to all appearance, is inhabited, why should not *Venus* be so too? You are so full of your Whys, and your Wherefores, *says she*, interrupting me, that I fancy you are sending Colonies to all the Planets. You may be certain, so I will, *I reply'd*, for I see no reason to the contrary; we find that all the Planets are of the same nature, all obscure Bodies, which receive no light but from the Sun, and then send it to one another; their motions are the same, so that hitherto they are alike; and yet if we are to believe that these vast Bodies are not inhabited, I think they were made but to little purpose; why should Nature be so partial as to except only the Earth? But let who will say the contrary, I must believe the Planets are peopled as well as the Earth. I find, *says the Countess*, with some concern, a Philosopher will never make a good Martyr, you can so quickly shift your Opinion, 'twas not many minutes since the Moon was a perfect Desart, now the rest of the Planets are inhabited. Why truly, Madam, *said I*, there is a time for all things, and your true Philosopher believes any thing, or nothing, as the Maggot bites. Had you taken me in the sceptical Vein, I would have as soon granted a Nation in a Mustard Ball, as a living Creature in the Moon; but the tide is turn'd, & all the Planets are Peopled like an Anthill; yet, Raillery apart, this is not so very improbable as you think it; for do you believe we discover, (as I may say) all the Inhabitants of the Earth? there be as many kinds of invisible as visible Creatures; we see from the Elephant to the very Hand-worm, beyond which our Sight fails us, & yet counting from that minute Creature, there are an infinity of lesser Animals, which were they perceptible, would be as little in comparison with a Mite, as a Mite is of an Ox.

• • •

THE FOURTH EVENING

. . . I find, *says the Countess*, it is easie enough to guess at the Inhabitants of *Venus;* they resemble what I have read of the *Moors* of *Granada*, who were a little black People, scorch'd with the Sun, witty, full of Fire, very Amorous, much inclin'd to Musick & Poetry, and ever inventing Masques & Turnaments in honour of their Mistresses. Pardon me, Madam, *said I*, you are little acquainted with the Planet; *Granada* in all its Glory, was a perfect *Greenland* to it; and your gallant *Moors*, in comparison with that People, were as stupid as so many *laplanders*.

But what do you think then of the Inhabitants of *Mercury?* They are yet nearer to the Sun, and are so full of Fire, that they are absolutely mad; I fancy, they have no memory at all, . . . that they make no reflections, and what they do is by sudden starts, and perfect hap-hazard; in short, *Mercury* is the Bedlam of the Universe; the Sun appears to them much greater than it does to us, because they are much nearer to it than we; it sends them so vast and strong a light, that the most glorious day here, would be no more with them than a declining twilight: I know not if they can distinguish Objects, but the heat to which they are accustom'd, is so excessive, that they would be starved with cold in the Torrid Zone; their year is but three Months, but we know not the exact length of their Day, because *Mercury* is so little, and so near the Sun; it is, (as it were) lost in his Rays, and is very hardly discover'd by the Astronomers; so that they cannot observe how it moves on its Centre, but because it is so little, fancy it compleats its motion in a little time; so that by consequence, the day there is very short, and the Sun appears to them like a vast fiery Furnace at a little distance, whose motion is prodigiously swift and rapid; and during their Night, *Venus* & the Earth (which must appear considerably big) give light to them; as for the other Planets which are beyond the Earth, towards the Firmament, they appear less to them in *Mercury*, than they do to us here, and they receive but little light from them, perhaps none at all; the fix'd Stars likewise seem less to them, and some of 'em totally disappear, which were I there, I should esteem a very great loss.

What signifies the loss of a few fix'd Stars, *says the Countess?* I pity 'em for the excessive heat they endure; let us give 'em some relief, and send *Mercury* a few of the refreshing

Showers they have sometimes four Months together in the hottest Countries during their greatest extremity. Your Fancy is good, Madam, *I reply'd,* but we will relieve 'em another way; In *China* there are Countries which are extreamly hot by their Situation: yet in *July* and *August* are so cold, that the Rivers are Frozen; the reason is, they are full of Salt-Petre, which being exhal'd in great abundance by the excessive heat of the Sun, makes a perfect Winter at Midsummer. We will fill the little Planet with Salt-petre, and let the Sun shine as hot as he pleases. And yet after all, who knows but the Inhabitants of *Mercury* may have no occasion either for Rain, or Salt-Petre. If it is a certain truth, that Nature never gives life to any Creature, but where that Creature may live; then thro' Custom, and ignorance of a better Life, those people may live happily.

• • •

Mars hath nothing curious that I know of, his Day is not quite an hour longer than ours, but his Year is twice as much as our Year; he is a little less than the Earth; and the Sun seems not altogether so large and so bright to him, as it appears to us; but let us leave *Mars,* he is not worth our stay: But what a pretty thing is *Jupiter,* with his four Moons, or Yeomen of the Guard; they are four little Planets that turn round him, as our Moon turns round us. But why, *says she, interrupting me,* must there be Planets to turn round other Planets, that are no better than themselves? I should think it would be more regular and uniform, that all the Planets, little and great, without any distinction, should have one and the same motion round the Sun.

Ah, Madam, *said I,* if you knew what were *Descartes's* Whirlpools or Vortex's, (whose Name is terrible, but their Idea pleasant) you would not talk as you do. Must my Head, *says she, smiling,* turn round to comprehend 'em, or must I become a perfect Fool to understand the misteries of Philosophy? Well, let the World say what it will, go on with your Whirlpools. I will, *said I,* and you shall see the Whirlpools are worthy of these transports: That then which we call a

Whirlpool or Vortex, is a Mass of Matter, whose parts are separated or detach'd one from another, yet have all one uniform motion, and at same time, every one is allow'd, or has a particular Motion of its own, provided follows the general Motion: Thus a Vortex of Wind, or Whirlwind, is an infinity of little particles of Air, which turn round all together, and involve what ever they meet with. You know the Planets are born up by the Celestial Matter, which is prodigiously subtile and active; so that this great Mass, or Ocean of Celestial Matter, which flows as far as from the Sun to the fix'd Stars, turns round, and bears the Planets along with it, making them all turn after the same manner round the Sun, who possesses the Centre, but in a longer or shorter time, according as they are farther or nearer in distance to it; there is nothing to the very Sun, which does not turn, but he turns on himself, because he is just in the middle of this Celestial Matter; and you must know by the way, that were the Earth in his place, it must turn on it self, as the Sun does. This is the great Vortex, of which the Sun is Lord; yet at the same time, the Planets make little particular Vortex's, in imitation of that of the Sun, each of 'em in turning round the Sun, doth at the same time turn round it self, and makes a certain quantity of Celestial Matter turn round it likewise, which is always prepar'd to follow the Motion the Planet gives it, provided it is not diverted from its general Motion; this then is the particular Vortex of the Planet, which pushes it as far as the strength of its Motion reaches; and if by chance, a lesser Planet falls into the Vortex of a greater Planet, it is immediately born away by the greater, and is indispensably forc'd to turn round it, tho' at the same time, the great Planet, the little Planet, and the Vortex which encloses 'em, all turn round the Sun: 'Twas thus at the beginning of the World, when we made the Moon follow us, because she was within reach of our Vortex, and therefore wholly at our dispose: *Jupiter* was stronger, or more fortunate than we, he had four little Planets in his neighbourhood, and he brought 'em all four under his subjection; and no doubt, we, tho' a principal Planet, had had the same Fate, had we been within the Sphere of his Activity; he is ninety times bigger than the Earth and would certainly have swallow'd us into his Vortex; we had then been no more than a Moon in his Family, when now we have one to wait on us; so that you see the advantage of Situation, decides often all our good Fortune. But pray, *says She*, who can assure us we shall still continue as we do now? if we should be such Fools as to go near *Jupiter*, or he so ambitious as to approach us, what will become of us? for if (as you say) the Celestial Matter is continually under this great Motion, it must needs agitate the Planets irregularly; sometimes drive 'em together, and sometimes separate 'em. Luck is all, *said I;* we may win as well as lose, and who knows, but we should bring *Mercury & Venus* under our Government; they are little Planets, and cannot resist us; but in this particular, Madam, we need not hope or fear; the Planets keep within their own bounds, and are oblig'd (as formerly the Kings of China were) not to undertake new Conquests. Have you not seen when you put Water and Oyl together, the Oyl swims a top; and if to these two Liquors, you add a very light Liquor, the Oyl bears it up, and it will not sink to the Water: But put an heavier Liquor, of a just weight, and it will pass through the Oyl, which is too weak to sustain it, and sink till it comes to the Water, which is strong enough to bear it up; so that in this Liquor, compos'd of two Liquors, which do not mingle, two Bodies of an unequal weight, will naturally assume two different Places; the one will never ascend, the other will never descend: Fancy then that the Celestial Matter which fills this great Vortex, hath several resting places, one by another, whose weight are different, like that of Oyl, Water, and other Liquors; the Planets too are of a different weight, and consequently, every Planet settles in that place which has a just strength to sustain and keep it equilibrate, so you see 'tis impossible it should, ever go beyond.

Would to God, *says the Countess*, our World were as well regulated, and every one among us knew their proper Place.

• • •

Pray tell me, if the Earth be so little in comparison with *Jupiter*, whether his Inhabitants do discover us? Indeed, *said I*, I believe not; for if we appear to him ninety times less than he appears to us; judge you if there be any possibility: Yet this we may reasonably conjecture, that there are Astronomers in *Jupiter*, that after they have made the most curious Telescopes, and taken the clearest Night for their observations, they may have discover'd a little Planet in the Heavens, which they never saw before; if they publish their discovery, most People know not what they mean, or laugh at 'em for Fools; nay, the Philosophers themselves will not believe 'em, for fear of destroying their own opinions; yet some few may be a little curious; they continue their observations, discover the little Planet again, and are now assur'd it is no Vision; then they conclude it hath a motion round the Sun, which it compleats in a year, and at last, (thanks to the Learned,) they know in *Jupiter* our Earth is a World, every body runs to see it at the end of the Telescope, tho' 'tis so little, 'tis hardly discover'd. It must be pleasant, *says she*, to see the Astronomers of both Planets, levelling their Tubes at one another, like two files of Musqueteers, and mutually asking, What World is that? What People inhabit it? Not so fast neither, *I reply'd*, for tho' they may from Jupiter discover our Earth, yet they may not know us; that is, they may not have the least suspicion it is inhabited; and should any one there chance to have such a fancy, he might be sufficiently ridicul'd, if not prosecuted for it; for my part, I believe they have work enough to make discoveries on their own Planet, not to trouble their Heads with ours; and had Sir *Francis Drake* and *Columbus* been in *Jupiter*, they might have had good employments; why, I warrant you, they have not yet discover'd the hundredth part of their Planet.

• • •

THE FIFTH EVENING

THE Countess was very impatient to know what would become of the fix'd Stars; are they inhabited, *says she*, as the Planets are, or are they not inhabited? What shall we do with 'em? You may soon guess, *said I;* the fix'd Stars cannot be less distant from the Earth than fifty millions of leagues; nay, if you anger an Astronomer, he will set 'em further. The distance from the Sun to the farthest Planet, is nothing in comparison of the distance from the Sun, or from the Earth, to the fix'd Stars, it is almost beyond Arithmetick. You see their light is bright and shining, and did they receive it from the Sun, it must needs be very weak after a passage of fifty millions of leagues; then judge how much it is wasted by reflection; for it comes back again as far to us; so that forwards and backwards, here are an hundred millions of leagues for it to pass; and it is impossible it should be so clear and strong as the light of a fix'd Star, which cannot but proceed from it self; so that in a word, all the fix'd Stars are so many Suns.

I perceive, *says the Countess,* where you would carry me; you are going to tell me that if the fix'd Stars are so many Suns, and our Sun the centre of a Vortex that turns round him, why may not every fix'd Star be the centre of a Vortex that turns round the fix'd Star? Our Sun enlightens the Planets; why may not every fix'd Star have Planets to which they give light? You have said it, *I reply'd,* and I will not contradict you.

You have made the Universe so large, *says she,* that I know not where I am, or what will become of me; what is it all to be divided into heaps confusedly, one among another? Is every Star the centre of a Vortex, as big as ours? Is that vast space which comprehends our Sun and Planets, but an inconsiderable part of the Universe? and are there as many such spaces, as there are fix'd Stars? I protest it is dreadful. Dreadful, Madam, *said I;* I think it very pleasant, when the Heavens were a little blue Arch, stuck with Stars, methought the Universe was too strait and close, I was almost stifled for want of Air; but now it is enlarg'd in height and breadth, and a thousand & a thousand Vortex's taken in; I begin to breath with more freedom, and think the Universe to be incomparably more magnificent than it was before. Nature hath spar'd no cost, even to profuseness, and nothing can be so glorious, as to see such a prodigious number of Vortex's, whose several centres are possess'd by a particular Sun, which makes the Planets turn round it. The Inhabitants of a Planet of one of these innumerable Vortex's, see on all sides those luminous centres of the Vortex, with which they are encompass'd; but perhaps they do not see the Planets, who receiving but a faint light from their Sun, cannot send it beyond their own World.

• • •

May not the Worlds, *reply'd the Countess,* notwithstanding this great resemblance between 'em, differ in a thousand other things; for tho' they may be alike in one particular, they may differ infinitely in others. It is certainly true, *said I;* but the difficulty is to know wherein they differ. One Vortex hath many Planets that turn round about its Sun; another Vortex hath but a few: In one Vortex, there are inferiour or less Planets, which turn about those that are greater; in another perhaps, there are no inferiour Planets; here, all the Planets are got round about their Sun, in form of a little Squadron; beyond which, is a great void space, which reacheth to the neighbouring Vortex's: In another place, the Planets take their course towards the out side of their Vortex, and leave the middle void. There may be Vortex's also quite void, without any Planets at all; others may have their Sun not exactly in their Centre; and that Sun may so move, as to carry its Planets along with it: others may have Planets which in regard of their Sun, ascend, and descend, according to the change of their Equilibration, which keeps them suspended. But I think I have said enough for a Man that was never out of his own Vortex.

• • •

We are arrived at the very roof and top of all the Heavens; and to tell you whether there be any Stars beyond it, you must have an abler Man than I am; you may place Worlds there, or no Worlds, as you please: 'Tis the Philosopher's Empire to describe those vast invisible Countries, which are, and are not, or are such as he pleases to make 'em: It is enough for me, to have carried your mind as far as you can see with your Eyes.

Well, *says the Countess,* I have now in my Head, the System of the Universe: How learned am I become? Indeed, Madam, *said I,* you are pretty knowing, and you are so with the advantage of believing, or not believing, any thing I have said: For all my pains, I only beg this favour, that when ever you see the Sun, the Heaven, or the Stars, you will think of me.

FINIS

CHRISTIAAN HUYGENS (1629–1695)

Perhaps never before, or ever since, has a scientist of the stature of Christiaan Huygens published an entire book devoted to the question of extraterrestrial life. Second only to Newton among the leading physicists of the second half of the seventeenth century and one of the most gifted observational astronomers of that period, this distinguished Dutch scientist is also known for his invention of the pendulum clock. Religiously, Huygens was a Protestant but with an attraction to both skeptical and stoic thought.[3] Huygens's treatise on extraterrestrials appeared posthumously in 1698 in Latin under the title *Κοσμοθεωρος, sive de terris coelestis earumque ornatu conjecturae*; it was translated and published in English in the same year as *The Celestial Worlds Discover'd: or, Conjectures Concerning the Inhabitants, Plants and Productions of the Worlds in the Planets.* Dutch, French, German, and Russian translations soon followed.

One of the least known but most important influences of Huygens's *Cosmotheoros* involved John Flamsteed, the director of England's Greenwich Observatory, who recommended the *Cosmotheoros* to the vicar of Greenwich, Archdeacon Thomas Plume (1630–1704). Plume's fascination with it led him to bequeath 1,902 pounds to Cambridge University to "erect an observatory and to maintain a professor of astronomy and experimental philosophy, and to buy or build a house with or near the same."[4] Thus came to be established the Plumian professorship of astronomy at Cambridge, a position held by some of the most distinguished astronomers of modern times. This is a striking example of how interest in extraterrestrials has affected the history of pure astronomy.

---

3. A. E. Bell, *Christiaan Huygens and the Development of Science in the Seventeenth Century* (London: Edward Arnold, 1947), 7, 94, and 202.

4. On this, see the article on Plume in the *Dictionary of National Biography* and J. Edleston, *Correspondence of Sir Isaac Newton and Professor Cotes* (London: F. Cass, 1969 reprinting of the 1850 original), lxxiv–lxxv.

Christiaan Huygens, *The Celestial Worlds Discover'd: or, Conjectures Concerning the Inhabitants, Plants and Productions of the Worlds in the Planets* (London, 1698), 1–11, 25–34, 36–39, 39–42, 43–44, 46–53, 56–58, 60–65, 73, 82–83, and 156. This selection is based in part on the selection from Huygens provided in Donald Goldsmith (ed.), *The Quest for Extraterrestrial Life: A Book of Readings* (Mill Valley, Calif.: University Science Books, 1980), 10–16.

A Man that is of *Copernicus*'s Opinion, that this Earth of ours is a Planet, carry'd round and enlighten'd by the Sun, like the rest of them, cannot but sometimes have a fancy, that it's not improbable that the rest of the Planets have their Dress and Furniture, nay and their Inhabitants too as well as this Earth of ours: Especially if he considers the later Discoveries made since *Copernicus*'s time of the Attendants of *Jupiter* and *Saturn*, and the Champain and hilly Countrys in the Moon, which are an Argument of a relation and kin between our Earth and them, as well as a proof of the Truth of that System. This has often been our talk, I remember, good Brother, over a large Telescope, when we have been viewing those Bodies, a study that your continual business and absence have interrupted for this many years. But we were always apt to conclude, that 'twas in vain to enquire after what Nature had been pleased to do there, seeing there was no likelihood of ever coming to an end of the Enquiry. Nor could I ever find that any Philosophers, those bold Heros, either antient or modern, ventur'd so far. At the very birth of Astronomy, when the Earth was first asserted to be Spherical, and to be surrounded with Air, even then there were some men so bold as to affirm, there were an innumerable company of Worlds in the Stars. But later Authors, such as Cardinal *Cusanus*, *Brunus*, *Kepler*, (and if we may believe him, *Tycho* was of that opinion too) have furnish'd the Planets with Inhabitants. Nay, *Cusanus* and *Brunus* have allow'd the Sun and fixed Stars theirs too. But this was the utmost of their boldness; nor has the ingenious French Author of the Dialogues about *the Plurality of Worlds* carry'd the business any farther. Only some of them have coined some pretty Fairy Stories of the Men in the Moon, just as probable as *Lucian*'s true History; among which I must count *Kepler*'s, which he has diverted us with in his Astronomical Dream. But a while ago thinking somewhat seriously of this matter (not that I count my self quicker sighted than those great Men, but that I had the happiness to live after most of them) methoughts the enquiry was not so impracticable, nor the way so stopt up with Difficulties, but that there was very good room left for probable Conjectures. As they came into my head, I clapt them down into common places, and shall now try to digest them into some tolerable Method for your better conception of them, and add somewhat of the Sun and Fixt Stars, and the Extent of that Universe of which our Earth is but an inconsiderable point. I know you have such an esteem and reverence for any thing that belongs to Heaven, that I perswade my self you will read what I have written without pain: I'm sure I writ it with a great deal of pleasure; but as often before, so now, I find the saying of *Archytas* true, even to the Letter, *That tho a Man were admitted into Heaven to view the wonderful Fabrick of the World, and the Beauty of the Stars, yet what would otherwise be Rapture and Extasie, would be but a melancholy Amazement if he had not a Friend to communicate it to*. I could wish indeed that all the World might not be my Judges, but that I might chuse my Readers, Men like you, not ignorant in Astronomy and true Philosophy; for with such I might promise my self a favourable hearing, and not need to make an Apology for daring to vent any thing new to the World. But because I am aware what other hands it's likely to fall into, and what a dreadful Sentence I may expect from those whose Ignorance or Zeal is too great, it may be worth the while to guard my self beforehand against the Assaults of those sort of People.

There's one sort who knowing nothing of Geometry or Mathematicks, will laugh at it as a whimsical and ridiculous undertaking. It's mere Conjuration to them to talk of measuring the Distance or Magnitude of the Stars: And for the Motion of the Earth, they count it, if not a false, at least a precarious Opinion; and no wonder then if they take what's built upon such a slippery Foundation for the Dreams of a fanciful

Head and a distemper'd Brain. What should we answer to these Men, but that their Ignorance is the cause of their Dislike, and that if they had more Sense they would have fewer Scruples? But few people having had an opportunity of prosecuting these Studies, either for want of Parts, Learning, or Leisure, we cannot blame their Ignorance; and if they resolve to find fault with us for spending time in such matters, because they do not understand the use of them, we must appeal to properer Judges.

The other sort, when they hear us talk of new Lands, and Animals endued with as much Reason as themselves, will be ready to fly out into religious Exclamations, that we set up our Conjectures against the Word of God, and broach Opinions directly opposite to Holy Writ. For we do not there read one word of the Production of such Creatures, no not so much as of their Existence; nay rather we read the quite contrary. For, That only mentions this Earth with its Animals and Plants, and Man the Lord of them; but as for Worlds in the Sky, 'tis wholly silent. Either these Men resolve not to understand, or they are very ignorant; For they have been answer'd so often, that I am almost asham'd to repeat it: That it's evident God had no design to make a particular Enumeration in the Holy Scriptures, of all the Works of his Creation. When therefore it is plain that under the general name of *Stars* or *Earth* are comprehended all the Heavenly Bodies, even the little Gentlemen round *Jupiter* and *Saturn*, why must all that multitude of Beings which the Almighty Creator has been pleased to place upon them, be excluded the Privilege, and not suffer'd to have a share in the Expression? And these Men themselves can't but know in what sense it is that all things are said to be made for the use of Man, not certainly for us to stare or peep through a Telescope at; for that's little better than nonsense. Since then the greatest part of God's Creation, that innumerable multitude of Stars, is plac'd out of the reach of any man's Eye; and many of them, it's likely, of the best Glasses, so that they don't seem to belong to us; is it such an unreasonable Opinion, that there are some reasonable Creatures who see and admire those glorious Bodies at a nearer distance?

But perhaps they'll say, it does not become us to be so curious and inquisitive in these things which the Supreme Creator seems to have kept for his own knowlege: For since he has not been pleased to make any farther Discovery or Revelation of them, it seems little better than presumption to make any inquiry into that which he has thought fit to hide. But these Gentlemen must be told, that they take too much upon themselves when they pretend to appoint how far and no farther Men shall go in their Searches, and to set bounds to other Mens Industry; just as if they had been of the Privy Council of Heaven: as if they knew the Marks that God has plac'd to Knowlege: or as if Men were able to pass those Marks. If our Forefathers had been at this rate scrupulous, we might have been ignorant still of the Magnitude and Figure of the Earth, or of such a place as *America*. The Moon might have shone with her own Light for all us, and we might have stood up to the ears in Water, like the *Indians* at every Eclipse: and a hundred other things brought to light by the late Discoveries in Astronomy had still been unknown to us. For what can a Man imagine more abstruse, or less likely to be known, than what is now as clear as the Sun? That vigorous Industry, and that piercing Wit were given Men to make advances in the search of Nature, and there's no reason to put any stop to such Enquiries. I must acknowlege still that what I here intend to treat of is not of that nature as to admit of a certain knowlege; I can't pretend to assert any thing as positively true (for that would be madness) but only to advance a probable guess, the truth of which every one is at his own liberty to examine. If any one therefore shall gravely tell me, that I have spent my time idly in a vain and fruitless enquiry after what by my own acknowlegement I can never come to be sure of; the answer is, that at this rate he would put down all Natural Philosophy as far as it concerns it self in searching into the Nature of things: In such noble and sublime Studies as these, 'tis a Glory to arrive at Probability, and the search it self rewards the pains. But there are many degrees of Probable, some nearer Truth than others, in the determining of which lies the chief exercise of our Judgment. But besides the Nobleness and Pleasure of the

Studies, may not we be so bold as to say, they are no small help to the advancement of Wisdom and Morality? so far are they from being of no use at all. For here we may mount from this dull Earth, and viewing it from on high, consider whether Nature has laid out all her cost and finery upon this small speck of Dirt. So, like Travellers into other distant Countrys, we shall be better able to judg of what's done at home, know how to make a true estimate of, and set its own value upon every thing. We shall be less apt to admire what this World calls great, shall nobly despise those Trifles the generality of Men set their Affections on, when we know that there are a multitude of such Earths inhabited and adorned as well as our own. And we shall worship and reverence that God the Maker of all these things; we shall admire and adore his Providence and wonderful Wisdom which is displayed and manifested all over the Universe, to the confusion of those who would have the Earth and all things formed by the shuffling Concourse of Atoms, or to be without beginning. But to come to our purpose.

In a section omitted here, Huygens explicates the Copernican system, noting that it constitutes one of the chief bases of his argument for extraterrestrial life. He also notes that he will make much use of the argument from analogy (11–25). He criticizes the Epicureans and Cartesians for viewing living beings as "haply jumbled together by a chance Motion of I don't know what little Particles" (21).

That the Planets are not without Water, is made not improbable by the late Observations: For about *Jupiter* are observ'd some spots of a darker hue than the rest of his Body, which by their continual change show themselves to be Clouds: For the Spots of *Jupiter* which belong to him, and never remove from him, are quite different from these, being sometimes for a long time not to be seen for these Clouds; and again, when these disappear, showing themselves. And at the going off of these Clouds, some spots have been taken notice of in him, much brighter than the rest of his Body, which remain'd but a little while, and then were hid from our sight.

These Monsieur *Cassini* thinks are only the Reflection from the Snow that covers the tops of the Hills in *Jupiter*: but I should rather think that it is only the colour of the Earth, which chances to be free from those Clouds that commonly darken it.

Mars too is found not to be without his dark spots, by means of which he has been observ'd to turn round his own Axis in 24 hours and 40 minutes; the length of his day: but whether he has Clouds or no, we have not had the same opportunity of observing as in *Jupiter*, as well because even when he is nearest the Earth, he appears to us much less than *Jupiter*; as that his Light not coming so long a Journey, is so brisk as to be an Impediment to exact Observations: And this Reason is as much stronger in *Venus* as its Light is. But since 'tis certain that the Earth and *Jupiter* have their Water and Clouds, there is no reason why the other Planets should be without them. I can't say that they are exactly of the same nature with our Water; but that they should be liquid their use requires, as their beauty does that they should be clear. For this Water of ours, in *Jupiter* or *Saturn*, would be frozen up instantly by reason of the vast distance of the Sun. Every Planet therefore must have its Waters of such a temper, as to be proportion'd to its heat: *Jupiter's* and *Saturn's* must be of such a nature as not to be liable to Frost; and *Venus's* and *Mercury's* of such, as not to be easily evaporated by the Sun. But in all of them, for a continual supply of Moisture, whatever Water is drawn up by the Heat of the Sun into Vapours, must necessarily return back again thither. And this it cannot do but in drops, which are caused as well there as with us, by their ascending into a higher and colder Region of the Air, out of that which, by reason of the Reflection of the Rays of the Sun from the Earth, is warmer and more temperate.

Here then we have found in these new Worlds Fields warm'd by the kindly Heat of the Sun, and water'd with fruitful Dews and Showers: That there must be Plants in them as well for Ornament as Use, we have shewn just now. And what Nourishment, what manner of Growth shall we allow them? Why, I think there can be no better, nay no other, than what we here experience; by having their Roots fastned into the Earth, and imbibing its nourishing Juices by their tender Fibres. And lest they should be only like so many bare Heaths, with nothing but creeping Shrubs and Bushes, we'll e'en send them some nobler and loftier Plants, Trees, or somewhat like them: These being the greatest, and, except Waters, the only Ornament that Nature has bestowed upon the Earth. For not to speak of those many uses that are made of their Wood, there's no one that is ignorant either of their Beauty or Pleasantness. Now what way can any one imagine for a continual Production and Succession of these Plants, but their bearing Seed? A Method so excellent that it's the only one that Nature has here made use of, and so wonderful, that it seems to be design'd not for this Earth alone. In fine, there's the same reason to think that this Method is observ'd in those distant Countries, as there was of its being follow'd in the remote Quarters of this same Earth.

'Tis much the same in Animals as 'tis in Plants, as to their manner of Nourishment, and Propagation of their kind. For since all the living Creatures of this Earth, whether Beasts, Birds, Fishes, Worms, or Insects, universally and inviolably follow the same constant and fixt Institution of Nature; all feed on Herbs, or Fruits, or the Flesh of other Animals that fed on them: since all Generation is perform'd by the impregnating of the Eggs, and the Copulation of Male and Female: Why may not the same rule be observ'd in the Planetary Worlds? For *'tis certain that the Herbs and Animals that are there would be lost, their whole Species destroy'd without some daily new Productions:* except there be no such thing there as Misfortune or Accident: except the Plants are not like other humid Bodies, but can bear Heat, Frost, and Age, without being dry'd up, kill'd or decay'd: except the Animals have Bodies as hard and durable as Marble; which I think are gross Absurdities. If we should invent some new way for their coming into the World, and make them drop like Soland Geese from Trees, how ridiculous would this be to any one that considers the vast difference between Wood and Flesh? Or suppose we should have new ones made every day out of some such fruitful Mud as that of *Nile*, who does not see how contrary this is to all that's reasonable? And that 'tis much more agreeable to the Wisdom of God, once for all to create of all sorts of Animals, and distribute them all over the Earth in such a wonderful and inconceivable way as he has, than to be continually obliged to new Productions out of the Earth? And what miserable, what helpless Creatures must these be, when there's no one that by his duty will be obliged, or by that strange natural fondness, which God has wisely made a necessary argument for all Animals to take care of their own, will be moved to assist, nurse or educate them?

As for what I have said concerning their Propagation, I cannot be so positive; but the other thing, namely, that they have Plants and Animals, I think I have fully proved. And by the same Argument, of their not being inferior to our Earth, they must have as great a variety of both as we have. What this is, will be best known to him that considers the different ways our Animals make use of in moving from one place to another. Which may be reduc'd, I think, to these; either that they walk upon two feet or four; or like Insects, upon six, nay sometimes hundreds; or that they fly in the Air bearing up, and wonderfully steering themselves with their Wings; or creep upon the Ground without feet; or by a violent Spring in their Bodies, or paddling with their feet, cut themselves a way in the Waters. I don't believe, nor can I conceive, that there should be any other way than these mention'd. The Animals then in the Planets must make use of one or more of these, like our amphibious Birds, which can swim in Water as well as walk on Land, or fly in the Air; or like our Crocodiles and Sea-Horses, must be Mongrels, between Land and Water. There can no other method be imagin'd but one of these. For where is it possible for Animals to live, except upon such a solid Body as our Earth, or a fluid one like the Water, or still a more fluid one than that, such as our Air is? The Air I confess may be much thicker and heavier than ours, and so, without any disadvantage to its Transparency, be fitter for the volatile Animals. There may be too many sorts of Fluids ranged over one another in rows as it were. The Sea perhaps may have such a fluid lying on it, which tho ten times lighter than Water, may be a hundred times heavier than Air; whose utmost Extent may not be so large as to cover the higher places of their Earth. But there's no reason to suspect or allow them this, since we have no such thing; and if we did, it would be of no advantage to them, for that the former ways of moving would not be hereby at all increas'd: But when we come to meddle with the Shape of these Creatures, and consider the incredible variety that is even in those of the different parts of this Earth, and that *America* has some which are no where else to be found, I must then confess that I think it beyond the force of Imagination to arrive at any knowlege in the matter, or reach probability concerning the figures of these Planetary Animals. Altho considering these ways of Motion we e'en now recounted, they may perhaps be no more different from ours than ours (those of ours I mean that are most unlike) are from one another.

In a section omitted here, Huygens describes the great variety of animals and plants seen on this Earth, which leads him to remark: "I make no doubt but that the Planetary Worlds have as wonderful a variety as we" (36).

But still the main and most diverting Point of the Enquiry is behind, which is the placing some Spectators in these new Discoveries, to enjoy these Creatures we have planted them with, and to admire their Beauty and Variety. And among all, that have never so slightly meddled with these matters, I don't find any that have scrupled to allow them their Inhabitants: not Men perhaps like ours, but some Creatures or other endued with Reason. For all this Furniture and Beauty the Planets are stock'd with seem to have been made in vain, without any design or end, unless there were some in them that might at the same time enjoy the Fruits, and adore the wise Creator of them. But this alone would be no prevailing Argument with me to allow them such Creatures. For what if we should say, that God made them for no other design, but that he himself might see (not as we do 'tis true; but that he that made the Eye sees, who can doubt?) and delight himself in the contemplation of them? For was not Man himself, and all that the whole World contains, made upon this very account? That which makes me of this opinion, that those Worlds are not without such a Creature endued with Reason, is, that otherwise our Earth would have too much the advantage of them, in being the only part of the Universe that could boast of such a Creature so far above, not only Plants and Trees, but all Animals whatsoever: a Creature that has a Divine somewhat within him, that knows, and understands, and remembers such an innumerable number of things; that deliberates, weighs and judges of the Truth: a Creature upon whose account, and for whose use, whatsoever the Earth brings forth seems to be provided. For every thing here he converts to his own ends. With the Trees, Stones, and Metals, he builds himself Houses: the Birds and Fishes he sustains himself with: and the Water and Winds he makes subservient to his Navigation; as he doth the sweet Smell and glorious Colours of the Flowers to his Delight. What can there be in the Planets that can make up for its Defects in the want of so noble an Animal? If we should allow *Jupiter* a greater variety of other Creatures, more Trees, Herbs and Metals, all these would not advantage or dignify that Planet so much as that one Animal doth ours by the admirable Productions of his penetrating Wit. If I am out in this, I do not know when to trust my Reason, and must allow my self to be but a poor Judg in the true estimate of things.

In a section omitted here, Huygens raises the question of the vices that beset inhabitants of the Earth, suggesting that these vices may have their uses.

For since it has so pleased God to order the Earth, and every thing in it as we see it is (for it's nonsense to say it happen'd against his Will or Knowlege) we must not think that those different Opinions, and that various multiplicity of Minds were plac'd in different Men to no end or purpose: but that this mixture of bad Men with good, and the Consequents of such a mixture, as Misfortunes, Wars, Afflictions, Poverty, and the like, were given us for this very good end, *viz.* the exercising our Wits, and sharpening our Inventions; by forcing us to provide for our own necessary defence against our Enemies. 'Tis to the fear of Poverty and Misery that we are beholden for all our Arts, and for that natural Knowlege which was the product of laborious Industry; and which makes us that we cannot but admire the Power and Wisdom of the Creator, which otherwise we might have passed by with the same indifference as Beasts. And if Men were to lead their whole Lives in an undisturb'd continual Peace, in no fear of Poverty, no danger of War, I don't doubt they would live little better than Brutes, without all knowlege or enjoyment of those Advantages that make our Lives pass on with pleasure and profit. We should want the wonderful Art of Writing, if its great use and necessity in Commerce and War had not forc'd out the Invention. 'Tis to these we owe our Art of Sailing, our Art of Sowing, and most of those Discoveries of which we are Masters; and almost all the Secrets in experimental Knowlege. So that those very things that make up their Indictment against Reason, are no small helps to its advancement and perfection. For those Virtues themselves, Fortitude and Constancy, would be of no use if there were no Dangers, no Adversity, no Afflictions for their exercise and trial.

If we should therefore imagine in the Planets some such reasonable Animal as Man is, adorn'd with the same Virtues, and infected with the same Vices, it would be so far from degrading or vilifying them, that while they want such a one, I must think them inferior to our Earth.

Well, but allowing these Planetarians some sort of Reason, must it needs be the same with ours? Why truly I think 'tis, and must be so; whether we consider it as applied to Justice and Morality, or exercised in the Principles and Foundations of Science. For Reason with us is that which gives us a true sense of Justice and Honesty, Praise, Kindness and Gratitude: 'tis that that teaches us to distinguish universally between Good and Bad; and renders us capable of Knowlege and Experience in it. And can there be any where a Reason contrary to this? or can what we call just and generous in *Jupiter* or *Mars* be thought unjust Villany? This is not at all, I don't say probable, but possible.

In a section omitted here, Huygens continues discussing similarities and differences between earthlings and extraterrestrials, remarking that "But allowing Morality and Passions with those Gentlemen to be somewhat different from ours, and supposing they may act by other principles in what belongs to Friendship, and Anger, Hatred, Honesty, Modesty, and Comeliness, yet still there would be no doubt, but that in the search after Truth, in judging of the Consequences of things, in reasoning, particularly in that sort which belongs to Magnitude or Quantity, about which their Geometry (if they have such a thing) is employ'd, there would be no doubt I say, but that their Reason here must be exactly the same, and go the same way to work with ours, and that what's true in one part will hold true over the whole Universe; so that all the difference must lie in the degrees of Knowlege, which will be proportional to the Genius and Capacity of the Inhabitants" (42–43).

But I perceive I am got a little too far: For till I have furnished them with Senses, neither will Life be any pleasure to them, nor Reason of any use. And I think it very probable, that all their Animals, as well their Beasts as rational Creatures, are like ours in all that relates to the Senses: For without the power of Seeing we should find it impossible for Animals to provide Food for themselves, or be forewarn'd of any approaching danger, so as to guard themselves from it. So that where-ever we plant any Animals, except we would have them lead the Life of Worms or Moles, we must allow them Sight; than which nothing can conduce more either to the preservation or pleasure of their Lives.

In a section omitted here, Huygens continues his discussion of sight, light, and the composition of eyes, from which he proceeds to generalize as follows.

It's likely then, and credible, that in these things the Planets have an exact correspondence with us, and that their Animals have the same Organs, and use the same way of sight that we do. Well then they have Eyes, and two at least we must grant them, otherwise they would not perceive some things close to them, and so could not avoid Mischiefs that take them on a blind side. And if we must allow them all Animals for the preservation of their Life, how much more must they that make more, and more noble uses of them, not be deprived of the Blessing of so advantageous Members? For by them we view the various Flowers, and the elegant Features of Beauty: with them we read, we write, we contemplate the Heavens and Stars, and measure their Distances, Magnitudes, and Journeys: which how far they are common to the Inhabitants of those Worlds with us, I shall strait examine. But first I shall enquire whether now we have given them one, we may not venture upon the other four Senses, to make them as good Men as our selves. And truly Hearing puts in hard, and almost perswades me to give it a share in the Animals of those new Countries. And 'tis of great consequence in defending us from sudden accidents; and, especially when Seeing is of no use to us, it supplies its place, and gives us seasonable warning of any imminent danger. Besides, we see many Animals call their fellow to them with their Voice, which Language may have more in it than we are aware of, tho we don't understand it. But if we do but consider the vast uses and necessary occasions of Speaking on the one side, and Hearing on the other, among those Creatures that make use of their Reason, it will scarce seem credible that two such useful, such excellent things were designed only for us. For how is it possible but that they that are without these, must be without many other Necessaries and Conveniencies of Life? Or what can they have to recompense this want? Then, if we go still farther, and do but meditate upon the neat and frugal Contrivance of Nature in making this same Air, by the drawing in of which we live, by whose Motion we sail, and by whose means Birds fly, for a conveyance of Sound to our ears; and this Sound for the conveyance of another man's Thoughts to our Minds: can we ever imagin that she has left those other Worlds destitute of so vast Advantages? That they don't want the means of them is certain, for their having Clouds in *Jupiter* puts it past doubt that they have Air too; that being mostly formed of the Particles of Water flying about, as the Clouds are of them gathered into small Drops. And another proof of it is, the necessity of breathing for the preservation of Life, a thing that seems to be as universal a Dictate of Nature, as feeding upon the Fruits of the Earth.

As for Feeling, it seems to be given upon necessity to all Creatures that are cover'd with a fine and sensible Skin, as a Caution against coming too near those things that may injure or incommode them: and without it they would be liable to continual Wounds, Blows and Bruises. Nature seems to have been so sensible of this, that she has not left the least place free from such a perception. Therefore it's probable that the Inhabitants of those Worlds are not without so necessary a Defence, and so fit a Preservative against Dangers and Mishaps.

And who is there that doth not see the inevitable necessity for all Creatures that live by feeding to have both Tast[e] and Smell, that they may distinguish those things that are good and nourishing, from those that are mischievous and harmful? If therefore we allow the Planetary Creatures to feed upon Herbs, Seeds, or Flesh, we must allow them a distinguishing Tast[e] and Smell too, that they may chuse or refuse any thing according as they find it likely to be advantageous or noxious to them.

I know that it hath been a question with many, whether there might not have been more Senses than these five. If we should allow this, it might nevertheless be reasonably doubted, whether the Senses of the Planetary Inhabitants are much different from ours. I must confess, I cannot deny but there might possibly have been more Senses; but when I consider the Uses of those we have, I cannot think but they would have been superfluous. The Eye was made to discern near and remote Objects, the Ear to give us notice of what our Eyes could not, either in the dark or behind our back: Then what neither the Eye nor the Ear could, the Nose was made (which in Dogs is wonderfully nice) to warn us of. And what escapes the notice of the other four Senses, we have Feeling to inform us of the too near approaches of, before it can do us any mischief. Thus has Nature so plentifully, so perfectly provided for the necessary preservation of her Creatures here, that I think she can give nothing more to those there, but what will be needless and superfluous. Yet the Senses were not wholly design'd for use: but Men from all, and all other Animals from some of them, reap Pleasure as well as Profit, as from the Tast[e] in delicious Meats; from the Smell in Flowers and Perfumes; from the Sight in the contemplation of beauteous Shapes and Colours; from the Hearing in the sweetness and harmony of Sounds; from the Feeling in Venery, unless you please to count that for a particular Sense by it self. Since it is thus, I think 'tis but reasonable to allow the Inhabitants of the Planets these same advantages that we have from them. For upon this consideration only, how much happier and easier a man's Life is render'd by the enjoyment of them, we must be obliged to grant them these Blessings, except we would engross every thing that is good to our selves, as if we were worthier and more deserving than any else. But moreover, that Pleasure which we perceive in eating or in copulation, seems to be a necessary and provident Command of Nature, whereby it tacitly compels us to the preservation and continuance of our Life and Kind. It is the same in Beasts. So that both for their happiness and preservation it's very probable the rest of the Planets are not without it. Certainly when I consider all these things, how great, noble, and useful they are; when I consider what an admirable Providence it is that there's such a Thing as Pleasure in the World, I can't but think that our Earth, the smallest part almost of the Universe, was never design'd to monopolize so great a Blessing. And thus much for those Pleasures which affect our bodily Senses, but have little or no relation to our Reason and Mind. But there are other Pleasures which Men enjoy, which their Soul only and Reason can relish: some airy and brisk, others grave and solid, and yet nevertheless Pleasures, as arising from the Satisfaction which we feel in Knowlege and Inventions, and searches after Truth, of which whether the Planetary Inhabitants are not partakers, we shall have an opportunity of enquiring by and by.

In a section omitted here, Huygens suggests that the four elements—earth, air, fire, and water—exist on other planets. He also discusses whether the inhabitants of large planets should be larger than humans, concluding in the negative.

There may arise another Question, whether there be in the Planets but one sort or more of rational Creatures possess'd of different degrees of Reason and Sense. There is something not unlike this to be observ'd among us. For to pass by those who have human Shape (altho some of them would very well bear that enquiry too) if we do but consider some sorts of Beasts, as the Dog, the Ape, the Beaver, the Elephant, nay some Birds and Bees, what Sense and Understanding they are masters of, we shall be forc'd to allow, that Man is not the only rational Animal. For we discover somewhat in them of Reason independent on, and prior to all teaching and practice.

But still no body can doubt, but that the Understanding and Reason of Man is to be prefer'd to theirs as being comprehensive of innumerable things, indued with an infinite memory of what's past, and capable of providing against what's to come. That there is some such rational Creatures in the other Planets, which is the Head and Sovereign of the rest, is very reasonable to believe: for otherwise, were many endued with the same Wisdom and Cunning, we should have them always doing mischief, always quarrelling and fighting one another for Empire and Sovereignty, a thing that we feel too much of where we have but one such Creature. But to let that pass, our next Enquiry shall be concerning those Animals in the Planets which are furnish'd with the greatest Reason, whether it's possible to know wherein they employ it, and whether they have made as great advances in Arts and Knowlege as we in our Planet. Which deserves most to be consider'd and examin'd of any thing belonging to their nature; and for the better performance of it we must take our rise somewhat higher, and nicely view the Lives and Studies of Men.

In a section omitted at this point, Huygens suggests that humans have no special preeminence over terrestrial animals in such abilities as protecting themselves or caring for their children.

What is it then after all that sets human Reason above all other, and makes us preferable to the rest of the Animal World? Nothing in my mind so much as the contemplation of the Works of God, and the study of Nature, and the improving those Sciences which may bring us to some knowlege in their Beauty and Variety. For without Knowlege what would be Contemplation? And what difference is there between a Man, who with a careless supine negligence views the Beauty and Use of the Sun, and the fine golden Furniture of the Heaven, and one who with a learned Niceness searches into their Courses; who understands wherein the Fixt Stars, as they are call'd, differ from the Planets, and what is the reason for the regular Vicissitude of the Seasons; who by sound reasoning can measure the magnitude and distance of the Sun and Planets? Or between such a one as admires perhaps the nimble Activity and strange Motions of some Animals, and one that knows their whole Structure, understands the whole Fabrick and Architecture of their Composition? If therefore the Principle we before laid down be true, that the other Planets are not inferior in dignity to ours, what follows but that they have Creatures not to stare and wonder at the Works of Nature only, but who employ their Reason in the examination and knowlege of them, and have made as great advances therein as we have? They do not only view the Stars, but they improve the Science of Astronomy: nor is there any thing can make us think this improbable, but that fond conceitedness of every thing that we call our own, and that pride that is too natural to us to be easily laid down. But I know some will say, we are a little too bold in these Assertions of the Planets, and that we mounted hither by many Probabilities, one of which, if it chance to be false, and contrary to our supposition, would, like a bad Foundation, ruin the whole Building, and make it fall to the ground. But I would have them to know, that all I have said of their Knowlege in Astronomy, has proofs enough, antecedent to those we now produc'd. For supposing the Earth, as we did, one of the Planets of equal dignity and honor with the rest, who would venture to say, that no where else were to be found any that enjoy'd the glorious sight of Nature's Opera? Or if there were any fellow-Spectators, yet we were the only ones that had dived deep into the secrets and knowlege of it? So then here's a proof not so far fetch'd for the Astronomy of the Planets, the same which we used for their having rational Creatures, and enjoying the other advantages we before talk'd of, which serves at the same time for the confirmation of our former Conjectures. But if Amazement and Fear at the Eclipses of the Moon and Sun gave the first occasion to the study of Astronomy, as they say it did, then it's almost impossible that *Jupiter* and *Saturn* should be without it; the Argument being of much greater force in them, by reason of the daily Eclipses of their Moons, and the frequent ones of the Sun to their Inhabitants. So that if a Person disinterested in his Judgment, and equally ignorant of the Affairs of all the Planets, were to give his Opinion in this matter, I don't doubt he would give the cause for Astronomy to those two Planets rather than us.

This supposition of their Knowlege and Use of Astronomy in the Planetary Worlds, will afford us many new Conjectures about their manner of life, and their state as to other things.

For, First: No Observations of the Stars that are necessary to the knowlege of their Motions, can be made without Instruments; nor can these be made without Metal, Wood, or some such solid Body. Here's a necessity of allowing them the Carpenters Tools, the Saw, the Ax, the Plane, the Mallet, the File: and the making of these requires the use of Iron, or some equally hard Metal. Again, these Instruments can't be without a Circle divided into equal Parts, or a streight Line into unequal. Here's a necessity for introducing Geometry and Arithmetick. Then the necessity in such Observations of marking down the Epochas or Accounts of Time, and of transmitting them to Posterity, will force us to grant them the Art of Writing; I won't say the same with ours which is commonly used, but I dare affirm not more ingenious or easy. For how much more ready and expeditious is our way, than by that multitude of Characters used in *China;* and how vastly preferable to Knots tied in Cords, or the Pictures in use among the barbarous People of *Mexico* and *Peru?* There's no Nation in the World but has some way or other of writing or marking down their Thoughts: So that it's no wonder if the Planetarians have been taught it by that great School-mistress

Necessity, and apply it to the study of Astronomy and other Sciences.

In a section omitted here, Huygens argues that planetarians have attained a level of sophistication in astronomy comparable to that of terrestrials, from which he infers that they must have the use of hands. He proceeds to note the advantages that hands confer in comparison with the proboscis of an elephant and the beak of a bird.

That they have Feet scarce any one can doubt, that does not consider what we said but just now of the different methods of Progression, which it's hard to imagin can be perform'd any other ways than what we there recounted. And of all those, there's none can agree so well with the state of the Planetarians, as that that we here make use of. Except (what is not very probable, if they live in Society, as I shall show they do) they have found out the art of flying in some of those Worlds.

In a section omitted here, Huygens continues his speculation about the nature of the intelligent extraterrestrials, suggesting that they will be upright in posture and must have organs adequate for the operation of instruments, but may differ greatly in shape. As he states: "For 'tis a very ridiculous opinion, that the common people have got among them, that it is impossible a rational Soul should dwell in any other shape than ours" (76). He also claims that they must live in societies and in houses, and probably, like us, are a blend of the good and the bad.

If their Globe is divided like ours, between Sea and Land, as it's evident it is (else whence could all those Vapours in *Jupiter* proceed?) we have great reason to allow them the Art of Navigation, and not proudly ingross so great, so useful a thing to our selves. Especially considering the great advantages *Jupiter* and *Saturn* have for sailing, in having so many Moons to direct their Course, by whose guidance they may attain easily to the Knowlege that we are not Masters of, of the Longitude of Places. And what a troop of other things follow from this allowance? If they have Ships, they must have Sails and Anchors, Ropes, Pullies, and Rudders, which are of particular use in directing a Ship's Course against the Wind, and in sailing different ways with the same Gale. And perhaps they may not be without the use of the Compass too, for the magnetical matter, which continually passes through the Pores of our Earth, is of such a nature, that it's very probable the Planets have something like it. But there's no doubt but that they must have the Mechanical Arts and Astronomy, without which Navigation can no more subsist, than they can without Geometry.

In a long section omitted here, Huygens discusses various topics.

- Regarding geometry, he states that it is "of such singular worth and dignity, so peculiarly imploys the Understanding, and gives it such a full comprehension and infallible certainty of Truth, as no other Knowlege can pretend to; it is moreover of such a nature, that its Principles and Foundations must be so immutably the same in all times and places, that we cannot without Injustice pretend to monopolize it, and rob the rest of the Universe of such an incomparable Study" (84).
- Regarding music: "It's the same with Musick as with Geometry, it's everywhere immutably the same, and always will be so" (86). His explication (86–91) of this claim makes it clear that by music he means not the various compositions in music, but rather the general principles that govern it.
- In the final portion of Part One, Huygens discusses the many advantages we get from such terrestrials sources as animals, plants, and metals. He also suggests that though planetarians may have different inventions from ours, their inventions will be just as useful as ours.

PART TWO

Early in Part Two of his book, Huygens, having made arguments for the habitability of the planets, remarks that "now that we have ventur'd to place Spectators in the Planets, let's take a Journey to each of them, and see what their Years, Days, and Astronomy are" (105–6).

- He notes that Jupiter and Saturn probably can observe each other, but not the other planets, the reason being that the orbits of Mercury, Venus, Earth, and Mars are so small and centered on the Sun that sight of those four planets will be lost in the Sun's glare to the inhabitants of Jupiter and Saturn (121–3).

- Turning to our Moon, Huygens reports that his observations indicate that it has no seas, rivers, or atmosphere, and he concludes against the existence of lunarians. Moreover, he expresses doubts about inhabitants of the moons of Jupiter and Saturn. His banishment of lunarians does not, however, preclude his describing how the heavens would look to them (128–38).
- Solarians fare no better than lunarians in Huygens's volume. He urges against their existence, the chief problem being the heat of the Sun. On the other hand, he notes that the Sun, even if without life, has an obvious purpose (141–44).
- Regarding the stars, Huygens argues against Kepler's notion that our Sun is larger than the other stars, stating: "But I must give my Vote . . . to have the Sun of the same nature with the Fix'd Stars. And this will give us a greater Idea of the world, than all those other Opinions. For then why may not every one of these stars or suns have as great a Retinue as our Sun, of Planets, with their Moons, to wait upon them?" (149).
- In explicating the last point, Huygens argues that the planets of other suns would be invisible to us because the stars are so extremely remote from us (149–50). This leads him into a famous attempt, important for pure astronomy, to estimate the distance of the stars.
- Huygens explains that his method was to place a screen in front of the Sun and to cut a tiny hole in the screen of sufficient size that the brightness of the hole would equal the brightness of the brightest star, Sirius. He finds that when he had in this way reduced the brightness of the Sun by 27,664 times, the brightness of the hole was equal to that of Sirius. From this he inferred that Sirius must be about 27,664 times further from us than our Sun, which certainly supported his claim for the remoteness of the stars. This in turn leads him to remark on the vastness of the number of stars.

Some of the Ancients, *and Jordanus Brunus* carry'd it further, in declaring the Number infinite: he would perswade us that he has prov'd it by many Arguments, tho in my opinion they are none of them conclusive. Not that I think the contrary can ever be made out. Indeed it seems to me certain, that the Universe is infinitely extended; but what God has bin pleas'd to place beyond the Region of the Stars, is as much above our Knowlege, as it is our Habitation.

In his concluding section (156–60), Huygens discusses the vortex cosmology worked out by Descartes, generally indicating his support for it.

It is interesting to compare and contrast these two highly gifted and widely read advocates of extraterrestrials. If the table below is correct, their differences were substantial.

Table 3. Comparing and contrasting Fontenelle and Huygens

Fontenelle	*Huygens*
Wrote his work when he was 29.	Wrote his book when he was near the end of his life; in fact, it appeared posthumously.
Although he did not attack religion, he was not friendly to it.	Showed himself quite friendly to theism. Mentioned God twelve times in our selection.
Stresses diversity among his extraterrestrials.	Stresses similarities among his extraterrestrials, e.g., they have the same math, virtues, senses.
Leaned toward life on the Moon.	Denied life to the Moon.
His gifts were chiefly literary.	His gifts were chiefly scientific.

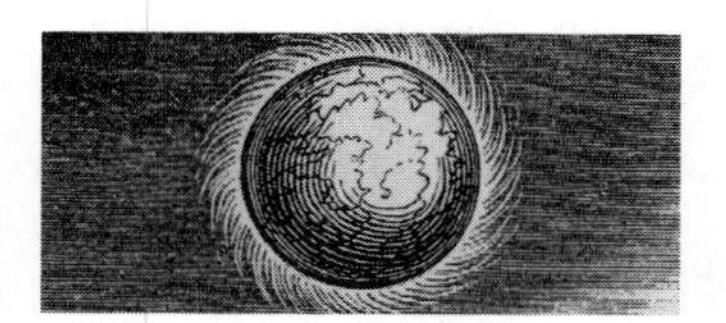

SIX

NEWTON, BENTLEY, AND DERHAM

ISAAC NEWTON (1642–1727)

Although Isaac Newton never published a sustained discussion of the question of extraterrestrial life, his published writings as well as his letters and manuscripts reveal his interest in this issue. Moreover, because of Newton's eminence as the most gifted scientist of the scientific revolution period, his contemporaries carefully scrutinized Newton's statements as they became known.

Newton's two most important books were his *Philosophiae naturalis principia mathematica* (*Mathematical Principles of Natural Philosophy*), which appeared in 1687, and his *Opticks* of 1704. Both contain relevant statements. It was in his *Principia* that Newton presented his theory of gravitational attraction, which had numerous implications concerning the question of life elsewhere in the universe. Newton, without directly mentioning this question, specified some of these in his first corollary to proposition 8 in book 3 of his *Principia.* Having developed a method by which the law of gravitation could be used to compare the masses of bodies that have satellites orbiting them, Newton applied this technique to compare the weights that an object (say, a human being) would have if placed on the four then-known solar system bodies that satisfied this condition. These four bodies were the Sun, Earth, Jupiter, and Saturn.

Isaac Newton, *Mathematical Principles of Natural Philosophy*, passage trans. William H. Donahue in Michael J. Crowe, *Mechanics from Aristotle to Einstein* (Santa Fe, N. Mex.: Green Lion Press, 2007), 191–93.

III.8 [BOOK 3, PROPOSITION 8] COROLLARY 1

Hence can be found and compared among each other the weights of bodies on different planets. For the weights of equal bodies revolving in circles about the planets are (by Book I Proposition 4 Corollary 2) as the diameters of the circles directly and the squares of the periodic times inversely; and the weights at the surfaces of the planets, or at any other distances from the planets you please, are greater or less (by this proposition) inversely in the duplicate ratio of the distances. Thus from the periodic times of Venus around the sun, 224 days and $16^3/_4$ hours, of the outermost satellite of Jupiter around Jupiter, 16 days and $16^8/_{15}$ hours, of the Huygenian satellite around Saturn, 15 days and $22^2/_3$ hours, and of the moon around the earth, 27 days 7 hours 43 minutes, compared with the mean distance of Venus from the sun and with the greatest heliocentric elongations of the outermost of Jupiter's satellites from the center of Jupiter, 8' 16", of the Huygenian satellite from the center of Saturn, 3' 4", and of the moon from the center of the earth, 10' 33", by entering into a computation I have found that the weights of equal bodies at equal distances from the center of the sun, Jupiter, Saturn, and earth, to the sun, Jupiter, Saturn, and earth, respectively, are as 1, $^1/_{1067}$, $^1/_{3021}$, and $^1/_{169282}$ respectively, and when the distances are increased or decreased, the weights are decreased or increased in the duplicate ratio. Thus the weights of equal bodies on the sun, Jupiter, Saturn, and earth, at distances of 10000, 997, 791, and 109 from their centers, and accordingly on their surfaces, will be as 10000, 943, 529, and 435, respectively. The magnitude of weights of bodies on the surface of the moon will be told in what follows.

Having shown that a 200 pound earthling, for example, would weigh more than twice that amount on Jupiter and weigh twenty-three times as much if transported to the Sun, Newton derived hardly less striking results when he employed his method to calculate the masses and densities of the Sun, Earth, Jupiter, and Saturn.

III.8 COROLLARY 2

The quantity of matter in the individual planets is also found. For the quantities of matter in the planets are as their forces at equal distances from their centers; that is, in the sun, Jupiter, Saturn, and the earth, they are as 1, $^1/_{1067}$, $^1/_{3021}$, and $^1/_{169282}$ respectively. If the sun's parallax be set at greater or less than 10" 30"', the quantity of matter in the earth will have to be increased or diminished in the triplicate ratio.

III.8 COROLLARY 3

The densities of the planets also become known. For, by Book I Prop. 72, at the surfaces of homogeneous spheres, the weights of equal and homogeneous bodies towards the spheres are as the diameters of the spheres; therefore, the densities of the homogeneous spheres are as those weights divided by the diameters of the spheres. Now the true diameters of the sun, Jupiter, Saturn, and earth were found to be to each other as 10,000, 997, 791, and 109, and the weights towards the same bodies are as 10,000, 943, 529, and 435, respectively, and therefore the densities are as 100, $94\,^1/_2$, 67, and 400. The density of the earth that results from this computation does not depend on the sun's parallax, but is determined by the moon's parallax, and accordingly is determined correctly here. The sun is therefore a little denser than Jupiter, and Jupiter a little denser than Saturn, and the earth is four times as dense as the sun. For the sun becomes rarefied through its tremendous heat. The moon, on the other hand, is denser than the earth, as will be made clear in what follows.

We can sum up Newton's findings in the following table:

Table 4. Newton's data comparing the Sun, Jupiter, Saturn, and Earth

	Sun	*Jupiter*	*Saturn*	*Earth*
Mass	1	1/1067	1/3021	1/169282
Density	100	94.5	67	400
Weight of person on	10000	943	529	435

One conclusion that could be drawn from this table is that although the Earth scarcely compares to the Sun, Jupiter, and Saturn in mass, it far surpasses each in density. Put differently, the fact that the Earth is six times more dense than Saturn eventually raised serious questions as to whether Saturn possesses a sufficiently solid surface to support inhabitants.

To the 1713 edition of his *Principia,* Newton added his famous "General Scholium" in which he discoursed on various subjects, including the relations of his system to theology. At one point, he stated:

> The six principal planets revolve about the sun in circles concentric upon the sun, in the same direction of motion, approximately in the same plane. Ten moons revolve about the earth, Jupiter, and Saturn, in concentric circles, in the same direction of motion, very nearly in the planes of the orbits of the planets. And all these regular motions do not have their origin from mechanical causes, inasmuch as comets are carried freely in highly eccentric orbits, and to all parts of the heavens. In this kind of motion, the comets pass through the orbits of the planets with greatest ease and swiftness, and at their aphelia, where they are slower and delay for a longer time, they are at the greatest distance from each other, so that they pull each other least. This most elegant arrangement of the sun, planets, and comets could not have arisen but by the plan and rule of an intelligent and powerful being. And if the fixed stars be centers of similar systems, all these, constructed by a similar plan, will be under the rule of One, especially because the light of the fixed stars is of the same nature as the light of the sun, and all the systems send light into all mutually. And so that the systems of the fixed stars should not fall into each other mutually, he will have placed this same immense distance among them.[1]

1. Isaac Newton, *Mathematical Principles of Natural Philosophy,* passage trans. William H. Donahue in Michael J. Crowe, *Mechanics from Aristotle to Einstein* (Santa Fe, N.Mex.: Green Lion Press, 2007), 191–93. For an explication of Newton's calculation, see Crowe, *Mechanics,* 190–91.

The fact that the last sentence in this paragraph, which has implications for the question of extraterrestrial life, is phrased in hypothetical terms suggests that Newton practiced caution in dealing with such a controversial topic. He showed similar caution when in a query added in 1706 to his *Opticks* he discussed the atomic theory of matter. To free it from the atheistic implications given it by the Epicureans, Newton was careful to begin his discussion by stating that God had created the atoms.[2] Later in his discussion, he suggested:

> And since Space is divisible *in infinitum*, and Matter is not necessarily in all places, it may be also allow'd that God is able to create Particles of Matter of several Sizes and Figures, and in several Proportions to Space, and perhaps of different Densities and Forces, and thereby to vary the Laws of Nature, and make Worlds of several sorts in several Parts of the Universe. At least I see nothing of Contradictions in all this.[3]

In 1725, two years before his death, Newton was visited by John Conduitt, who was the husband of Newton's niece. In the conversation that resulted, Newton revealed to Conduitt a number of his speculations about extraterrestrials and about other features of the universe and its history. Conduitt's record of that conversation was first published in 1806.

John Conduitt's Conversation with Newton as recorded in Edmund Turnor, *Collections for the History of the Town and Soke of Grantham: Containing Authentic Memoirs of Sir Isaac Newton* (London: William Miller, 1806), 172–73.

> I was on Sunday night, the 7th of March, 1724–5, at Kensington with Sir Isaac Newton, in his lodgings, just after he was come out of a fit of the gout, which he had had in both his feet, for the first time, in the eighty-third year of his age; he was better after it, and his head clearer, and memory stronger than I had known them for some time. He then repeated to me, by way of discourse, very distinctly, though rather in

2. Isaac Newton, *Opticks*, based on the 4th edition (New York: Dover, 1952), 400.
3. Newton, *Opticks*, 403–4.

answer to my queries, than in one continued narration, what he had often hinted to me before, viz. that it was his conjecture (he would affirm nothing) that there was a sort of revolution in the heavenly bodies; that the vapours and light emitted by the sun, which had their sediment as water, and other matter had, gathered themselves by degrees, into a body, and attracted more matter from the planets; and at last made a secondary planet (viz. one of those that go round another planet), and then by gathering to them and attracting more matter, became a primary planet; and then by increasing still, became a comet, which after certain revolutions, by coming nearer and nearer to the sun, had all its volatile parts condensed, and became a matter fit to recruit, and replenish the sun (which must waste by the constant heat and light it emitted), as a faggot would this fire, if put into it (we were sitting by a wood fire), and that that would probably be the effect of the comet of 1680 sooner or later, for by the observations made upon it, it appeared, before it came near the sun, with a tail only two or three degrees long, but by the heat it contracted in going so near the sun, it seemed to have a tail of thirty or forty degrees, when it went from it; that he could not say when this comet would drop into the sun; it might perhaps have five or six revolutions more first; but whenever it did, it would so much increase the heat of the sun, that this earth would be burnt, and no animals in it could live. That he took the three phenomena seen by Hipparchus, Tycho Brahe, and Kepler's disciples to have been of this kind, for he could not otherwise account for an extraordinary light as those were, appearing all at once among the fixed stars (all which he took to be suns enlightening other planets, as our sun does ours) as big as Mercury, or Venus, seems to us; and gradually diminishing for sixteen months, and then sinking into nothing. He seemed to doubt whether there were not intelligent beings superior to us, who superintended these revolutions of the heavenly bodies, by the direction of the Supreme Being. He appeared also to be very clearly of [the] opinion, that the inhabitants of this world were of a short date, and alledged as one reason for that opinion, that all arts, as letters, ships, printing, needle, &c.

> were discovered within the memory of history; which could not have happened, if the world had been eternal; and that there were visible marks of ruin upon it, which could not be effected by a flood only. When I asked him how this earth could have been repeopled, if ever it had undergone the same fate it was threatened with hereafter by the comet of 1680; he answered, that required the power of a creator. He said, he took all the planets to be composed of the same matter with this earth, viz. earth, water, stones, &c, but variously concocted.

RICHARD BENTLEY (1662–1742)

One of Newton's earliest and most gifted disciples was the Anglican divine Richard Bentley, who eventually became Master of Trinity College, Cambridge. Bentley's contemporaries recognized his abilities by selecting him, when he was only thirty, to inaugurate in 1692 a series of lectures funded by the famous chemist Robert Boyle, who specified that they should be devoted to "proving the Christian religion."[4] Before composing the last two of the eight lectures or sermons he delivered, Bentley wrote to Newton to ask whether Newton saw the scientific system presented in his *Principia* as being supportive of religion. Newton responded by writing four famous letters, which letters reveal many of his views about science and religion. Bentley in turn incorporated some of Newton's ideas into his seventh and eighth sermons.

One of the topics treated by Bentley in his seventh and especially his eighth sermon was the question of the plurality of worlds. In his seventh sermon, he made clear that he favored a positive response to that question by stating that "because every Fixt Star is supposed by Astronomers to be of the same Nature with our Sun; and each may very possibly have Planets about them, though by reason of their vast distance they be invisible to Us: we will assume this reasonable supposition."[5]

In his eighth sermon, Bentley dealt with the question of extraterrestrial life more fully. The context in which he did this deserves special attention. Bentley faced the problem of trying to prove God's beneficent design not in the limited world of

4. As quoted in R. C. Hebb, *Bentley* (New York: Harper, [1882, n.d.]), 20.

5. Richard Bentley, "A Confutation of Atheism," in *Isaac Newton's Papers and Letters on Natural Philosophy*, ed. I. Bernard Cohen (Cambridge, Mass.: Harvard University Press, 1958), 326.

medieval thought but rather in the far larger universe brought into view both by the Copernican revolution and by Newtonian mechanics. Moreover, Bentley's views are especially important because he was the first author to assess the question of extraterrestrial life from within the framework of the system that Newton had created five years earlier in his *Principia*.

Richard Bentley, "A Confutation of Atheism," in I. Bernard Cohen (ed.) *Isaac Newton's Papers and Letters on Natural Philosophy* (Cambridge, Mass.: Harvard University Press, 1958), 355–60.

Having abundantly proved in our Last Exercise, That the Frame of the present World could neither be made nor preserved without the *Power* of God; we shall now consider the structure and motions of our own System, if any characters of Divine *Wisdom* and *Goodness* may be discoverable by us. And even at the first and general View it very evidently appears to us (which is our FOURTH and Last Proposition,) That the Order and Beauty of the Systematical Parts of the World, the Discernible Ends and Final Causes of them, the τὸ βελτίον or Meliority above what was necessary to be, do evince by a reflex Argument, that it could not be produced by Mechanism or Chance, but by an Intelligent and Benign Agent, *that by his excellent Wisdom made the Heavens.*

But before we engage in this Disquisition, we must offer one necessary Caution; that we need not nor do not confine and determin the purposes of God in creating all Mundane Bodies, merely to Human Ends and Uses. Not that we believe it laborious and painfull to Omnipotence to create a World out of Nothing; or more laborious to create a great World, than a small one: so as we might think it disagreeable to the Majesty and Tranquillity of the Divine Nature to take so much pains for our sakes. Nor do we count it any absurdity, that such a vast and immense Universe should be made for the sole use of such mean and unworthy Creatures as the Children of Men. For if we consider the Dignity of an Intelligent Being, and put that in the scales against brute inanimate Matter; we may affirm, without over-valuing Humane Nature, that the Soul of one vertuous and religious Man is of greater worth and excellency than the Sun and his Planets and all the Starrs in the World. If therefore it could appear, that all the Mundane Bodies are some way conducible to the service of Man; if all were as beneficial to us, as the Polar Starrs were formerly for Navigation: as the Moon is for the flowing and ebbing of Tides, by which an inestimable advantage accrues to the World; for her officious Courtesy on dark Winter nightes, especially to the more Northern Nations, who in a continual Night it may be of a whole month are So pretty well accommodated by the Light of the Moon reflected from frozen Snow, that they do not much envy their *Antipodes* a month's presence of the Sun; if all the Heavenly Bodies were thus serviceable to us, who should not be backward to assign their usefulness to Mankind, as the sole end of their Creation. But we dare not undertake to shew, what advantage is brought to Us by those innumerable Starrs in the Galaxy and other parts of the Firmament, not discernible by naked eyes, and yet each many thousand times bigger than the whole body of the Earth: If you say, they beget in us a great Idea and Veneration of the mighty Author and Governor of such stupendious Bodies, and excite and elevate our minds to his adoration and praise; you say very truly and well. But would it not raise in us a higher apprehension of the infinite Majesty and boundless Beneficence of God, to suppose that those remote and vast Bodies were formed, not merely upon Our account to be peept as through an Optick Glass but for different ends and nobler purposes? And yet who will deny, but that there are great multitudes of lucid Starrs even beyond the reach of the best Telescopes; and that every visible Starr may have opake Planets revolve about them, which we cannot discover? Now if they were not created for Our sakes; it is certain and evident, that they were not made for their own. For Matter hath no life nor perception, is not conscious of its own existence, nor capable of happiness, nor gives the Sacrifice of Praise and Worship to the Author of its Being. It remains therefore, that all Bodies were formed for the sake of Intelligent Minds: and as the Earth was principally designed for the Being and Service and Contemplation of Men; why may not all other

Planets be created for the like Uses, each for their own Inhabitants which have Life and Understanding? If any man will indulge himself in this Speculation, he need not quarrel with revealed Religion upon such an account. The Holy Scriptures do not forbid him to suppose as great a Multitude of Systems and as much inhabited, as he pleases. 'Tis true; there is no mention in *Moses*'s Narrative of the Creation, of any People in other Planets. But it plainly appears, that the Sacred Historian doth only treat of the Origins of Terrestrial Animals: he hath given us no account of God's creating the Angels; and yet the same Author in the ensuing parts of the Pentateuch makes not unfrequent mention of the *Angels of God*. Neither need we be sollicitous about the condition of those Planetary People, nor raise frivolous Disputes, how far they may participate in the miseries of *Adam*'s Fall, or in the benefits of *Christ*'s Incarnation. As if, because they are supposed to be *Rational* they must needs be concluded to be Men? For what is Man? Not a *Reasonable Animal* merely, for that is not an adequate and distinguishing Definition; but a Rational Mind of such particular Faculties, united to an Organical Body of such a certain Structure and Form, in such peculiar Laws of Connexion between the Operations and Affections of the Mind and the Motions of the Body. Now God Almighty by the inexhausted fecundity of his creative Power may have made innumerable Orders and Classes of Rational Minds; some higher in natural perfections, others inferior to Human Souls. But a Mind of superior or meaner capacities than Human would constitute a different Species, though united to a Human Body in the same Laws of Connexion: and a Mind of Human Capacities would make another Species, if united to a different Body in different Laws of Connexion: For this Sympathetical Union of a Rational Soul with Matter, so as to produce a Vital communication between them, is an arbitrary institution of the Divine Wisdom: there is no reason nor foundation in the separate natures of either substance, why any Motion in the Body should produce any Sensation at all in the Soul; or why This motion should produce That particular Sensation, rather than any other. God therefore may have join'd Immaterial Souls, even of the same Class and Capacities in their separate State, to other kinds of Bodies and in other Laws of Union; and from those different Laws of Union there will arise quite different affections and natures and species of the compound Beings. So that we ought not upon any account to conclude, that if there be Rational Inhabitants in the *Moon* or *Mars* or any unknown Planets of other Systems, they must therefore have Human Nature, or be involved in the Circumstances of Our World. And thus much was necessary to be here inculcated (which will obviate and preclude the most considerable objections of our Adversaries) that we do not determin the Final Causes and Usefulness of the Systematical parts of the World, merely as they have respect to the Exigencies or Conveniencies of Human Life.

Later in his eighth sermon, Bentley discoursed on the very favorable location of our Earth in the solar system, taking this to be evidence of God's beneficent design. Were we located nearer the Sun, our weather would be too warm, whereas if more distant, we would suffer from excessive cold. This arrangement, Bentley suggested, indicates that our Earth was placed "where it is, by the Wisdom of some voluntary Agent; and not by the blind motions of Fortune or Fate."[6] This, however, creates a problem, which Bentley attempted to solve in a manner that jeopardized not only the notion that matter is identical throughout the universe, but also the even more fundamental notion that the laws of nature are uniform and identical everywhere, even in distant regions of space.

If any one shall think with himself, How then can any thing live in *Mercury* and *Saturn* in such intense degree of Heat and Cold? Let him only consider, that the Matter of each Planet may have a different density and texture and form, which will dispose and qualifie it to be acted on by greater and less degrees of Heat according to their several Situations; and that the Laws of Vegetation and Life and Sustenance and Propagation are the arbitrary pleasure of God, and may vary in all Planets according to the Divine Appointment and the Exigencies of Things, in manners incomprehensible to our Imagination.[7]

6. Bentley, "Confutation," 368.
7. Bentley, "Confutation," 368.

Although Newton's view of the relation of God to the material has often been described as voluntaristic, one suspects that Bentley had in this claim proceeded further than Newton would have favored.

WILLIAM DERHAM (1657–1735)

The historical relations between ideas of extraterrestrial life and religion are complex. Although it is certainly true that religious concerns led many persons in the sixteenth and seventeenth centuries to express reservations about accepting extraterrestrials, it is also true that various religious writers who saw ways in which extraterrestrials could be marshaled in support of religion encouraged their acceptance. Such an author was Rev. William Derham, an Oxford graduate and Anglican clergyman deeply interested in astronomy who in 1714 became chaplain to the future King George II. Derham was not only well connected in religious circles; he also knew Edmond Halley and Isaac Newton, whose system he supported. Moreover, Derham made various observations with telescopes that had belonged to Christiaan Huygens, whose *Cosmotheoros* may be the book that most influenced Derham's *Astro-Theology*. Despite his debt to Huygens, Derham did not hesitate to disagree with some of Huygens's ideas of extraterrestrial life. For example, Derham opposed Huygens's claim that our Moon lacks an atmosphere and also seas and oceans on its surface.

In 1714, Derham published his *Astro-Theology, or a Demonstration of the Being and Attributes of God from a Survey of the Heavens*, a book that by 1777 had gone through fourteen English and six German editions. This publication was part of a genre called natural theology or physico-theology, which was popular during that period not only in Britain but also on the Continent. The spirit animating such writings was to show God's existence, wisdom, and power by examining aspects of nature. The range of writings in natural theology was quite wide, extending even to books on ichthyo-theology and insecto-theology. Derham was a major contributor to this natural theology, having edited some of the relevant writings of John Ray and having himself published in 1713 his *Physico-Theology*, which showed God's beneficence as manifested in organic nature.

At least two features of Derham's *Astro-Theology* merit attention. First, his book nicely illustrates how an author could adapt ideas of extraterrestrial life to foster religious sentiments. In fact, Derham may be the first author who devoted a whole volume to showing how astronomical ideas could be used in support of religion. Second, Derham's volume is especially memorable as marking a stage through which astronomy passed in the period around 1700. Although much could be said in support of the idea that the history of astronomy can be divided into two periods, the period before Copernicus and the period after him, Derham suggested in the "Preliminary Discourse" to his *Astro-Theology* an interesting tri-partite periodization. After describing his first system, the Ptolemaic, and his second, the Copernican system, he stresses that astronomy has now moved beyond Copernicanism to a third system.

William Derham, *Astro-Theology, or a Demonstration of the Being and Attributes of God from a Survey of the Heavens*, 2nd ed. (London: W. Innys, 1715), xl–xlvii, xlix–l, and lvi.

OF THE NEW SYSTEME

And now I pass from the Second Systeme to the Third, which is calculated the *New Systeme*; which extends the Universe to a far more immense compass, than any of the other Systemes do, even to an indefinite Space; and replenishes it with a far more grand Retinue than ever was before ascribed unto it.

This *New Systeme* is the same with the *Copernican*, as to the Systeme of the Sun and its Planets; as may be seen by the Scheme of it in Fig. 3 [in Fig. 4 in this edition]. But then whereas the *Copernican* Hypothesis supposeth the Firmament of the Fixt Stars to be the Bounds of the Universe, and to be placed at equal Distance from its Center the Sun; the *New Systeme* supposeth there are many other Systemes of *Suns* and *Planets*, besides that in which we have our residence: namely, that every Fixt Star is a Sun, and encompassed with a Systeme of Planets, both Primary and Secondary, as well as ours.

These several Systemes of the Fixt Stars, as they are at a great and sufficient distance from the Sun and us; so they are imagined to be at as due and regular distances from one another. By which means it is, that those multitudes of Fixt Stars appear to us of different Magnitudes, the nearest to us large; those farther and farther less and less.

Of those Systemes of the Fixt Stars I have given a rude representation in Fig. 3 [in Fig. 4 in this edition] together with

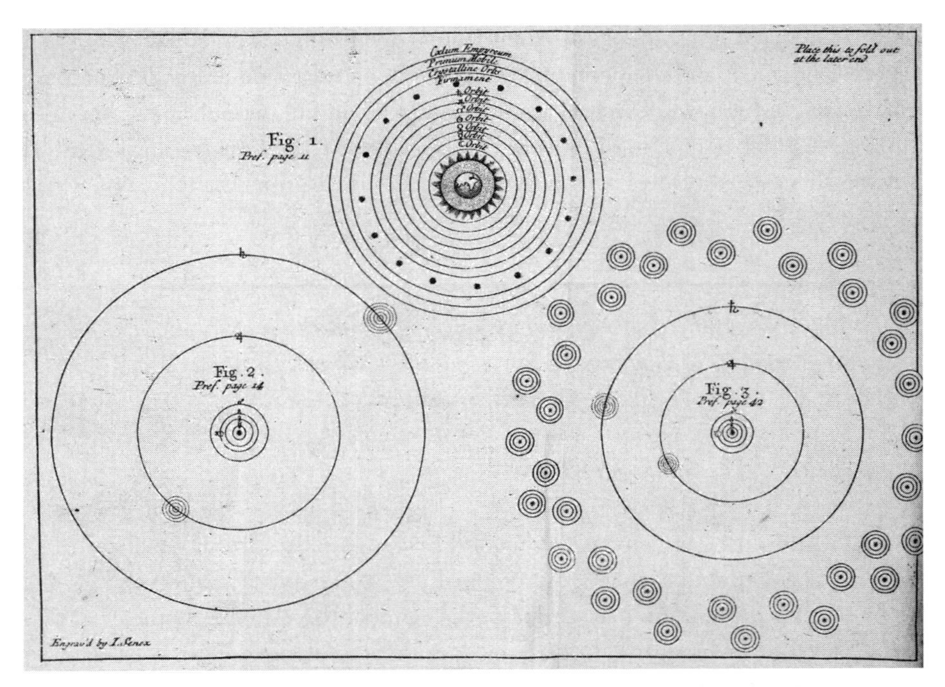

Fig. 4. Derham's diagram of his third system. (Reproduced from the original held by the Department of Special Collections of the University Libraries of Notre Dame)

that of the Sun; which may serve to give an unskilful Reader some conception of the state of the Universe, altho' there be but little likeness in it, for want of room to lay out all the several Systemes in due proportion; which is necessary to a true representation of the matter.

In this 3d Fig. the Fixt Stars with their Systemes (represented by little Circles about those Stars, which Circles signify the Orbits of their respective Planets) are placed without the limits of the Solar Systeme, and the Solar Systeme is set in the Center of the Universe, and figured as a more grand and magnificent part thereof. And so it may be looked upon by us, by reason of its proximity and relation to us. But whether it be really so, whether it be in the Center of the Universe, and whether among all the noble Train of Fixt Stars, there be no Systeme exceeding ours in its magnificent Retinue of Planets both Primary and Secondary, and other admirable Contrivances, is a difficulty as out of the reach of our Glasses, so consequently above our ability to fathom, although not at all improbable. But be the various Systemes of the Universe as they will as to their Dignity, it is sufficient that in all probability there are many of them, even as many as there are Fixt Stars, which are without number.

This *Systeme* of the *Universe,* as it is physically demonstrable, so is what, for the most part, I have followed in the ensuing Book, but not so rigorously and obstinately, as utterly to exclude or oppugne any other Systeme; because as the *Works of GOD* are truly great, and sufficiently manifest their excellence and magnificence in any Systeme; so I was willing to shew the same in such Systemes as I had occasion to speak of them in; because I would not offend, and consequently not bar the force of my arguments upon such Readers, as might happen to be wedded to the *Aristotelian Principles,* or prejudiced to the *Ptolemaick,* or any other *Systeme*: not that I had my self any doubts about this *New Systeme,* but think it to be far the most rational and probable of any, for these reasons.

1. Because it is far the most magnificent of any; and worthy of an infinite CREATOR: whose *Power* and *Wisdom* as they are without bounds and measure, so may in all probability exert themselves in the Creation of many Systemes, as well as one. And as Myriads of Systemes are more for the *Glory* of GOD, and more demonstrate his *Attributes* than one, so it is no less probable than possible, there may be many besides this which we have the privilege of living in. But it is very highly probable the matter is so, by reason[.]

2. We see it is really so, as far as it is possible it can be discerned by us, at such immense distances as those Systemes of the Fixt Stars are from us. Our Glasses are indeed too weak so to reach those Systemes, as to give us any assurance of our seeing the Planets themselves, that encompass any of the Fixt Stars. We cannot say we see them actually moving round their respective Suns or Stars, as I have made probable in Book 2. Chap. 2. As also that there are something very like unto Planets, which sometimes appear and disappear in the regions of the Fixt Stars; as I have shewn in my discourse of *New Stars,* Book 2. Chap. 3.

But besides what I have said there, I have this farther to add from some late observations I have made since my writing that part of my Book; and that is, That the *Galaxy* being well known to be the fertile place of *New Stars*, the region in which they commonly appear, I am much inclined to be of opinion, that the *Whiteness* there is not caused by the bare Light of the great number of Fixt Stars in that place, as hath commonly been thought, but partly by their Light, and partly (if not chiefly) by the Reflections of their Planets; which stop and reflect, intermix and blend the Light of their respective Stars or Suns, and so cause that Whiteness the *Galaxy* presents us with; which hath rather the colour of the Reflected Light of our Moon, then the Primary Light of our Sun.

• • •

Having thus represented the State of the *Universe* according to the *New Systeme* of it, the usual Question is, what is the use of so many Planets as we see about the Sun, and so many as are imagined to be about the Fixt Stars? To which the answer is, That they are *Worlds*, or places of *Habitation*, which is concluded from their being *habitable*, and well provided for Habitation. This is pretty manifest in our Solar Planets, from their being opake Bodies as our Earth is, consisting in all probability of Land and Waters, Hills and Valleys, having Atmospheres about them, Moons ministering unto them, and being enlightened, warmed and influenced by the Sun, whose yearly Visits they receive for Seasons, and frequent Returns for Days and Nights. All which particulars are fully treated of in the following Book, and need not therefore to be anticipated here.

After giving his arguments in opposition of Huygens's claim that the Moon lacks seas and an atmosphere and providing a map of the Moon, Derham turns directly to the question of the habitability of the Moon and planets.

And now considering how accomplished the Moon, and all the other Planets are for Habitation, how solemn an Apparatus is in them for this service: and considering also that these Accoutrements relate to their respective Planets only, and in all probability are of little or no use to our Earth; with great reason therefore the Maintainers of the new Systeme conclude those Planets, yea all the Planets of the Sun and of the Fixt Stars also to be *habitable Worlds*; places as accommodated for Habitation, so stocked with proper Inhabitants.

But now the next Question commonly put is what Creatures are they inhabited with? But this is a difficulty not to be resolved without a Revelation, or far better Instruments than the World hath hitherto been acquainted with.

PART TWO

THE EIGHTEENTH CENTURY

Fig. 5. William Herschel's 48-inch aperture reflecting telescope

SEVEN

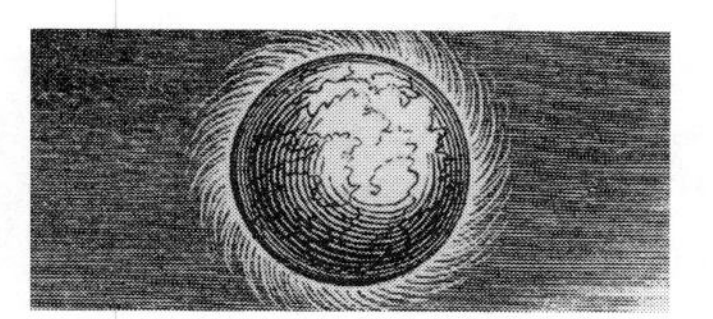

ASTRONOMERS AND EXTRATERRESTRIALS

The interactions between the science of astronomy and ideas of extraterrestrial life have been numerous and complex. It is certainly true that developments in astronomy have deeply influenced ideas of extraterrestrial life. But it is also true that biological, physical, metaphysical, and theological beliefs have had an impact on our conceptions of life beyond the Earth. It is less widely recognized that changing conceptions regarding extraterrestrial life have influenced astronomy itself. A central thesis of this chapter is the suggestion that in the period from 1750 to about 1800 aspects of the growing conviction that intelligent life is widespread in the universe had important effects on astronomy. In particular, a claim examined in this chapter is that astronomy in this period began to go through a major transformation, which transformation was significantly influenced by convictions concerning extraterrestrials. In 1750, astronomers were above all concerned with the solar system and especially with positional and mathematical astronomy. They saw the chief goals of astronomy as precise observation of the planetary motions and as development of methods for predicting and explaining these motions. The stars within this perspective provided a backdrop for observing and charting the motions of planets and comets, but beyond this the stellar realm was of minor interest. We shall see that in the period after 1750 interest in stars and in the dim nebulous patches scattered throughout the heavens gradually began to increase. Questions were raised about the arrangement and possible motions of the stars and also about their nature. To a first approximation, it is correct to say that between 1750 and 1800 astronomy began to be transformed from being the science of the solar system to being the science of the universe.

Who were the pioneers of the development of stellar astronomy, which is now such a central area in astronomy, and what new ideas facilitated the transformation? The consensus among experts is that the eighteenth-century pioneers of stellar astronomy were Thomas Wright, Immanuel Kant, Johann Lambert, and above all William Herschel. It is to these authors that credit is accorded for the idea that the appearance known as the Milky Way is due to the stars in our region forming a giant, disk-shaped structure. Similarly these authors explored the claim that the nebulous patches seen in the heavens are nothing less than island universes, massive and remote structures comparable to our Milky Way.

It is a striking fact, evident from the readings that follow, that these four authors were also deeply concerned about extraterrestrial life. This, it is suggested, is no accident. For example, they adopted the notion strongly associated with ideas of extraterrestrial life that stars, rather than being just points of light scattered on the inside of a giant sphere with the Sun at the center, are themselves suns, possibly surrounded by inhabited planets. This notion greatly increased the significance of the stars, raised questions about their arrangement and possible motions, and inspired efforts to discover their nature. In the case of William Herschel, the pioneer of stellar astronomy whose observational contribution was paramount, it can plausibly be suggested that his efforts to build the largest telescopes that humans had ever constructed were at first motivated not chiefly by a desire to observe the nebulae in which objects his contemporaries took little interest, but rather by a desire to achieve the most sought after telescopic detection of the modern times: the discovery of direct evidence of extraterrestrial intelligent life. Herschel not only believed this was possible, but compelling evidence (included in this chapter) shows that for a period he believed that he had succeeded in this quest.

THOMAS WRIGHT (1711–1786)

Among many books that played a role in drawing attention to the stellar region, one of the earliest was a volume published in 1750 by Thomas Wright. His *An Original Theory or New Hypothesis of the Universe* was possibly the first astronomy book to concentrate on the structure of the stellar region. Moreover, no earlier book had extended the claim for extraterrestrial intelligent life so vigorously into the stellar heavens. It is, however, a curious fact that what has given Wright's volume its celebrity has been the claim, now recognized as erroneous, that it contains the first published presentation of the disk theory of the Milky Way. Possibly the recognition that Wright fell short of attaining this idea, which is central to modern astronomy, makes it easier to understand his book for what it was: an elaborate attempt to delineate in both astronomical *and* religious terms the overall structure of the universe, its relationship to God, and, not least, the place of extraterrestrials within it.

Wright launched his foray into the stellar region as early as his title page, which featured a quotation from Edward Young's recently published poem *Night Thoughts*:

> One Sun by Day, by Night ten Thousand shine,
> And light us deep into the Deity.

Presenting his ideas in the form of nine letters, nearly all of which were filled with quotations and striking illustrations, Wright revealed as early as his preface that his universe contained a "vast Infinity of Worlds . . . crouded full of Beings, all tending through their various States to a final Perfection."[1] He devoted much of his first letter to quoting such authors as Bruno, Milton, Huygens, Newton, Pope, Derham, and Young, in support of the claims that stars are suns and that these suns are surrounded by inhabited planets. In his second letter, he attempted to justify the argument from analogy, which was central in many pluralist writings. Wright summarized the chief claim in this letter by asserting: "All then that I pretend to argue for, is a Universality of rational Creatures to people Infinity, or rather such Parts of the Creation, as from the Analogy and Nature of Things, we judge to be habitable Seats for Beings, not unlike the mortal human."[2] In his third letter, Wright presented a case for planets being comparable to our Earth and like it inhabited.

Wright devoted his "Letter the Fourth" to making his most sustained case for the idea that stars are suns. After citing various authorities on this issue, he stressed that if stars are comparable to our Sun, then they should, like our Sun, be encircled by planets. The religious associations Wright saw in these claims became evident when he unfolded his vision of a richly populated universe.

---

1. Thomas Wright, *An Original Theory or New Hypothesis of the Universe*, ed. Michael Hoskin (New York: Science History Publications, 1971), viii.

2. Wright, *Original Theory*, 12.

Thomas Wright, *An Original Theory or New Hypothesis of the Universe*, ed. by Michael Hoskin (New York: Science History Publications, 1971), 33–36.

May we not with the greatest Confidence imagine, that Nature as justly abhors a *Vacuum* in Place, as much as Virtue does in Time? Surely yes: And by supposing the Infinity of Stars, all centers to as many Systems of innumerable Worlds, all alike unknown to us; how naturally do we open to ourselves a vast Field of Probation, and an endless Scene of Hope to ground our Expectation of an *ever*-future Happiness upon, suitable to the native Dignity of the awful Mind, which made and comprehends it; and whose Works are all as the Business of an Eternity?

If the Stars were ordained merely for the Use of us, why so much Extravagance and Ostentation in their Number, Nature, and Make? For a much less Quantity, and smaller Bodies, placed nearer to us, would every Way answer the vain End we put them to; and besides, in all Things else, Nature is most frugal, and takes the nearest Way, through all her Works, to operate and effect the Will of God. It scarce can be reckoned more irrational, to suppose Animals with Eyes, destined to live in eternal Darkness, or without Eyes to live in perpetual Day, than to imagine Space illuminated, where there is nothing to be acted upon, or brought to Light; therefore we may justly suppose, that so many radiant Bodies were not created barely to enlighten an infinite Void, but to make their much more numerous Attendants visible; and instead of discovering a vast unbounded desolate Negation of Beings, display an infinite shapeless Universe, crowded with Myriads of glorious Worlds, all variously revolving round them. . . .

That the sidereal Planets are not visible to us, can be no Objection to their actual Existence, and being there, is plain from this; it is well known, that the Stars themselves, which are their Centeral, and only radiant Bodies, are little more to us at the Earth, than mathematical Points. How ridiculous then is it to expect, that any of their small opaque Attendance, should ever be perceived so far as the Earth by us; and besides, to show the impossibility of such a Discovery, we need only consider, what is, and what is not to be expected, or known in our own home System. All the Planets in this our sensible Region, every Astronomer knows, [are] far from being visible to one another, in every individual Sphere; for to an Eye at the Orb of *Saturn*, this Earth we live upon, which requires Years to circumscribe, and Ages to be made acquainted with, and is far from being yet all known, cannot possibly from the above Planet be seen: And further, since *Saturn* and *Jupiter*, two of the most material and considerable Globes we know of, except the Sun himself, are Bodies apparently of the same kind, and are observed to have each a Number of lesser Planets moving round them; why may we not expect with equal Certainty and Propriety, that all other Bodies, under the same Circumstances, are in like manner attended; that is, seeing the Sun is found to be the Center of a System of Bodies, all variously volving round him? where lies the Improbability of his fellow Luminaries, the Stars, being surrounded in like sort, with more or less of such Attendance.

After discussing a number of diagrams presented in support of the claims made in these paragraphs, Wright quoted Joseph Addison:

"When I consider . . . that infinite Host of Stars . . . with those innumerable Sets of Planets or Worlds, which were then moving round their respective Suns; when I still enlarge the Idea, and supposed another Heaven of Suns and Worlds rising still above this which we discovered; and these still enlightened by a superior Firmament of Luminaries, which are planted at so great a Distance, that they may appear to the Inhabitants of the former as the Stars do to us; in short, whilst I pursued this Thought, I could not but reflect on that little insignificant Figure which I myself bore amongst the Immensity of God's Works. . . ."

In his fifth letter, Wright discussed the Milky Way, nebulae, clusters of stars, and various types of stars. Regarding the Milky Way or what he called the "Via Lactea," he asserted:

> Now admitting the Breadth of the Via Lactea to be at a Mean but nine degrees, and supposing only twelve hundred Stars in every square Degree, there will be nearly in the whole orbicular Area 3,888,000 Stars, and all these in a very minute Portion of the great Expanse of Heaven. What! a vast Idea of endless Beings must this produce and generate in our Minds; and when we consider them all as flaming Suns, Progenitors, and Primum Mobiles of a still greater Number of peopled Worlds, what less than an Infinity can circumscribe them, less than an Eternity comprehend them, and less than Omnipotence produce and support them, and where can our Wonder cease?[3]

With no less enthusiasm for the principle of plenitude, Wright later in the letter stated:

> What an amazing Scene does this display to us! what inconceivable Vastness and Magnificence of Power does such a Frame unfold! Suns crowding upon Suns, to our weak Sense, indefinitely distant from each other; and Miriads of Miriads of Mansions, like our own, peopling Infinity, all subject to the same Creator's Will; a Universe of Worlds, all deck'd with Mountains, Lakes, and Seas, Herbs, Animals, and Rivers, Rocks, Caves, and Trees; and all the Produce of indulgent Wisdom, to chear Infinity with endless Beings, to whom his Omnipotence may give a variegated eternal Life.[4]

Wright, in his sixth letter, discussed stars and their motions. As part of his exposition, he presented a diagram of over a dozen stars, all of which he portrayed as surrounded by an array of planets. This and other portions of the book indicate that Wright did not hold that *some* stars are attended by satellites, but rather that *all* have such accompaniment.

In his seventh letter, Wright provided the insightful suggestion that were our solar system located within a vast planar array of stars, we would perceive this area much as we actually see the Milky Way. He moved on from this idea, however, to the incorrect conclusion that the appearance of the Milky Way is due to our being

3. Wright, *Original Theory,* 42–43.
4. Wright, *Original Theory,* 46.

located either in one portion of a vast spherical shell or within a giant ring comparable in form to Saturn's.

Employing a quantified approach in his eighth letter, Wright effectively sketched the vast measures of space and time needed to describe his universe. Rather than excluding religious concerns, such quantification in Wright's approach naturally led into it, for example, by showing how insignificant in size our Earth is within the vast universe. In the same spirit, the final section of his eighth letter contains a passage blending the quantified and the religious.

Wright, *Original Theory*, 76.

> Of these habitable Worlds, such as the Earth, all which we may suppose to be also of a terrestrial or terraqueous Nature, and filled with Beings of the human Species, subject to Mortality, it may not be amiss in this Place to compute how many may be conceived within our finite View every clear Starlight Night. It has already been made appear, that there cannot possibly be less than 10,000,000 Suns, or Stars, within the Radius of the visible Creation; and admitting them all to have each but an equal Number of primary Planets moving round them, it follows that there must be within the whole celestial Area 60,000,000 planetary Worlds like ours. And if to these we add those of the secondary Class, such as the Moon, which we may naturally suppose to attend particular primary ones, and every System more or less of them as well as here; such Satellites may amount in the Whole perhaps to 100,000,000, or more, in all together then we may safely reckon 170,000,000, and yet be much within Compass, exclusive of the Comets which I judge to be by far the most numerous Part of the Creation.
>
> In this great Celestial Creation, the Catastrophy of a World, such as ours, or even the total Dissolution of a System of Worlds, may possibly be no more to the great Author of Nature, than the most common Accident in Life with us, and in all Probability such final and general Doom-Days may be as frequent there, as even Birth-Days, or Mortality with us upon the Earth.

> This Idea has something so chearful in it, that I own I can never look upon the Stars without wondering why the whole World does not become Astronomers; and that Men endowed with Sense and Reason, should neglect a Science they are naturally so much interested in, and so capable of inlarging the Understanding, as next to a Demonstration, must convince them of their Immortality, and reconcile them to all those little Difficulties incident to human Nature, without the least Anxiety.

Religious concerns dominated in Wright's ninth and final letter. For example, he speculated that God resides at the center of his universe and suggested various locations for the Paradise promised by religion. Within this context, he assigned humans their significance: "Man may be of a very inferior Class; the second, third, or fourth perhaps, and scarce allowed to be a rational Creature."[5] Moreover, not content with limiting God's beneficence to the universe visible to us, Wright at the end of his ninth letter expanded yet again the region of God's creation by postulating (and indeed drawing; see Fig. 6) other universes, each with the "Eye of Providence" at the center. Ironically, this passage, so filled with sentiments of religion and seemingly so remote from twentieth-century astronomy, contains within its second paragraph what some scholars take to be the first statement of the "island universe" theory, that is, the correct and astronomically fundamental notion that some of the nebulous patches seen in the heavens are actually other universes comparable to our Milky Way.

Wright, *Original Theory*, 83–84.

> [S]ince as the Creation is, so is the Creator also magnified, we may conclude in Consequence of an Infinity, and an infinite all-active Power; that as the visible Creation is supposed to be full of siderial Systems and planetary Worlds, so on, in like similar Manner, the endless Immensity is an unlimited

5. Wright, *Original Theory*, 81.

Fig. 6. Thomas Wright's multiverses

Plenum of Creations not unlike the known Universe. See . . . *Plate* XXXII [Fig. 6], which represents their Sections, if all may be a proper Term for an infinite or indefinite Number, we may justly imagine to be the Object of that incomprehensible Being, which alone or in himself comprehends and constitutes supreme Perfection.

That this in all Probability may be the real Case, is in some Degree made evident by the many cloudy Spots, just perceivable by us, as far without our starry Regions, in which tho' visibly luminous Spaces, no one Star or particular constituent Body can possibly be distinguished; those in all likelyhood may be external Creation, bordering upon the known one, too remote for even our Telescopes to reach.

With the raptur'd Poet may we not justly say
O, what a Root! O what a Branch is here!
O what a Father! What a Family!
Worlds! Systems! and Creations! . . .

So many varied Seats where every Element may have its proper Beings and all adapted to partake of everything suited to their Natures, argue such Maturity of Wisdom, and the vast Production such mysterious Power; 'tis hardly possible for Mortals not to see divine Intelligence preside, and that every Being somewhere must be happy.

IMMANUEL KANT (1724–1804)

Ranked as the most creative philosopher of modern times, Kant is above all famous for the publication in 1781 of his *Kritik der reinen Vernunft* (*Critique of Pure Reason*). Kant was also significantly involved with astronomy and with the question of extraterrestrial intelligent life, an issue he engaged in many of his writings. In the history of astronomy, Kant is best known for having been the first to publish the claim that the Milky Way is a disk-shaped structure. Ironically, Kant believed that he had derived this idea from Thomas Wright's *Original Theory or New Hypothesis of the Universe*, which Kant knew only through a lengthy book review. Kant is moreover credited with being either the first or the second author (Wright is at times assigned priority) to publish the claim that various nebulous appearing patches in the heavens are in fact other universes comparable to our Milky Way. It was in 1755 that Kant published these two theories in his *Allgemeine Naturgeschichte und Theorie des Himmels* (*Universal Natural History and Theory of the Heavens*), which Kant published anonymously. Because this book presents what are now recognized as two keystone theories in stellar astronomy, the book has become a classic.

One feature of Kant's *Universal Natural History* that is not well known is the attention he bestowed on the issue of a plurality of worlds. Kant devoted the third of the three parts of his book precisely to this subject. References to it also appear elsewhere in the book. For example, in the second part Kant sets out a vigorous formulation of the principle of plenitude:

> Now, it would be senseless to set Godhead in motion with an infinitely small part of his creative ability and to imagine his infinite force, the wealth of a true inexhaustibility of nature and worlds to be inactive and locked up in an eternal absence of exercise. Is it not much more proper, or to say better, is it not necessary to represent the very essence of creation as it ought to be, to be a witness of that power which can be measured by no yardstick? For this reason the field of the manifestation of divine attributes is just as infinite as these themselves are.[6]

Kant employs this principle to support his claim that "the cosmic space will be enlivened by worlds without number and without end."[7] Moreover, it seems plausible that Kant's interest in stellar astronomy (an area of astronomy far from the mainstream of mid-eighteenth-century astronomy) arose from his conviction that stars are suns surrounded by planets inhabited by intelligent beings. Support for this suggestion comes from the fact that Wright, Lambert, and William Herschel (all recognized along with Kant as *the* pioneers of stellar astronomy) shared this conviction with Kant.

One reason why the third part of Kant's book is not widely known is that Kant himself removed the third part (along with some other sections) from it when in 1791 the book was republished. Moreover, when in 1900 William Hastie published the first English translation of Kant's book, Hastie excised the third part. Nor was this part restored when between 1968 and 1970 three different editions of the Hastie translation were published. English readers first gained access to the entire book

---

6. Immanuel Kant, *Universal Natural History and Theory of the Heavens*, trans. Stanley L. Jaki (Edinburgh: Scottish Academic Press, 1981), 151.

7. Kant, *Universal Natural History*, 152.

only in 1981 when S. L. Jaki brought out a newly translated and extensively referenced translation of the entire work. The key portions of the third part are provided below.

Immanuel Kant, *Universal Natural History and Theory of the Heavens*, trans. Stanley L. Jaki (Edinburgh: Scottish Academic Press, 1981), 183–96.

THIRD PART,
which contains an essay on a comparison, based on the analogies of nature, between the inhabitants of the various planets. . . .

Since I hold that the character of philosophy is dishonored when one uses it to assert with levity [and] with some appearance [of plausibility][8] free flights of fancy, if [unless] one also declares that all this is merely to entertain; therefore, I will in the present essay adduce no other proposition than such which can truly contribute to the enlargement of our knowledge and the probability of which is so well grounded that one can hardly refuse to let them be valid.

It may seem though that in this kind of project, the freedom has no proper limits to poetize and that one, in judging the characteristics of the inhabitants of distant worlds, can with greater abandon give free rein to phantasy than would a painter in depicting the plants or animals of undiscovered lands, and that such thoughts can neither be well proven or refuted: still one must admit that the distances of celestial bodies [planets] from the sun embody certain relationships, which in turn entail a decisive influence on the various characteristics of thinking natures [beings] that are found there; whose manner of operating and feeling is bound to the condition of the material with which they are connected and [also]

8. [The square brackets in this selection are found in the Jaki translation. MJC]

depends on the measure [intensity] of impressions which the [external] world evokes in them according to the properties of the relation of their habitat to [the sun,] the center of attraction and heat.

I am of the opinion that it is not even necessary to assert that all planets must be inhabited, although it would be sheer madness to deny this in respect to all, or even to most of them. In the richness of nature, where worlds and [world-] systems are but specks of sun-dust in respect to the whole of creation, there may very well exist barren and uninhabited regions that are not useful in the slightest for the purpose of nature, namely, for [making possible her] contemplation by intelligent beings. It would have to be admitted, even if one wanted to consider things on the basis of God's wisdom, that sandy and uninhabited deserts cover large tracts of the earth's surface and that there are in the ocean abandoned islands on which no man can be found. At any rate, a planet is much smaller in respect to the whole of creation than a desert or island in respect to the earth's surface.

It may well be that not all celestial bodies are yet fully developed; hundreds or perhaps thousands of years are needed before a great celestial body obtains a firm state of its material. Jupiter seems still to be in that phase. The notable changes in its form at different times have long ago made astronomers suspect that it must undergo great convulsions and that it is not at all so quiet on its surface, as [this] should be [with] a habitable planet. If it has no inhabitant, or if it never should have one would this not be still an infinitely smaller waste of nature in respect to the immensity of the whole creation? And would it not be a sign of nature's poverty much rather than an evidence of her abundance, if she had to display with diligence all her richness at every point of space?

But it is more satisfactory to imagine that although Jupiter is not inhabited, it will be at a time when the period of its development is completed. Perhaps our earth had been around a thousand or even more years before it found itself in the condition to support men, animals, and plants. It does not disrupt the purposefulness of a planet's existence that it should reach such [stage of] perfection only a few thousand years later. Precisely, because of this a planet will in the future stay longer in its state of perfection once it has reached it; for there is a certain law in nature: everything that has a beginning steadily approaches its decline and is all the closer to it, the farther it gets from its starting point.

One cannot indeed help agreeing with the satirical portrayal by that witty author from The Hague who, after listing the general news from the r[egister] of sciences, knew how to present the humorous side of the idea about all celestial bodies being necessarily inhabited: "Those creatures," says he, "which live in the forests on the head of a beggar, had long since considered their location as an immense ball, and themselves as the masterpiece of creation, when one of them, endowed by Heaven with a more refined spirit, a small Fontenelle of his species, unexpectedly became familiar with the head of a nobleman. Immediately he called together all the witty heads of his quarters and told them with excitement: 'We are not the only living beings in nature; see, there, that new land, more lice live there.'" When the unfolding of this conclusion provokes a laughter, it does not happen because it departs, to wit, very far from man's ways, but because the same error, which with man has a similar cause for its basis, seems to deserve more excuse with these [ways of his].

Let us judge without prejudice. This insect, which both in respect to its way of life and insignificance well expresses the condition of most men, can with good reason be used for such a comparison. Because, according to its belief, nature is infinitely well adapted to its existence, the insect holds as irrelevant the entire remainder of nature which does not imply a direct reference to its species as the center of her aims. Man, standing immensely removed from the uppermost rank of beings, is indeed bold to flatter himself in a similar delusion about the necessity of his own existence. [But] the infinity of nature includes within herself with the same necessity all beings which display her overwhelming richness. From the highest class of thinking beings to the most abject insect, no member [of that chain] is indifferent to nature; and nothing can be missing [from it] without breaking up the beauty of the whole, which consists in interconnectedness. There everything is determined by universal laws which nature makes operative through the connection of their originally implanted forces. Because in her procedure nature displays only aptitude and order, no particular purpose should disturb and interrupt her course. In its first stage the formation of a planet was but an infinitely small effect of nature's fruitfulness; and it would now be senseless that her well-established laws should be subservient to the particular aims of that atom [which man is]. If the condition of a celestial body sets obstacles to its being inhabited, then it will be uninhabited, although in and for itself it would be more beautiful that it should have inhabitants of its own. The splendor of creation loses nothing thereby: for the infinite is among all magnitudes the one which by the subtraction of a finite part is not diminished. One would [however] imply this by complaining that the space between Jupiter and Mars is unnecessarily empty, and that there are comets which are not populated. In fact, that insect may appear to us as insignificant as possible, still nature is more interested in maintaining its whole species than in supporting a small number of prominent creatures of which an infinite number would remain, even though an entire region or locality would be berobbed of them. Because nature is inexhaustible in producing both [the highest and the lowest], one sees both left mercilessly, in their preservation and decay, to [the course of] universal laws. Did the owner of those populated forests on the beggar's head ever make greater devastations among the species of that colony [of lice], than did the son of Philip in the species of his fellow citizens when his evil genius put it in his head that the world has been produced only for his sake?

At any rate, most planets are certainly inhabited, and those that are not, will be one day. What relationship will then be produced among the different kinds of those inhabitants through the relation of their place[s] in the world-edifice to the center, out of which diffuses the heat which keeps all alive? For it is certain that this heat produces specific relationships in the materials of those celestial bodies in proportion to their distance [from the sun].

In a section omitted here, Kant discusses how humanity's place in the solar system affects the way the body functions, and in turn how this affects the mind and the ability to think.

It becomes evidently clear from all this that the forces of the human soul become hemmed in and impeded by the obstacles of a crude matter to which they are most intimately bound; but is even more noteworthy that this specific condition of the stuff has a fundamental relation to the degree of influence by which the sun in the measure of its distance enlivens them and renders them adapted to the maintenance of animal economy [regimen]. This necessary relation to the fire, which spreads out from the center of the world-system to keep matter in the necessary [degree of] excitation, is the basis of an analogy which will be firmly stated in respect to the different inhabitants of planets; and in virtue of that relationship each and any class of theirs is tied through the necessity [necessary structure] of their nature to the place which has been assigned to it [that nature] in the universe.

The inhabitants of the earth and Venus cannot exchange their habitats without mutual destruction. The former, whose [bodily] building stuff is proportioned to the measure of heat of his distance [from the sun] and is therefore too light and volatile for a greater heat, would in a hotter sphere suffer enormous upheavals and a collapse of his nature, which would arise from the dissipation and evaporation of his fluids and from the violent tension of his elastic fibers; the latter, whose grosser build and the sluggishness of the elements of his structure needed a greater influence of the sun, would in a cooler celestial region grow numb and perish in lifelessness. By the same token, much lighter and more volatile ought to be the matter of which the body of Jupiter's inhabitant is composed, so that the weak excitation which the sun can produce at that distance may move those [bodily] machines as powerfully as it performs this [function] in the lower regions [closer to the sun], and thus all would be summed up in a general notion: *The stuff, out of which the inhabitants of different planets as well as the animals and plants on them, are built, should in general be all the lighter and of finer kind, and the elasticity of the fibers together with the principal disposition of their build should be all the more perfect, the farther they stand from the sun.*

This relationship is so natural and well grounded that not only the motivations of final purpose, which in natural science are generally looked upon as only weak reasons, lead to it, but also the proportion of the specific conditions of materials composing the planets (which are established from Newton's calculations as well as from the foundations of cosmogony) confirm the same [relationship] according to which the stuff of which the celestial bodies are built is always of a lighter kind in those that are more distant than in those which are closer [to the sun], which [circumstance] should entail a similar relationship in the creatures that produce and sustain themselves on them.

We have established a comparison between the condition[s] of the material with which rational creatures on the planets are essentially united: and it may also easily be seen after the introduction of this consideration that these relations would entail a consequence also in respect of their spiritual faculties. If, accordingly, these spiritual faculties have a necessary dependence on the stuff of the [bodily] machines which they inhabit, then we can conclude with more than probable confidence: *That the excellence of thinking natures, the promptness in their reflections, the clarity and vivacity of the notions that come to them through external impression, together with the[ir] ability to put them together, finally also the skill in their actual use, in short, the whole range of their perfection, stand[s] under a certain rule, according to which these [natures] become more excellent and perfect in proportion to the distance of their habitats from the sun.*

Since this relationship has a measure of credibility, which is not far from demonstrated certainty, we find an open field for pleasant speculations that stem from the comparison of the characteristics of these various [planetary] inhabitants. Human nature, which on the ladder of beings occupies exactly the middle rung, sees itself in the middle between the two extreme limits of perfection, from which both ends it is equally distant. When the consideration of the most elevated classes of rational creatures, which inhabit Jupiter or Saturn, hurts its pride and humiliates it through the knowledge of its lowliness, then a look at the lower rungs would bring it satisfaction and peace, [because] those on the planets Venus and Mercury are lowered far beneath the perfection of human nature. What an outlook worthy of wonderment! From one side we saw thinking creatures among whom a man from Greenland or a Hottentot would be a Newton, and on the other side some others who would admire him as [if he were] an ape:

> When recently the superior sages saw
> What not long ago truly wonderfully
> A mortal among us has done,
> And how he unfolds nature's law, they wondered
> That it was possible for such a thing
> To happen through an earthly creature,
> *And looked upon our Newton as we look upon an ape.*
>
> Pope[9]

What advances of knowledge should not be achieved by the insight of those happy beings of the uppermost spheres of the heavens! What beautiful consequences would not this brightness of insights have on their ethical disposition! The insights of intellect, when they possess the proper degree of perfection and clarity, have much more vivid stirrings than do sensory allurements, and are able to overcome these and rule them victoriously. With what majesty would not God, who depicts himself in all creatures, depict himself in these thinking natures, which as an ocean undisturbed by the storms of passions would quietly receive and reflect his image! We do not wish to stretch such speculations beyond the limits prescribed

9. [These verses are Jaki's translation into English of the German translation made by B. H. Brockes of part of Alexander Pope's *Essay on Man.* Pope's lines in their original English are: "Superior beings when of late they saw / A mortal Man unfold all Nature's law, / Admired such wisdom in an earthly shape / And showed a NEWTON as we show an Ape." MJC]

to a physical treatise; we merely wish to recall once more the analogy set forth above: *That the perfection of the spiritual world, just as well as the material [world] in [the realm of] the planets from Mercury to Saturn, or perhaps even beyond (insofar as there are still other planets), increases and progresses in a straight-forward gradation according to the measure of their distance from the sun.*

In a section omitted here, Kant discusses potential objections to the idea of those farther from the Sun being more perfect than those closer to the source of light and heat. He also defends the notion that the faster rotation of Jupiter is suited to the sort of beings that would live there.

Finally, the excellence of natures in these higher celestial regions seems to be tied through a physical connection to a durability which is most proper to it. Decay and death cannot affect those excellent creatures to the extent to which they affect us lower beings. For the very sluggishness of matter and crudeness of the stuff, which in the lower stages [regions] are the specific principle of their debasement are also the cause of that propensity which they have for [their] perishing. When the fluids, which nourish and make grow the animal, or man, incorporate themselves amidst its small fibers and add to its mass, can no longer enlarge those vessels and canals in spatial extension once the growth has been completed, these additional fluids of nourishment must therefore, owing to the mechanical drive which is expended for the nourishment of the animal, constrict and block the cavities of its vessels, and destroy the build of the whole machine through a gradually increasing numbness. It is believable that although decay affects even the most perfect natures, nevertheless the advantage in the refinement of the stuff in the elasticity of vessels, and in the lightness and efficiency of fluids of which those perfect beings that inhabit the more distant planets are composed, held up far longer this frailty, which is a consequence of the sluggishness of the cruder matter and secure for those creatures an endurance the length of which is proportional to their perfection just as the frailty of the lives of men has a direct relation to their unworthiness.

I cannot leave these considerations without facing up to a doubt which may naturally arise from the comparison of these ideas with our previous statements. In respect to the abodes in the world-edifice, we have recognized in the greater number of satellites, which illuminate the planets of the most distant orbits, in the speed of their axial rotations, and in the composition of their stuff proportioned against the working of the sun, the wisdom of God that disposed so fittingly everything for the benefit of intelligent beings who inhabit them. But how would now one reconcile with the doctrine of purposiveness a mechanistic doctrine, so that what the highest Wisdom itself planned is entrusted to raw matter, and the rule of providence to a nature left to herself for its execution? Is the former [or the doctrine of purposiveness] not rather an understanding that the orderly disposition of the world-edifice could not have developed through the latter [or the mechanistic doctrine]?

One will quickly dissipate this doubt if one only recalls what was previously set forth in a similar connection. Should not the mechanism of all natural motions have a fundamental propensity for only such consequences that fittingly correspond to the project of the Highest Reason in the whole realm of interconnections?

In a section omitted here, Kant argues that a knowledge of nature fosters an understanding that the laws of motion have been arranged by the Creator, and that they have been arranged so that more perfect beings develop farther away from the Sun. He also endorses the idea of a Great Chain of Being, urging that "in the entire span of nature all is tied together into an uninterrupted gradation through the eternal harmony which makes all members related to one another," in support of which notion Kant cites a passage from Alexander Pope.

We have set forth the foregoing considerations faithfully to the directive of physical relationships, which has kept them on the path of rational credibility. Should we permit ourselves one more escapade from these tracks into the field of phantasy? Who shows us the limits where the well-founded probability ends and arbitrary fictions begin? Who is so bold as to dare an answer to the question whether sin would exercise its dominion also on the other globes of the world-edifice, or virtue alone has her regime set up there?

> The stars are perhaps abodes of glorious souls,
> As vice rules here, there virtue is the lord.
> von Haller

Does not a certain middle position between wisdom and unreason belong to the unfortunate faculty of being able to sin? Who knows, are not also the inhabitants of those distant celestial bodies too noble and wise to degrade themselves to [the level of] that stupidity which is inherent in sin, those, however, who inhabit the lower planets are grafted too fast to matter and endowed with all too weak faculties to be obligated to carry the responsibility of their actions before the judgment seat of justice? In such a way, only the earth and perhaps Mars (so that we would not be deprived of the miserable comfort of having companions in misery) would alone be in the dangerous middle road, where temptations of sensible stirrings against the domination of the spirit would possess a strong potential for seduction, [and] yet, this [spirit] cannot deny that it has the ability by which it is in a position to put up resistance to them, provided its sluggishness would not rather take pleasure in being carried away by them; [in that position] where the dangerous middle point is between weakness and ability, there precisely those advantages, which put him above the lower classes, place him at a height from which he may sink again infinitely deeper below them. In fact, both planets, the earth and Mars, are the middle members of the planetary system, and perhaps not without probability an intermediate physical as well as moral constitution between the two extremes may be assumed about their inhabitants; however, I will readily leave these considerations to those who feel they can muster more repose in the face of undemonstrable considerations and more readiness in providing answer.

In the concluding section, Kant speculates that humanity may have the capacity to improve itself. Then his final paragraph suggests how strongly the views laid out in his book impacted on his own life and thought.

> In fact, when man has filled his soul with such considerations and with the foregoing ones, then the spectacle of a starry heaven in a clear night gives a kind of pleasure which only noble souls can absorb. In the universal quiet of nature and in the tranquillity of mind there speaks the hidden capacity for knowledge of the immortal soul in unspecifiable language and offers undeveloped concepts that can be grasped but not described. If there are among the thinking creatures of this planet lowly beings who, unmindful of the stirrings through which such a great vision can captivate them, are in the position of fastening themselves to the servitude of vanity, then how unfortunate is this globe to have been able to generate such miserable creatures! On the other hand, how fortunate is that same globe, since a road is open for them under most desirable conditions to reach a happiness and nobility which are infinitely far above those advantages which nature's most exceptional dispositions can achieve on all celestial bodies.

This selection shows clearly the young Kant's heavy commitment to the idea of extraterrestrial life. As he matured and became more critical regarding speculative ideas, Kant may have regretted that he made such speculative claims in his *Universal Natural History*. Nonetheless, three indications suggest that he grew no less committed to the idea of extraterrestrial life. One of these consists of a remarkable passage from late in his masterwork, his *Critique of Pure Reason*: "I should not hesitate to stake my all on the truth of the proposition—if there were any possibility of bringing it to the test of experience—that, at least, some one of the planets, which we see, is inhabited. Hence I say that I have not merely the opinion, but the strong belief, on the correctness of which I would stake even many of the advantages of life, that there are inhabitants in other worlds."[10] Secondly, a plausible case can be made for the suggestion that in formulating his ethics and his epistemology, he was careful to do so in such a manner as would allow its applicability not only to earthlings but to rational beings throughout the universe. Third, Kant's

10. Immanuel Kant, *Critique of Pure Reason*, trans. J. M. D. Meiklejohn (London: E. P. Dutton, 1956), 468.

grand vision of a universe filled with extraterrestrials roaming planets orbiting stars suggests why he placed the following passage at the conclusion of his *Critique of Practical Reason* (1788):

> Two things fill the mind with ever new and increasing admiration and awe, the oftener and more steadily they are reflected on: the starry heavens above me and the moral law within me. . . . The former . . . broadens the connection in which I stand into an unbounded magnitude of worlds beyond worlds and systems of systems. . . . The former view of a countless multitude of worlds annihilates, as it were, my importance as an animal creature, which must give back to the planet (a mere speck in the universe) the matter from which it came.[11]

It is significant also to note that the first sentence of this famous passage, so revealing of Kant's core concerns, also appears on his tombstone.

JOHANN HEINRICH LAMBERT (1728–1777)

That Lambert was one of the most remarkable figures of the eighteenth century is evident in many ways. One such way consisted of his efforts to gain entrance to the German intellectual elite.

> On an evening in March 1764, a prospective member of the Berlin Academy of Sciences was introduced to Frederick the Great, its royal patron. The scene could not have been more unusual. The candidate was seated and almost all the candles were extinguished shortly before the king entered. "Would you do me the favor of telling me in what sciences you are specialized?" the king asked the visitor of whom he could at best see a dark silhouette. "In all of them," came the answer from the man in the dark. "Are you a skilful mathematician?" the king asked again. "Yes," the visitor answered. "Which professor taught you mathematics?" the king pressed on. "I myself," went the reply as curt as before. "Are you therefore another Pascal?" "Yes, your majesty."[12]

11. Immanuel Kant, *Critique of Practical Reason and Other Writings in Moral Philosophy*, trans. Lewis White Beck (Chicago: University of Chicago Press, 1949), 258–59.

12. Stanley L. Jaki, "Introduction," *Cosmological Letters on the Arrangement of the World Edifice*, Johan Heinrich Lambert, trans. S. L. Jaki (New York: Science History Publications, 1976), 1.

A skillful and creative mathematician, Lambert also made important contributions to physics, cartography, philosophy, and astronomy, particularly cosmology, in which his most important publication appeared in 1761, his *Cosmologische Briefe über die Einrichtung des Weltbaues* (*Cosmological Letters on the Arrangement of the World-Edifice*).

The best known idea from that book is Lambert's suggestion that the Milky Way is a flattened disk-shaped structure composed of millions of stars. It is of course true that because Kant had published this idea in 1755, he achieved priority of publication. Nonetheless, sound evidence indicates not only that Lambert did not derive this idea from Kant, whose book he had not seen when he wrote his *Cosmological Letters*, but also that he attained this idea as early as 1749. This evidence centers on a letter Lambert wrote Kant after he had learned of Kant's 1755 book. Lambert explained:

> What gave occasion to my *Cosmological Letters* . . . was this: that in the year 1749, on a certain occasion immediately after supper, . . . I . . . went into another room. I there wrote down my thoughts on a quarto page, and in the year 1760, when I wrote the *Cosmological Letters*, I had still nothing further on the subject in hand. In the year 1761, I was told at Nürnberg that some years previously an Englishman had printed similar thoughts in letters to certain persons, but that he had not had much success, and that the translation of his Letters, begun at Nürnberg, had not been completed. I answered that I believed my *Cosmological Letters* would not make a great impression, but that perhaps in the future an Astronomer would discover something in the Heavens which could not be otherwise explained.[13]

A comparison of Kant's universe with Lambert's reveals interesting differences and similarities. Regarding differences, Kant viewed the universe as infinite and evolving, whereas Lambert argued for a finite and more static cosmos. On the other hand, Kant and Lambert shared a fundamentally Newtonian orientation and both were heavily involved (although in somewhat different ways) with ideas of extraterrestrial intelligent life.

Lambert's advocacy of extraterrestrials was such that he informed King Frederick the Great that the public considered his *Cosmological Letters* as a second volume "to the one on the plurality of worlds by Fontenelle."[14] Lambert's special pas-

---

13. As quoted in *Kant's Cosmogony*, ed. and trans. W. Hastie (Glasgow: James Maclehose and Sons, 1900), lxx.

14. As quoted in Jaki, "Introduction," 24.

sion was postulating inhabitants of comets. Never before, and probably never since, has an astronomer shown such enthusiasm for life on comets, millions of which, he claims, circulate in our own as well as other solar systems.[15] Most remarkable is his assertion that comets are peopled by extraterrestrial astronomers "created for the purpose of viewing the edifice of the heavens."[16] Lambert writes of the astronomers,

> Their route proceeds from suns to suns as we go from city to city on earth, and when in our case a few days go by they count in myriads of our years. They are destined to admire the ground plan of the world-edifice, and understand its foundation and order the series of divine counsels about its structure. . . . They know the warmth and brightness of each sun, and with a single conclusion they determine the general characteristics of the inhabitants of each planet which orbits around it at a given distance. Their year is the time [which it takes to go] from one sun to another. Their winter falls in the middle of the intervening space, or of the journey which they make to another sun, and they celebrate the moment when their former course turns into a new one. The perihelion of each course is their summer. Their habitat is suited for each distance from the suns, and each degree of heat conditions their habitat for the growth of such plants that serve for them as specimens of those which occur on planets and comets at a similar distance from the suns. Each entry into a new solar system is their spring, and they celebrate the fall when they leave it again.[17]

Three major sources of Lambert's extraterrestrials are his Copernicanism, his acceptance of the principle of plenitude, that is, the notion that God's goodness needs to be expressed in a multiplicity of forms, and his teleological orientation, in other words, his belief that God designed every part of the universe with a purpose. He repeatedly asks in the course of his book whether we are yet sufficiently Copernican, whether we have fully recognized all that is entailed by the Copernican assertion that the Earth is a planet and the associated claim that the Sun is a star and stars are suns. Regarding the fullness of nature as a result of the appropriateness of God creating life in all its forms, he remarks that "all possible varieties which are permitted by general law ought to be realized, because perfection becomes greater thereby."[18] His

15. Lambert, *Cosmological Letters*, 72.
16. Lambert, *Cosmological Letters*, 73.
17. Lambert, *Cosmological Letters*, 79.
18. Lambert, *Cosmological Letters*, 89.

teleological orientation not only motivates his conviction that life must be spread throughout the universe, but also at times leads him, as is evident in passages that follow, to structure the *physical* universe in a particular manner.

Lambert's *Cosmological Letters* consists of a preface and twenty letters exchanged between the author (even numbered letters) and an interlocutor (odd numbered letters). The book's epistolary structure as well as the fact that Lambert wrote in a somewhat playful manner and allowed both writers to express his views make it difficult at times to determine his precise intentions, but his overall message is clear. The following selection is drawn from letters eight through twelve.

J. H. Lambert, *Cosmological Letters on the Arrangement of the World Edifice*, trans. Stanley L. Jaki (New York: Science History Publications, 1976), 95–98, 100, 102–4, 105–6, 109–10, 110–12, 112–113, 113–14, 120, and 124–25.

EIGHTH LETTER

Lambert begins by pointing out that although the teleological character of his approach to the issue of extraterrestrial life precludes him from offering proofs with geometrical exactness, nonetheless none of his claims is contradicted by empirical information. He also suggests that observational evidences supportive of extraterrestrials may be forthcoming in the future. Regarding comets, he admits that persons who doubt the habitability of planets will be even more hesitant about claims for cometarians. Lambert also presents an analysis of the number of comets, concluding that comets far exceed planets in number. He then offers an answer to the question of why comets move in ellipses or hyperbolas of high eccentricity.

One will just as well concede to me that elongated ellipses which comets describe are more suited to have their number increased, and that thereby the variety and greater diversity in our solar system will become far greater. One will admit that as soon as the number of these bodies, through the safeguarding of the law of gravity and of their orbits, ought to be in all cases the greatest, the very oblong ellipses become necessary and ought to occur in greater quantities. Why do they, however, actually exist? It is very conceivable that the law of gravity, the habitability of the solar system, the variety and periodic changes in the celestial bodies stand in such a mutual connection that taken together they constitute a maximum. One is not able yet to show why the Newtonian law of gravitation rather than any other law was brought about in the world. About various other laws Newton investigated the consequences and their unsuitability. If, however, a rigorous demonstration is required, I am much inclined to think that the habitability of the solar system and the variety of orbits will be counted among the basic reasons.

In a section omitted at this point, Lambert explores the possibility of spiral orbits, concluding that such orbits would reduce the chances for habitable worlds.

On the contrary, how much more suited is the Newtonian law of gravity for [securing] habitability and variety in each solar system! How much more harmonious is the return of orderly disposition in each periodic orbit! How many varieties similar to one another [are there] among the celestial bodies that stay around the sun! It is always ellipses, well-defined periods, succession of seasons of such duration, which bring about on each celestial body the changes ascribed to them.

Should, however, even higher viewpoints be reached, and should the distance from sun to sun be sized up, and should the world-edifice, the system of fixed stars in its entirety, the ground plan of the world and its arrangement be considered, then this law of gravity still will do and will give the shortest way to it. If you transform the ellipses into hyperbolas, the celestial body which moves along it does not stay with any sun. It hardly bends its path to turn toward other suns. For this change of direction it needs but the shortest time, because it moves at the highest speed. Then it approaches its asymptote and enters along a straight line into the [gravitational] area of another sun where its velocity again increases only to move soon again toward new [solar] systems. Could one here think of a more convenient route than hyperbolas and of a law of gravity more capable in every respect than the one which is demanded by conic sections, the simplest of all curved lines of which one half is periodic, the other, however, gives ever new varieties, and with both of which all possible variations can occur?

I believe that one can here conclude that these changes in the world do not remain merely possible. The [principle of] variety has to say much too much in the doctrine about perfection than to permit me to omit to a large extent those changes from the world-edifice. This would happen if I limited the number of comets to a few hundred in spite of the fact that millions of them are possible without disturbing one another. About comets whose perihelia are equally distant from the sun some can stay in the outer regions much longer than others, because the longer axis of the ellipse does not depend on the focal length. Again, different comets can have similar periods but very different seasons, because the longer axis alone determines the time of their revolution but the focus can be at any point of that axis. How many changes and varieties are here, too, possible! Should I exclude by far their greatest portion? For this I have no reason; on the contrary, the greatest perfection also demands the greatest fullness. It must have all that it can have. How much can the distance of the real from the possible still amount to and what else will be lacking to my system than the complete register of all comets, if one does not consider these conclusions as adequate prior to experience?

In a few paragraphs omitted at this point, Lambert cautions that he is presenting his cometary system "only as probable" (98). He also distinguishes comets moving in elliptical paths from those that move in hyperbolic orbits and suggests that the latter type are more numerous. Lambert speculates that the habitability of the solar system is enhanced by the planets all moving in the same orbital plane. Such an arrangement brings greater order and a more uniform distribution of heat than a system of planets in dissimilar orbits. From here, Lambert passes to a discussion of the proper limits to speculating about the characteristics of extraterrestrials.

. . . About the inhabitants of celestial bodies I was not much concerned, because I saw it well that our concepts were not rich enough to describe them each according to their circumstances and according to the conditions of their habitats. The remarks which you, Sir, have made in that connection pleased me very much. I see with you the usefulness of light as a very universal one, and would the ability to absorb it be granted to the inhabitants of each celestial body, it might be transplanted into their souls either through the eyes or through some other means. Here again I state nothing definitively, because we cannot form a notion of any other means.

He also speculates on the nature and function of the materials composing the tails of comets, suggesting that they provide protection for the extraterrestrials living on the comet. In the final paragraph of his Eighth Letter, Lambert makes the theological claim that "the Creator spreads out through the world-edifice the choicest features from the riches of his omniscience" (101).

NINTH LETTER

In the first two paragraphs of this letter, Lambert's interlocutor urges that inhabitants of comets that move in hyperbolic orbits, which entails that some comets move from one solar system to another, gain a fuller view of the universe than persons on celestial bodies that orbit a single sun. Lambert's interlocutor suggests that similarly Lambert himself has a fuller view of the universe than persons whose convictions concerning extraterrestrials derive from empirical evidence. The interlocutor then praises Lambert for his idea that the celestial bodies in the universe may be linked up by comets moving through space.

How much more the world becomes inhabited than one would have thought only a short time ago! We find worlds in each speck of dust, in each droplet, and soon the specks of dust will not be as numerous as the celestial bodies in the firmament. It is unquestionable, Sir, that he who grants inhabitants on the planets has no more new steps to make to surround them with as many celestial bodies as you wish. For he will grant that the world should not remain barren and that there ought to be in it more live than dead masses. The Creator, the eternal source of all life, is much too efficient not to imprint life, forces, and activity on each speck of dust. How should one then consider your enterprise mistaken when you, Sir, do nothing else than show that, if one is to form a correct notion of the world, one should set as a basis God's intention in its true extent to make the whole world inhabited and leave no part, no side [section] of it out of consideration! The only thing that one can object to the universality of this viewpoint is the concern that some higher and to us unknown reasons may prevent letting so many celestial bodies wander around each sun. He who is concerned in that respect will certainly wait for the complete register of all comets, and that will take many centuries, and for all that one shall never draw it up in a complete form.

I have made for myself a quick analysis of this register. From 1500 until 1600 one observed over 40 comets of which it seems none reappeared for a second time during these hundred years, because I still consider the one of 1759 as the speediest and I assign to the rest more than 100 years but to most of them many centuries for the completion of their orbits. According to that calculation I shall have in 400 years 4 times 40, or 160 comets. I set at 60 [the number of] those that among these return twice or several more times, so there still remain 100. For these 100 I must, however, take at least 300, because so many hindrances stand in the way of the visibility of a comet that I can assume that we see of those which can be seen hardly a third. Thus I had at least 300 comets that descend to our sphere of visibility. This sphere is, however, as you, Sir, remark in a previous letter, 40 times smaller than the sphere of Saturn, [and] consequently I had to take these 300 comets another 40 times to account for all those that come within the sphere of Saturn. The calculation gives 12,000 comets.

After extending and expanding this estimate of the number of (presumedly inhabited) comets associated with our solar system, Lambert's interlocutor continues.

These 12,000 comets all still come closer to the sun than Saturn. Nothing prevents, however, that there should not be such comets which stay out 10 times farther, and the habitability of the world demands that they should be all the more densely spaced, the closer they are to the sun. In such a way I had to multiply by 100 these 12,000 and I will come up with 1,200,000 comets even when the register of those that can be seen by us does not stretch beyond 300.

Comets that come close to the sun still leave so much space outside Saturn that I would not hesitate at all to place the greatest comets in that outer region and add satellites to them. For I do not believe that a comet with satellites would ever come into our view. The closer they come to the sun, the narrower will be the space and the more sparingly ought it be used. A comet that has satellites around it has a very large sphere of influence, but it still must leave room for other comets. This is, however, much more possible if the comet stays farther away from the sun. I think, as you do, Sir, that order, diversity, habitability, and the law of gravity determine one another in each solar system, and this is necessarily the reason why you consider only the general, because the individual things must be discovered through experimentation.

In the long paragraph omitted at this point, Lambert's correspondent supports the author's denial that comets can become planets or satellites. The point is put forward that planets move nearly in circles on the basis of the idea that variety demands that there exist locations for inhabitants who need a stable environment. Regarding life on comets, it is claimed that "The inhabitants of comets seem to be unaffected by heat and cold, and longer or shorter winter should not mean so much to them as to us" (105).

How excellent the Newtonian law of gravity becomes through the considerations which you, Sir, have presented on it, and how precisely you connect them with other purposes of creation, in terms of which you assess the arrangement of the whole solar system and similarly predict what only posterity will carry into the realm of experience! I wished I could determine the maximum which you have mentioned in that connection. You connect the law of gravity with the habitability and diversity of celestial bodies in such a way that these general purposes of creation certainly seem to require that law, which anyhow is the simplest and facilitates so much for us the determination of the orbit of each celestial body. How admirable become thereby the arrangement and order in the whole solar system, and what a manifold diversity and variety are spread through that law across the whole world! Thus light is shed on the choicest of all orderings, which should necessarily appear to us as disorder if we, unaware of that law, could survey the solar system at a glance in such a way that each planet and comet would come into our view. In fact, it would present itself to us in no other way than as an entangled and disorderly scattered heap of globes which were unrelated with respect to size. We would find no reason why their distances from one another are so unsimilar and in all appearance so clumsy. I cannot sufficiently exert my imagination to present to myself the position of these celestial bodies, as it is, for instance, at the present moment, and to track down each comet in its orbit. This much, however, I think that it would appear to me as if they all had gone astray and many of them were at one place and almost none at another.

But I take the law of gravity and the resultant order in the course of celestial bodies that are around our sun, and this erratic picture at once vanishes. For this law teaches me that I must not consider the celestial bodies in themselves but also their orbit and the law of their motion, and then I certainly find the exquisite harmony which you, Sir, point out so worthily in your letters. How clear it becomes from all this the great tenet: the disorder in the world is only apparent, and where it appears to be the greatest, there the true order is even more excellent though only more hidden to us! We shall feel, when we consider the members of the solar system only with respect to space and position, as if one were in a well-ordered library where one only considered the place to find each book simultaneously. This order would be much too simple in the world. Here space and time must be connected with one another, and the order ought to be extended to both. How perfectly harmoniously this occurs in the motion of celestial bodies around the sun, which, from the viewpoint of space, do not seem to have any order!

At this point, Lambert's interlocutor raises and develops the question: "Do you think, Sir, that these considerations may not be extended to the fixed stars as well?" (106). He states that in examining the realm of stars, he can see no "definite symmetry" (106) and elaborates on this apparent disorderliness. A specific case he cites is the Milky Way, which seems to show no order. In this context, he remarks: "Here I became confused and I must tell you, Sir, that soon I began to doubt again the universality of your basic tenet that the world is as habitable as possible and ought to be inhabited" (106–7). This leads him to request illumination from Lambert. Regarding the Milky Way, he asks such questions as "Do you believe that the stars in this luminous streak are in fact more densely together, or do they merely lie in endless long rows behind one another?" And more generally, referring to the stellar regions, which become from this point on the central concern of the book, he asks: "Can you, Sir, find harmony and order in this apparent irregularity?" (107).

TENTH LETTER

In the introductory paragraph to this letter, Lambert admits that the theories that follow may lack observational evidence, but that they are at least probable and consistent with experience.

Lambert then turns to the question of the arrangement of the entire stellar realm, noting first of all that Newtonian gravitation governs in this domain. This, however, creates a problem. Were gravitation the only force operative, the system would eventually collapse upon itself. Because this does not occur, there must be a counteracting force, which Lambert identifies as a centrifugal force, which he portrays as comparable to the centrifugal force that keeps the planets in our system from falling into the Sun. This explanation leads to the further conclusion: that the so-called fixed stars must be in motion. Lambert admits that as yet empirical confirmation of this motion has not come forth. He also stresses that stars not only move, but also that the distances between them change.

This motion of fixed stars can be considered in a twofold manner. Since there, too, the law of gravity obtains, they move, [as] they belong so much together, around their common center of gravity just as this takes place also in our solar system. The question now merely becomes whether one leaves this center completely void, or whether one should put there a body very large compared with the fixed stars which revolve around it, which is exactly what the sun is in comparison with the celestial bodies that orbit around it. Were this center quite empty, the motion of fixed stars ought to be very slow, because they would have with respect to that center no other centripetal force than the one arising from the fact that they gravitate toward one another. This gravity would, however, be very small because of the great distance, and therefore the centrifugal force and consequently also the [orbital] velocity would not be great. The planets and comets of our system would therefore move around their common center of gravity even if the sun were not there. But their velocity would be incomparably slower because [otherwise] they would soon be scattered.

Should I, however, put in the common center of the fixed stars, which together constitute a system, a body toward which they all gravitate, then I must give that body an enormous size and I must enhance its mass so much that even the most distant stars of the system would have considerable gravity toward it, because this is always proportional to the mass. Were I to write here a novel, I would then state that this body has either no proper light or only a very weak one. I would so arrange the world that the smaller dark bodies, like, for instance, the planets would move around bright suns, but these again would orbit around dark bodies. For the suns would need no other light because they themselves possess so great a brightness, while the dark body can still sufficiently be illuminated by the suns which move closest around it. But for such an arrangement I can give no reason other than the mere possibility. You know, however, Sir, that possibility is considered satisfactory only in the poetical but not in the philosophical world. I leave it therefore undecided whether a system of fixed stars revolves only around a central point, or whether at that point there is in fact a body of enormous mass toward which the fixed stars gravitate.

Two paragraphs dealing with the Sun's rotation and revolution have been omitted at this point. This brings him to his now famous presentation of his disk theory of the Milky Way.

What you, Sir, write to me about the Milky Way, has already astonished me repeatedly. It seems unquestionable that this band is much behind the other fixed stars which we see outside it, and the bigger fixed stars in that circle ought to be incomparably closer to us because only with the aid of telescopes do we see more distinctly those which specifically form the Milky Way. Since, however, they are so immeasurably distant, it seems certain that they do not yield to our sun in size and brightness, and therefore they must have a considerable distance from one another.

Therefore start out in thought from our sun straight toward a fixed star of first magnitude, from that to a more distant one, from that to those that are farther according to the order [of magnitude], and you will perhaps sooner come to the outermost [star] if you had taken the road outside the Milky Way rather than the one going through it. I set the nearest star, which properly belongs to the Milky Way, many times farther away from us than the farthermost among those that do not belong to it. For since I assume that the stars in that band are as much separated from one another as any of the nearest stars are from our sun, I must therefore necessarily put them in inconceivably long [many] rows behind one another, and thus I conclude that the whole system of fixed stars visible to us is not spherical but flat, roughly like a disk whose diameter is many times longer than its thickness. For I must take here a physical plane which has a certain thickness. In that plane lies the Milky Way and all visible stars outside it; that plane is also the ecliptic where all fixed stars move.

But this is far from enough for me. The Milky Way distinguishes itself clearly from the rest of the sky. When I take together all the other fixed stars, I must completely separate from them the Milky Way and I also must divide that band into innumerable smaller parts. Many of these parts show themselves to us through the fact that they appear separated from the rest. The others cover one another because one lies behind the other. Each of these parts I see as a separate system of fixed stars. We find ourselves in one such system and I count into it all stars that are visible to us and lie outside the Milky Way as well as the larger ones that cover that arch of the sky. I set each such system similar to our solar system in the sense that all fixed stars or suns which belong to it move it around a common center, and I would be inclined to believe that all these systems, or the whole Milky Way, have a common center around which they move.

You will see from this, Sir, that I make my conclusions on the basis of analogy. Thus, for instance, the satellites belong to the planets, these to the sun, the sun to its system, and this to the system of the whole Milky Way. Farther our eyes do not reach and I leave it undecided whether the Milky Way visible to us still belongs to uncounted others and forms with these a whole system. Perhaps the light of this so immeasurably distant [super] Milky Way is so weak that we are unable to see it. For the nearest fixed stars still can spread a weak light through our air, as we can see at night when the sky is clear. This so weak a light can unquestionably obscure and make insensible a still weaker light, and I conclude from the fact that the Milky Way is still visible that in that streak there ought to be innumerably many stars. To the naked eye the stars of the seventh and subsequent magnitudes vanish and they become visible to us only when many of them are densely together. We see this in the so-called nebulous stars. The telescopes teach us that they are but a heap of stars which the distance makes too small for us to distinguish with the naked eye. When, however, the light from many of them fuses together, it becomes stronger and therefore perceptible to our eyes. This is the reason why we see the Milky Way.

Omitted at this point is a paragraph in which Lambert again stresses the speculative character of these ideas and also points out the centrality of ideas of motion to his system.

Sir, you have helped me to extend the Newtonian law of gravity through the whole world and also through each system of fixed stars. Just as far do I extend the circles and ellipses in which the fixed stars move around the center of their system. The closer a star is to that center, the faster it moves in its orbit and the shorter time it takes in moving around. I have already remarked above that I take the stars which are outside the Milky Way together into one system, great as their number may be. It is natural and consonant with the [principle of] analogy that number, space, and time grow together with the system. The earth has only one satellite, Jupiter 4, Saturn 5, the sun many millions of celestial bodies around itself, because its mass is [so much] greater and its heat and light are of more general benefit. Should I put together accordingly a system of suns, millions of them are still far from being sufficient. Their number must be incomparably greater if it should increase in the proportion which the satellites of Saturn have to the number of bodies in the solar system, which I have set at millions in one of my previous letters.

In two paragraphs omitted at this point, Lambert discusses the question of whether "our sun is near the center of the system to which it belongs, or whether farther away from that center." He builds a case for it being near the center and also somewhat off the plane of the Milky Way's disk.

I have not yet pursued these considerations any further and they cannot therefore be presented here in a more elaborate way. I also submit them only as arbitrary and I have perhaps described them unclearly and without a proper order. But you know, Sir, how difficult it is to extend order over a matter to which most pieces are still lacking. . . .

ELEVENTH LETTER

Now I see in its full extent why you, Sir, have always said that we are still far from thinking in a truly Copernican way. It would not be enough to stir the earth from its rest; rather, not a single body in the whole firmament should remain at rest. The sun may forever be in the center of its system and let the planets and comets wander around itself. The sun is no more than Jupiter and Saturn are in respect to their satellites. But that the sun should be in the center of the whole world-edifice is far from evident; and should the sun once happen to be there, it would soon be removed. In short, resting is banned from the world-edifice because that would make it too uniform. Variety demands changes and these cannot occur without motion. Motion becomes therefore essential, and the general law of gravity will suffice to show that everything is alive and moving. No point of the whole world-edifice remains, even for a moment, in absolute rest. . . .

For purposes of space, most of the eleventh letter has been omitted. Lambert's correspondent reiterates the features of the system that the author outlines in the previous letter. The correspondent adds that the Milky Way is not the largest system of stars in the universe. He begins to discuss the distance between stars, ruling out the possibility that the stars are closely packed. A better model, he suggests, pictures the stars as arranged like a series of lamps in "indescribably long [many] rows" (115). The interlocutor also states that "You readily speak to me of still other Milky Ways which are outside the one visible to us and which taken together with ours must constitute a greater system. I suspect that you would look at the faint light one sees in Orion as such a Milky Way which would be the one closest to ours" (119). Lambert's correspondent ends by returning to the idea with which he began the letter, that Lambert's system is a hypothesis that will be eventually proven and that is defensible in the meantime because it is an extension of Copernicanism.

Even though you offer your thoughts on this point only as a hypothesis, the effort to derive such consequences in greater amounts will always have its rewards. A hypothesis will in the end become a truth, when all phenomena let themselves be derived from it in a natural and in an obvious manner, when all these consequences are connected with one another and with the general reasons, in short, when that hypothesis is consistent in all its parts with itself. When in the edifice of truths there remains somewhere a gap and one finds a system of thought which fits exactly in that gap, then it becomes very likely even though one cannot quickly connect it with all the remaining truths. One comes closest to the truth when new phenomena can be inferred and foreseen from it. You know, Sir, that it so happened with the Copernican world-edifice which no astronomer doubts any more. Since you try to make the whole world-edifice completely Copernican, you build on similar grounds and I see it as a good omen that you are having such success in the process. It will satisfy me if I can only provide one handyman to your edifice. . . .

TWELFTH LETTER

After some introductory comments, Lambert recounts how he came to his disk theory of the Milky Way and explicitly states that the Milky Way is flat. He then predicts that his theory will stand or fall on the basis of evidence as to the distance of the stars. He notes that light can be dimmed by the ether and that not all stars are equally bright. He stresses that the stars are in motion, that they orbit, and are unevenly distant. He notes again that a sun with its surrounding planets and comets needs substantial space to preserve the order of the system. He says the Milky Way is made up of an indescribably large number of systems.

Then as I separate our system of fixed stars from those that appear in the Milky Way, I leave between these a similar intervening space. They lie in about one plane. Therefore I cannot assume more than six that are immediately around us if I set their distances from one another [to be] about the same. All the rest I must remove farther and farther out. The apparent diameter of the next systems of the Milky Way can hardly be equal to its mean width and therefore not more than 10 degrees. If I take the true diameter of such a system as a yardstick, then it has to be laid off six or more times to find the distance of such a system from the next bordering on it. I think, however, that the apparent diameter amounts to less than 10 degrees and perhaps hardly to 5 or 6 degrees, and I conclude from this that a system will be removed from the nearest 10 to 12 times more than it is wide. Since, however, the

Milky Way is quite compact, uncounted such systems must therefore lie behind one another to cover each intervening space which the next ones would leave empty.

Yet their number is definite and the Milky Way must have its limits somewhere. I have derived this in my last letter from [the principle of] analogy, since I have posited still uncounted other Milky Ways outside that streak, because it does not seem likely to me that the world-edifice is so confined, though on the contrary just as little do I grant that it is infinite. A great deal could be said if you, Sir, ask, whether the pale light in Orion, which Derham considered to be an opening in the *coelum empyreum,* is not the nearest of such Milky Ways, and I will not now address myself to the point that a change has been observed in its visible shape. It appears to be too far removed for one to be able to see it always equally clearly through our air and through any telescope.

One can perhaps conclude most correctly from the apparent form of the Milky Way that it has narrow confines. Were it immeasurably extended, it should appear to us as a great circle of the sphere. But this is not so. Its apparent form is rather an oval. Its center stands from the North Pole 35, from the South Pole only 25 degrees away. On the other hand, it cuts the equator into two fairly equal parts. So does a ring appear to the eyes when one is located both outside it and away from its axis. Our system of fixed stars appears to lie not only somewhat outside the plane of the Milky Way but also closer to its periphery than to its center.

Several paragraphs are omitted from the end of this letter, in which Lambert suggests that the revolution of the Sun around the center of its system of stars is detectable, in principle, through the detection of perturbations in this orbit caused by the orbit of nearby stars. Lambert concludes the letter by admitting that his theory is still in development and that it will become more convincing with further observation and reflection.

The selections from Kant and Lambert (and to a lesser extent Wright's materials) are the founding documents of stellar astronomy, now the dominant area of astronomy. Each presented a form of the disk theory of the Milky Way and each put forth the hypothesis that the nebulous patches seen in the heavens are other universes, comparable in structure to our Milky Way. These core ideas of modern stellar astronomy are not all that Kant and Lambert contributed; they also raised questions that in some cases today's astronomers continue to seek to resolve. Most importantly, they sought to draw the attention of astronomers to the vast regions beyond our solar system, an area that nearly all eighteenth-century astronomers neglected.

It is a striking point not to be missed that extraterrestrials make numerous appearances in the pages of their respective books. A plausible explanation of why Kant and Lambert, far more than earlier astronomers or their contemporaries, took the stellar realm seriously is that they saw it as filled with possible places of habitation, as regions abounding in extraterrestrials, and therefore worthy of study.

Similar as Kant and Lambert were in many aspects, they differed in significant ways, as the following table contrasting and comparing their views suggests.

Table 5. Comparing and contrasting Kant and Lambert

Kant	*Lambert*
Disk theory of the Milky Way.	Disk theory of the Milky Way.
Passionate about extraterrestrials, centered on planets.	Passionate about extraterrestrials, centered on comets.
Lucretius and Newton were his masters.	Newton was his master.
Evolving universe.	Changing universe.
Limited knowledge of science.	Extensive knowledge of science.
Boldly speculative.	Boldly speculative, but with numerous admissions of the fallibility of his formulations.
Universe infinite.	Universe finite.

WILLIAM HERSCHEL (1738–1822)

As background for the Herschel readings and because James Ferguson (1710–1776) is important in his own right, it will be useful briefly to consider the career of this shepherd turned scientist and astronomical popularizer. Ferguson's *Astronomy Explained upon Sir Isaac Newton's Principles*, first published in 1756, went through at least seventeen editions. Among its readers was William Herschel, who may also have heard Ferguson on one of the latter's lecture tours of England. Thomas Paine certainly heard Ferguson's lectures; moreover, that itinerant astronomer significantly influenced both Herschel and Paine, although in very different ways, as will soon become clear. Throughout Ferguson's book, he enriched its presentations of astronomy by including ideas of extraterrestrial life as well as religious themes, the two frequently in combination. He followed the same practice in his *Easy Introduction to Astronomy for Young Gentlemen and Ladies* (1768), which went through a dozen editions. When the *Encyclopaedia Britannica* first appeared in 1776, the unsigned article "Astronomy" consisted wholly or in largest part of materials from Ferguson's 1756 book. The selection that follows consists of the first chapter of the article "Astronomy," which is essentially identical to chapter 1 of Ferguson's book.

James Ferguson, "Astronomy," *Encyclopaedia Britannica* (Edinburgh, 1771), 1:434–35.

ASTRONOMY

ASTRONOMY is the science which treats of the nature and properties of the heavenly bodies

CHAP. I. OF ASTRONOMY in general.

By astronomy we discover that the earth is at so great a distance from the sun that if seen from thence it would appear no bigger than a point, although its circumference is known to be 25,020 miles. Yet that distance is so small, compared with the earth's distance from the fixed stars that if the orbit in which the earth moves round the sun were solid and seen from the nearest star, it would likewise appear no bigger than a point although it is at least 162 millions of miles in diameter.[19] For the earth, in going round the sun, is 162 millions of miles nearer to some of the stars at one time of the year than at another; and yet their apparent magnitudes, situations

19. [The modern value for the distance of the Earth from the Sun is about 93,000,000 miles, which makes the diameter of the Earth's orbit 186,000,000 miles. Hence Ferguson's value for the diameter of the Earth's orbit is about 24,000,000 miles too low. MJC]

and distances from one another still remain the same; and a telescope which magnifies above 200 times does not sensibly magnify them; which proves them to be at least 400 thousands times farther from us than we are from the sun.

It is not to be imagined that all the stars are placed in one concave surface, so as to be equally distant from us; but that they are scattered at immense distances from one another through unlimited space. So that there may be as great a distance between any two neighbouring stars, as between our sun and those which are nearest to him. Therefore an observer, who is nearest any fixed star, will look upon it alone as a real sun; and consider the rest as so many shining points, placed at equal distances from him in the firmament.

By the help of telescopes we discover thousands of stars which are invisible to the naked eye; and the better our glasses are, still the more become visible; so that no limits can be set either to their number or their distances.

The sun appears very bright and large in comparison of the fixed stars, because we keep constantly near the sun, in comparison of our immense distance from the stars. For a spectator, placed as near to any star as we are to the sun, would see that star a body as large and bright as the sun appears to us: and a spectator, as far distant from the sun as we are from the stars, would see the sun as small as we see a star, divested of all its circumvolving planets; and would reckon it one of the stars in numbering them.

The stars, being at such immense distances from the sun, cannot possibly receive from him so strong a light as they seem to have nor any brightness sufficient to make them visible to us. For the sun's rays must be so scattered and dissipated before they reach such remote objects, that they can never be transmitted back to our eyes, so as to render these objects visible by reflexion. The stars therefore shine with their own native and unborrowed lustre, as the sun does; and since each particular star, as well as the sun, is confined to a particular portion of space, it is plain that the stars are of the same nature with the sun.

It is noways probable that the Almighty, who always acts with infinite wisdom, and does nothing in vain, should create so many glorious suns, fit for so many important purposes, and place them at such distances from one another, without proper objects near enough to be benefited by their influences. Whoever imagines they were created only to give a faint glimmering light to the inhabitants of this globe, must have a very superficial knowledge of astronomy, and a mean opinion of the Divine Wisdom; since, by an infinitely less exertion of creating power, the Deity could have given our earth much more light by one single additional moon.

Instead then of one sun and one world only in the universe, astronomy discovers to us such an inconceivable number of suns, systems, and worlds, dispersed through boundless space, that if our sun, with all the planets, moons, and comets belonging to it, were annihilated, they would be no more missed, by an eye that could take in the whole creation, than a grain of sand from the sea-shore: The space they possess being comparatively so small, that it would scarce be a sensible blank in the universe, although Saturn, the outermost of our planets, revolves about the sun in an orbit of 4884 millions of miles in circumference, and some of our comets make excursions upwards of ten thousand millions of miles beyond Saturn's orbit; and yet, at that amazing distance, they are incomparably nearer to the sun than to any of the stars; as is evident from their keeping clear of the attractive power of all the stars, and returning periodically by virtue of the sun's attraction.

From what we know of our own system, it may be reasonably concluded, that all the rest are with equal wisdom contrived, situated, and provided with accommodations for rational inhabitants. Let us therefore take a survey of the system to which we belong; the only one accessible to us; and from thence we shall be the better enabled to judge of the nature and end of the other systems of the universe. For although there is almost an infinite variety in the parts of the creation which we have opportunities of examining; yet there is a general analogy running through, and connecting all the parts into one great and universal system.

To an attentive considerer, it will appear highly probable, that the planets of our system, together with their attendants called satellites or moons, are much of the same nature with our earth, and destined for the like purposes. For they are solid opaque globes, capable of supporting animals and vegetable. Some of them are larger, some less, and some much about the size of our earth. They all circulate round the sun, as the earth does, in a shorter or longer time, according to their respective distances from him; and have, where it would not be inconvenient, regular returns of summer and winter, spring and autumn. They have warmer and colder climates, as the various productions of our earth require: And, in such as afford a possibility of discovering it, we observe a regular motion round their axes like that of our earth, causing an alternate return of day and night; which is necessary for labour, rest, and vegetation, and that all parts of their surfaces may be exposed to the rays of the sun.

Such of the planets as are farthest from the sun, and therefore enjoy least of his light, have that deficiency made up by several moons, which constantly accompany and revolve about them, as our moon revolves about the earth. The remotest planet has, over and above, a broad ring encompassing it; which like a lucid zone in the heavens reflects the sun's light very copiously on that planet; so that if the remoter planets have the sun's light fainter by day than we, they have an addition made to it morning and evening by one or more of their moons, and a greater quantity of light in the night-time.

On the surface of the moon, because it is nearer us than any other of the celestial bodies are, we discover a nearer resemblance of our earth. For, by the assistance of telescopes we observe the moon to be full of high mountains, large valleys, and deep cavities. These similarities leave us no room to doubt, but that all the planets and moons in the system are designed as commodious habitations for creatures endued with capacities of knowing and adoring their beneficent Creator.

Since the fixed stars are prodigious spheres of fire like our sun, and at inconceivable distances from one another as well as from us, it is reasonable to conclude they are made for the same purposes that the sun is; each to bestow light, heat, and vegetation, on a certain number of inhabited planets, kept by gravitation within the sphere of its activity.

One significance of this short selection is not to be missed. The fact that Ferguson's presentation appeared in the *Encyclopaedia Britannica* indicates not only its widespread dissemination but also the acceptability of the ideas contained in it to eighteenth-century readers. A contrast may highlight this point. Were a person of the twenty-first century possessing a good knowledge of today's astronomy to read the selections from Kant and Lambert they would probably be impressed by the quality of these authors' insights regarding stellar astronomy but put off by their speculations about extraterrestrials. The reading by Ferguson in effect suggests that were a late eighteenth-century astronomer to read Kant and Lambert, their reaction might be just the reverse of their twenty-first century counterpart. They might well have taken the astronomical sections as wildly speculative, and the extraterrestrial discussions as sufficiently mainline that they would have been acceptable even in such a source as the *Encyclopaedia Britannica.*

There is hardly any story of the acceptance of ideas of extraterrestrial life more striking than that of Sir William Herschel, who was not only the most important astronomer of the latter half of the eighteenth century, but possibly of the entire post-Newtonian period. Although Herschel is most widely known as the discoverer of the planet Uranus (1781), his greatest achievement was his pioneering contributions to stellar and extragalactic astronomy.

Born in Hanover, Herschel lacked a university education but, after immigrating to England in the 1750s, established himself as a musician, composer, and music teacher, working in various locations, especially in Bath, where he settled with his sister Caroline, who also contributed to astronomy. Until well into his forties, Herschel worked as a professional musician. One of the key factors that led Herschel to turn from music to astronomy occurred in 1773, when he purchased and began reading Ferguson's *Astronomy Explained upon Sir Isaac Newton's Principles,* which (according to Caroline's diary) he took to bed "with a bason [*sic*] of milk or a glass of water"[20] for a number of months. During this period and influenced by such readings, Herschel began to construct telescopes and to observe the heavens.

Herschel's first main appearance before the scientific community came in 1780 with the publication of two of his papers in the *Philosophical Transactions of the Royal Society.* The longer of these, which reported on Herschel's efforts to measure the heights of various mountains on the Moon, included the comment that "knowledge of the construction of the Moon leads us insensibly to several consequences . . .

---

20. As quoted in Mrs. John Herschel, *Memoir and Correspondence of Caroline Herschel,* 2nd ed. (London: John Murray, 1879), 35.

such as the great probability, not to say almost absolute certainty, of her being inhabited." Before Herschel's lunar paper was published, Nevil Maskelyne, England's Astronomer Royal, wrote to request details on Herschel's methods of measurement as well as to ask about his comment on the habitability of the Moon, a somewhat surprising claim because at that time astronomers, including Herschel himself, had solid evidence that the Moon lacks an appreciable atmosphere. In particular, Maskelyne asked: "Does not Mr. Herschall [*sic*] go too far in saying that there is almost an absolute certainty of the moon being inhabited, most especially as many philosophers think she is destitute of three out of four of our elements; i.e. air, water, and fire?" Maskelyne's query was no doubt directed at suggesting to this neophyte astronomer that unsupported claims for lunar life were not appropriate for inclusion in a formal scientific paper. A selection from Herschel's remarkable response to Maskelyne is given next. It was first published in 1913.

William Herschel, "[Is the Moon Inhabited?]" in *The Collected Scientific Papers of Sir William Herschel*, ed. J. L. E. Dreyer (London: Royal Society and Royal Astronomical Society, 1913), 1:xc.

[IS THE MOON INHABITED?]
Letter from Mr. Herschel to the
Reverend Dr. Maskelyne, Astronomer Royal.

Some time ago I delivered to this Society a series of Observations upon the height of the lunar Mountains, which paper, having since been presented to the Royal Society by Dr. Watson, I have been desired, in a Letter from the Revd. Dr. Maskelyne, Astronomer Royal, to Dr. Watson, to deliver a further explanation of my method of measuring the projection of the moon's Mountains, and in answer to that Letter, I have drawn up the inclosed Memorandum in a Letter to Dr. Maskelyne. Now, as the contents of that Memorandum, are in some respect necessary to compleat the Theory of my former paper, I take the liberty of presenting a copy of the Letter to this Society.

I beg leave to observe Sir, that my saying there is almost an absolute certainty of the Moon's being inhabited, may perhaps be ascribed to a certain Enthusiasm which an observer, but young in the Science of Astronomy can hardly divest himself of when he sees such wonders before him; And if you will promise not to call me a Lunatic I will transcribe a passage (from a series of observations on the Moon of a different nature I begun about 18 months ago) which will shew my real sentiments on the subject.

"Perhaps conclusions from the analogy of things may be exceedingly different from truth; but as in things beyond the reach of Observation we have no other way to come at knowledge, the imperfection of these Arguments may in some measure be excused; And I may venture to say that if we do not go so far as to conclude a perfect resemblance, we must allow great weight to inferences taken from this source. For instance, seeing that our Earth is inhabited and comparing the Moon with this planet: finding that in such a satellite there is a provision of light and heat: also, in all appearance a soil proper for habitation full as good as ours, if not perhaps better — who can say that it is not extremely probable, nay beyond doubt, that there must be inhabitants on the Moon of some kind or other? Moreover it is perhaps not altogether so certain that the moon is out of the reach of observation in this respect. I hope, and am convinced, that some time or other very evident signs of life will be discovered on the moon."

"When we call the Earth by way of distinction a planet and the moon a satellite or attendant, we should consider whether we do not perhaps, in a certain sense, mistake the matter. Perhaps — and not unlikely — the moon is the planet and the earth the satellite! Are we not a larger moon to the moon than she is to us? Does it not appear that there is a much more uninterrupted, even temperature there than here? What a glorious View of the heavens from the moon! How beautifully diversified with hills and valleys! No large oceans to take up immense plains, fit for pasture &c: Uninterrupted day on one half, and on the other a day and night of a noble length, equal to many of ours! Do not all the elements seem at war here when we compare the earth with the moon? Air, Water, Fire, Clouds, Tempests, Vulcano[e]s &c: all these are either not on the moon, or at least kept in much greater subjection than here. If as a prerogative we assign the size of the earth, and its motion since it carries the moon along with it; I answer, And is not the sun still larger than the earth, yet we allow him to be only the servant as he yields us light and heat. And tho' by his attraction the earth is made to revolve round him, yet we allow the earth to be the nobler of the two. Even so I say it is with moon. The Earth acts the part of a Carriage, a heavenly waggon to carry about the more delicate moon, to whom it is destined to give a glorious light in the absence of the sun; whereas we as it were travel on foot and have but a small lamp to give us light in our dark nights and that too, often enough extinguished by clouds. For my part, were I to chuse between the Earth and Moon I should not hesitate a moment to fix upon the moon for my habitation."

Colorful as this letter is, it is even more remarkable that it must have taken great restraint on Herschel's part not to have made still bolder claims. This at least is the natural inference to be drawn from Herschel's unpublished lunar observation book. The next selection is a transcription of his entry for 28 May 1776. The handwriting, a sample of which is given in figure 7, is Caroline's, but the observations are William Herschel's and the underlinings are in the original. Some of the spellings are such as one might expect to find in persons whose native language was German, writing only for their own records. This selection has never been published before.

William Herschel, "Lunar Notebooks," as given in the Royal Astronomical Society microfilm (Reel 17) of the Herschel MSS, W.3/1.4, pp. 1–2.

This Evening I tryed a new ten foot Reflector on the Moon with a power of 240. The moment I saw the Moon I was struck with the appearance of something I had never observed before, which I ascribed to the power and distinctness of my Instrument, but which may perhaps be an optical fallacy. But in the first place I will argue and describe the Phenomena, as if those appearances I saw had been founded

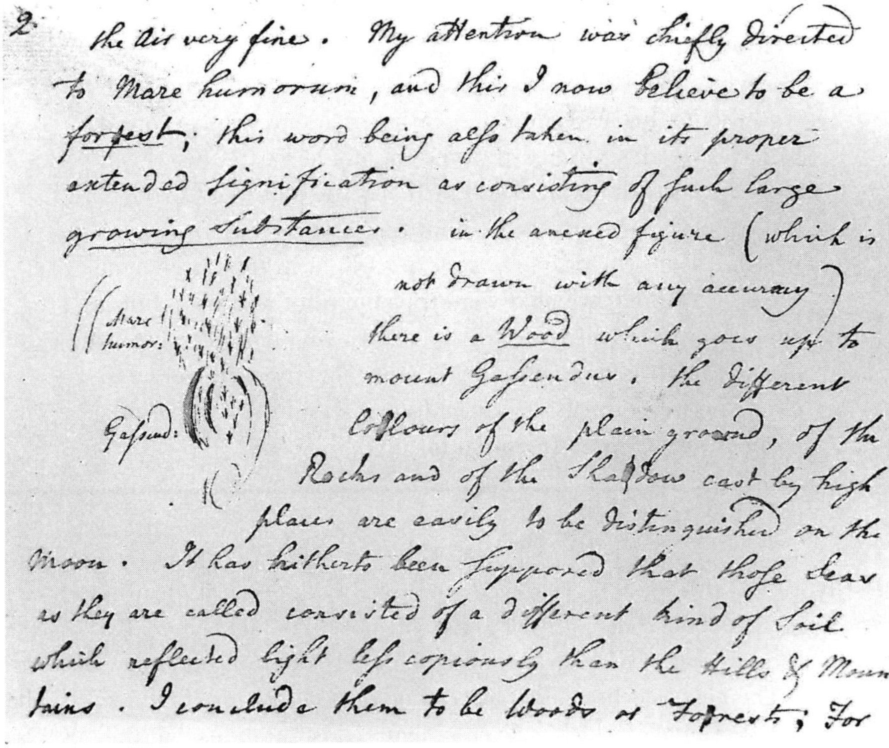

Fig. 7. A section from William Herschel's lunar observation notebook (Courtesy of SPL/Royal Astronomical Society)

upon reality. I believed to perceive some thing which I immediately took to be growing substances, I will not call them Trees as from their size they can hardly come under that denomination, or if I do, it must be understood in that extended signification so as to take in any Size how great soever. The Moon was gibbous, being 12 Days old, also in and near the Meridian; And the air very fine. My attention was chiefly directed to Mare humorum, and this I now believe to be a forest, the word being also taken in its proper extended significance as consisting of such large growing substances. in the anexed figure [Fig. 7 in this edition] (which is not drawn with any accuracy) there is a Wood which goes up to Mount Gassendus. The different Colours of the plain ground, of the Rocks and of the Shadow cast by high places are easily to be distinguished on the Moon. It has hitherto been supposed that those Seas as they are called consisted of a different kind of Soil, which reflected light less copiously than the Hills & and Mountains. I conclude them to be Woods or Forests; For which Opinion I shall bring the following argumts.

Herschel then provided an analysis of these observations, which led him to the following concluding statement.

However, not to lay too much stress on these appearances till they have been better confirmed since I can hardly imagine that any growing Substance could be long enough to be visible from the Earth to the Moon. Our tallest Trees would vanish at that distance. It is not impossible but that the vegetable Creation (and indeed the Animal too) may be of a much larger size on the Moon than it is here, tho' perhaps not very likely. And I suppose that the borders of forests, to be visible, would require Trees at least 4, 5 or 6 times the height of ours.

But the thought of Forrests or Lawn and Pastures still remains exceedingly probable with me; As that will much better account for the different Colours, than different colour'd soils can do.

William Herschel continued to observe the Moon over the next decade, spotting at various times features that he labeled "roads," "canals," "pyramids," and "vegetation." At one point, referring to variously sized circular features of the Moon, he stated: "As it appears probable to me that those round holes of certain size with a raised edge about them are the works of Art, perhaps in some of art and Nature together, I shall distinguish them by the names Metropolis, Cities, Villages and believe there will be less impropriety in these names than there is in calling those larger darkish coloured places Seas which evidently contain no water." Then, thinking better of the matter, he crossed out the words "Metropolis, Cities, Villages" and substituted "Large places, Middling places, Small places"! Nonetheless, extensive as Herschel's lunar observations were and as large as he built his telescopes, he never secured observations of lunar life that satisfied him to the point that he published them. Not all astronomers from that era managed to muster such restraint.

Herschel entered the 1780s essentially unknown in the astronomical community, though a few knew that he had begun building reflecting telescopes of uncommon quality. By 1782, he was world famous, this being due to the fact that in 1781 he had discovered the first new planet detected in historical time, the planet now known as Uranus. This, however, was not his greatest achievement, which consisted in developing both the observational and the conceptual basis of stellar astronomy. One example: when Herschel began observing nebulae, slightly over a hundred of these mysterious objects were known. By the time Herschel put his telescopes aside, he had increased this number by a factor of approximately 25. This extraordinary increase was made possible by the superlative reflecting telescopes that Herschel constructed, the largest of which had a mirror forty-eight inches in diameter. A careful study of Herschel's interests and activities during the 1770s and early 1780s has led me to the hypothesis that what chiefly motivated Herschel's efforts to build giant telescopes was not his desire to discover more nebulae, but rather his quest to detect evidence of extraterrestrial life, a discovery that the previous document shows he believed for a period he had already attained.[21]

One final example of William Herschel's enthusiasm for extraterrestrial life comes from his published writings, in particular, a 1795 paper in the *Philosophical Transactions of the Royal Society* in which he presented a new theory of the structure of the Sun. A particularly striking feature of his new model for the Sun (and for stars in general) was that it entailed the possibility that the Sun (and stars) may be inhabited.

William Herschel, "On the Nature and Construction of the Sun and Fixed Stars," in Herschel, *Scientific Papers*, 1:479.

> The sun, viewed in this light, appears to be nothing else than a very eminent, large, and lucid planet, evidently the first, or in strictness of speaking, the only primary one of our system; all others being truly secondary to it. Its similarity to the other globes of the solar system with regard to its solidity, its atmosphere, and its diversified surface, the rotation upon its axis, and the fall of heavy bodies, leads us on to suppose that it is most probably also inhabited, like the rest of the planets,

21. For details, see M. J. Crowe, *The Extraterrestrial Life Debate, 1750–1900: The Idea of a Plurality of Worlds from Kant to Lowell* (Cambridge: Cambridge University Press, 1986; new ed., Mineola, N.Y.: Dover Publications, 1999), 61–70.

by beings whose organs are adapted to the peculiar circumstances of that vast globe.

Whatever fanciful poets might say, in making the sun the abode of blessed spirits, or angry moralists devise, in pointing it out as a fit place for the punishment of the wicked, it does not appear that they had any other foundation for their assertions than mere opinion and vague surmise; but now I think myself authorized, *upon astronomical principles*, to propose the sun as an inhabitable world, and am persuaded that the foregoing observations, with the conclusions I have drawn from them, are fully sufficient to answer every objection that may be made against it.

EIGHT

EXTRATERRESTRIALS AND THE ENLIGHTENMENT

THE CELESTIAL CITIES OF THE EIGHTEENTH-CENTURY PHILOSOPHERS

What was the Enlightenment? Franklin L. Baumer described it by stating that "The Enlightenment. . . , as Ernst Troeltsch and many others were later to say, was the hinge on which the European nations turned from the Middle Ages to 'modern' times, marking the passage from a supernaturalistic-mystical-authoritative to a naturalistic-scientific-individualistic type of thinking."[1] Another well known historian, Carl Becker, did not discuss ideas of extraterrestrial life in his classic book *The Heavenly Cities of the Eighteenth-Century Philosophers,* in which he maintained that the Enlightenment "*Philosophes* demolished the Heavenly City of St. Augustine only to rebuild it with more up-to-date materials."[2] In that book, Becker also sug-

1. Franklin L. Baumer, *Modern European Thought* (New York: Macmillan, 1977), 141.

2. Carl Becker, *The Heavenly Cities of the Eighteenth-Century Philosophers* (New Haven: Yale University Press, 1932), 31.

gested that some of the differences between medieval and Enlightenment thought were not as great as sometimes assumed.

It is certainly true that during the Enlightenment a major shift occurred regarding the idea of a plurality of worlds, which went from being seen as a somewhat bizarre and dangerous notion to being viewed by many as fully reconcilable with both astronomy and religion. Nonetheless, careful reading of a number of the authors included in this and the previous chapter suggests that at least some Enlightenment intellectuals, although intent on destroying the celestial spheres and angelic hierarchies of medieval cosmology, proceeded to erect new celestial cities concerning which they debated whether the superbeings of our solar system live on its outer or inner planets or on the Sun itself. Moreover, although some Enlightenment figures analyzed the issue of extraterrestrial life in properly scientific terms, the majority were heavily influenced in the positions they adopted by religious or metaphysical suppositions. This suggests that although a shift from a "supernaturalistic-mythical" to a "naturalistic-scientific" mode of thought does describe the program of many Enlightenment pluralists, it may be less descriptive of their practice.

VOLTAIRE (1694–1778)

François-Marie Arouet, who in 1717 assumed the name Voltaire, dealt with extraterrestrials in many of his writings, most notably in his *Micromégas*, which belongs, along with Voltaire's famous *Candide*, to the genre "les contes philosophiques" (philosophical stories). One source of Voltaire's *Micromégas* was his reading of Fontenelle's *Conversations*; in fact, Fontenelle in one form appears in that short story (selected below) as the secretary of the Academy of Saturn. Another source was Voltaire's irritation at a calculation made by the German philosopher Christian Wolff, who was in some ways a follower of Leibniz. In his *Elementa matheseos universae*, Wolff had attempted to calculate the height of inhabitants of Jupiter by assuming that the diminished intensity of the Sun's rays when they reach Jupiter would entail that Jupiterians would have enlarged eyes to compensate for this. Assuming that bodily height and eye-size would be proportional, Wolff assigned Jupiterians a height far beyond that of earthlings. Voltaire expressed his outrage at Wolff's readiness to indulge in such speculation and more generally his distaste for the approach of Leibniz in a 1741 letter to the scientist Maupertuis, who had recently returned from an expedition that provided good empirical evidence for Newton's theory that the Earth bulges at the equator. Sending Maupertuis some verses honoring him, Voltaire stated that

> I will not place such quatrains, my dear flattener of planets . . . , at the base of the portrait of Christianus Volfius. For a long time I have viewed with the stupor of a monad the height which that Germanic babbler assigns to the inhabitants of Jupiter. He judges by the size of our eyes and by the distance of the sun . . . from the earth. . . . That man has brought back to Germany all the horrors of scholasticism surcharged with *sufficient reasons, monads, indiscernibles,* and all the scientific absurdities which Leibniz has in his vanity put on the world, and which the Germans study because they are Germans.[3]

Eleven years later Voltaire made his distaste for Wolff's approach more public when he published his *Micromégas,* which is sometimes seen as a delightful effort to instill humility in earthlings, and at other times as the message of a despairing misanthrope, intent on ridiculing leading thinkers by means of a literary device long available.

Voltaire, *Micromegas,* in *Favorite Works of Voltaire* (Garden City, N.Y.: De Luxe Editions Club, 1900?), 227–28, 229–30, 232–40.

In one of the planets that revolve round the star known by the name of Sirius, was a certain young gentleman of promising parts, whom I had the honor to be acquainted with in his last voyage to this our little ant-hill. His name was Micromegas, an appellation admirably suited to all great men, and his stature amounted to eight leagues in height, that is, twenty-four thousand geometrical paces of five feet each.

Some of your mathematicians, a set of people always useful to the public, will, perhaps, instantly seize the pen, and calculate that Mr. Micromegas, inhabitant of the country of Sirius, being from head to foot four and twenty thousand paces in length, making one hundred and twenty thousand royal feet, that we, denizens of this earth, being at a medium little more than five feet high, and our globe nine thousand leagues in circumference: these things being premised, they will then conclude that the periphery of the globe which produced him

3. *Oeuvres complètes de Voltaire* (Paris: Baudouin Frères, 1825–1834), 72:471–72.

must be exactly one and twenty millions six hundred thousand times greater than that of this our tiny ball. Nothing in nature is more simple and common. The dominions of some sovereigns of Germany or Italy, which may be compassed in half an hour, when compared with the empires of Ottoman, Russia, or China, are no other than faint instances of the prodigious difference that nature hath made in the scale of beings. The stature of his excellency being of these extraordinary dimensions, all our artists will agree that the measure around his body might amount to fifty thousand royal feet,—a very agreeable and just proportion.

His nose being equal in length to one-third of his face, and his jolly countenance engrossing one-seventh part of his height, it must be owned that the nose of this same Sirian was six thousand three hundred and thirty-three royal feet to a hair, which was to be demonstrated. With regard to his understanding, it is one of the best cultivated I have known. He is perfectly well acquainted with abundance of things, some of which are of his own invention; for, when his age did not exceed two hundred and fifty years, he studied, according to the custom of the country, at the most celebrated university of the whole planet, and by the force of his genius discovered upwards of fifty propositions of Euclid, having the advantage by more than eighteen of Blaise Pascal, who, (as we are told by his own sister,) demonstrated two and thirty for his amusement and then left off, choosing rather to be an indifferent philosopher than a great mathematician.

About the four hundred and fiftieth year of his age, or latter end of his childhood, he dissected a great number of small insects not more than one hundred feet in diameter, which are not perceivable by ordinary microscopes, of which he composed a very curious treatise, which involved him in some trouble. The mufti of the nation, though very old and very ignorant, made shift to discover in his book certain lemmas that were suspicious, unseemly rash, heretic, and unsound, and prosecuted him with great animosity; for the subject of the author's inquiry was whether, in the world of Sirius, there was any difference between the substantial forms of a flea and a snail.

Micromegas defended his philosophy with such spirit as made all the female sex his proselytes; and the process lasted two hundred and twenty years; at the end of which time, in consequence of the mufti's interest, the book was condemned by judges who had never read it, and the author expelled from court for the term of eight hundred years.

After being banished, Micromegas decides to spend the time hopping comets from planet to planet, learning about the inhabitants. Eventually he reaches Saturn, where he makes the acquaintance of one of the inhabitants of that planet, who is "only about a thousand fathoms tall" (415) and, for this reason is alternately referred to as "the dwarf" throughout the rest of the story. The story picks up again with Micromegas and the Saturnian discussing philosophy. Micromegas is speaking.

". . . begin, therefore, without further preamble, and tell me how many senses the people of this world enjoy."

"We have seventy and two," said the academician, "but we are daily complaining of the small number, as our imagination transcends our wants, for, with the seventy-two senses, our five moons and ring, we find ourselves very much restricted; and notwithstanding our curiosity, and the no small number of those passions that result from these few senses, we have still time enough to be tired of idleness."

"I sincerely believe what you say," cried Micromegas, "for, though we Sirians have near a thousand different senses, there still remains a certain vague desire, an unaccountable inquietude incessantly admonishing us of our own unimportance, and giving us to understand that there are other beings who are much our superiors in point of perfection. I have traveled a little, and seen mortals both above and below myself in the scale of being, but I have met with none who had not more desire than necessity, and more want than gratification. Perhaps I shall one day arrive in some country where nought is wanting but hitherto I have had no certain information of such a happy land."

The Saturnian and his guest exhausted themselves in conjectures upon this subject, and after abundance of argumentation equally ingenious and uncertain, were fain to return to matter of fact.

"To what age do you commonly live?" said the Sirian.

"Lack-a-day! a mere trifle," replied the little gentleman.

"It is the very same case with us," resumed the other, "the shortness of life is our daily complaint, so that this must be an universal law in nature."

"Alas!" cried the Saturnian. "few, very few on this globe outlive five hundred to about fifteen thousand years. So, you see, we in a manner begin to die the very moment we are born: our existence is no more than a point, our duration an instant, and our globe an atom. Scarce do we begin to learn a little, when death intervenes before we can profit by experience. For my own part, I am deterred from laying schemes when I consider myself as a single drop in the midst of an immense ocean. I am particularly ashamed, in your presence, of the ridiculous figure I make among my fellow-creatures."

To this declaration, Micromegas replied:

"If you were not a philosopher, I should be afraid of mortifying your pride by telling you that the term of our lives is seven hundred times longer than the date of your existence: but you are very sensible that when the texture of the body is resolved, in order to reanimate nature in another form, which is the consequence of what we call death—when that moment of change arrives, there is not the least difference betwixt having lived a whole eternity, or a single day. I have been in some countries where the people live a thousand times longer than with us, and yet they murmured at the shortness of their time. But one will find everywhere some few persons of good sense, who know how to make the best of their portion, and thank the author of nature for his bounty.

The Saturnian accompanies Micromegas to Jupiter and then Mars, the latter of which is "so small that they feared they might not have enough room to lay themselves down" (421). They continue on to the Earth, where they land on the northern shore of the Baltic Sea. They set off exploring the Earth, at first concluding that it is uninhabited because they can see no living beings. This sets off a discussion between them of the value of our planet.

"But this globe," said the dwarf, "is ill contrived; and so irregular in its form as to be quite ridiculous. The whole together looks like a chaos. Do but observe these little rivulets; not one of them runs in a straight line: and these ponds which are neither round, square, nor oval, nor indeed of any regular figure; together with these little sharp pebbles (meaning the mountains) that roughen the whole surface of the globe, and have torn all the skin from my feet. Besides, pray take notice of the shape of the whole, how it flattens at the poles, and turns round the sun in an awkward oblique manner, so as that the polar circles cannot possibly be cultivated. Truly, what makes me believe there is no inhabitant on this sphere, is a full persuasion that no sensible being would live in such a disagreeable place."

"What then?" said Micromegas, "perhaps the beings that inhabit it come not under that denomination; but, to all appearance, it was not made for nothing. Everything here seems to you irregular; because you fetch all your comparisons from Jupiter or Saturn. Perhaps this is the very reason of the seeming confusion which you condemn; have I not told you, that in the course of my travels I have always met with variety?"

The Saturnian replied to all these arguments; and perhaps the dispute would have known no end, if Micromegas, in the heat of the contest, had not luckily broken the string of his diamond necklace, so that the jewels fell to the ground; they consisted of pretty small unequal karats, the largest of which weighed four hundred pounds, and the smallest fifty. The dwarf, in helping to pick them up, perceived, as they approached his eye, that every single diamond was cut in such a manner as to answer the purpose of an excellent microscope. He therefore took up a small one, about one hundred and sixty feet in diameter, and applied it to his eye, while Micromegas chose another of two thousand five hundred feet. Though they were of excellent powers, the observers could perceive nothing by their assistance, so they were altered and adjusted. At length, the inhabitant of Saturn discerned something almost imperceptible moving between two waves in the Baltic. This was no other than a whale, which, in a dexterous manner, he caught with his little finger, and, placing it on the nail of his thumb, showed it to the Syrian, who laughed heartily at the excessive smallness peculiar to the inhabitants of this our globe. The Saturnian, by this time convinced that our world was inhabited, began to imagine we had no other animals than whales; and being a mighty debater, he forthwith set about investigating the origin and motion of this small atom, curious to know whether or not it was furnished with ideas, judgment, and free will. Micromegas was very much perplexed upon this subject. He examined the animal with the most patient attention, and the result of his inquiry was, that he could see no reason to believe a soul was lodged in such a body. The two travelers were actually inclined to think there was no such thing as mind in this our habitation, when, by the help of their microscope, they perceived something as large as a whale floating upon the surface of the sea. It is well known that, at this period, a flight of philosophers were upon their return from the polar circle, where they had been making observations, for which nobody has hitherto been the wiser. The gazettes record that their vessel ran ashore on the coast of Bothnia and that they with great difficulty saved their lives; but in this world one can never dive to the bottom of things. For my own part, I will ingenuously recount the transaction just as it happened, without any addition of my own; and this is no small effort in a modern historian.

Micromegas stretched out his hand gently toward the place where the object appeared, and advanced two fingers, which he instantly pulled back, for fear of being disappointed, then opening softly and shutting them all at once, he very dexterously seized the ship that contained those gentlemen, and placed it on his nail, avoiding too much pressure, which might have crushed the whole in pieces.

"This," said the Saturnian dwarf, "is a creature very different from the former."

Upon which the Sirian placing the supposed animal in the hollow of his hand, the passengers and crew, who believed themselves thrown by a hurricane upon some rock, began to put themselves in motion. The sailors having hoisted out some casks of wine, jumped after them into the hand of Micromegas: the mathematicians having secured their quadrants, sectors, and Lapland servants, went overboard at a different place, and made such a bustle in their descent, that the Sirian at length felt his fingers tickled by something that seemed to move. An iron bar chanced to penetrate about a foot deep into his forefinger; and from this prick he concluded that something had issued from the little animal he held in his hand; but at first he suspected nothing more: for the microscope, that scarce rendered a whale and a ship visible, had no effect upon an object so imperceptible as man.

I do not intend to shock the vanity of any person whatever; but here I am obliged to beg your people of importance to consider that, supposing the stature of a man to be about five feet, we mortals make just such a figure upon the earth, as an animal the sixty thousandth part of a foot in height would exhibit upon a bowl ten feet in circumference. When you reflect upon a being who could hold this whole earth in the palm of his hand, and is provided with organs proportioned to those we possess, you will easily conceive that there must be a great variety of created substances; and pray, what must such beings think of those battles by which a conqueror gains a small village, to lose it again in the sequel?

I do not at all doubt, but if some captain of grenadiers should chance to read this work, he would add two large feet at least to the caps of his company; but I assure him his labor will be in vain; for, do what he will, he and his soldiers will never be other than infinitely diminutive and inconsiderable.

What wonderful address must have been inherent in our Sirian philosopher that enabled him to perceive those atoms of which we have been speaking. When Leuwenhoek and Hartsoecker observed the first rudiments of which we are formed, they did not make such an astonishing discovery. What pleasure, therefore, was the portion of Micromegas, in observing the motion of those little machines, in examining all their pranks, and following them in all their operations! With what joy did he put his microscope into his companion's hand; and with what transport did they both at once exclaim:

"I see them distinctly,—don't you see them carrying burdens, lying down and rising up again?"

So saying, their hands shook with eagerness to see, and apprehension to lose such uncommon objects. The Saturnian, making a sudden transition from the most cautious distrust to the most excessive credulity, imagined he saw them engaged in their devotions and cried aloud in astonishment.

Nevertheless, he was deceived by appearances: a case too common, whether we do or do not make use of microscopes.

Micromegas being a much better observer than the dwarf, perceived distinctly that those atoms spoke; and made the remark to his companion, who was so much ashamed of being mistaken in his first suggestions that he would not believe such a puny species could possibly communicate their ideas: for, though he had the gift of tongues, as well as his companions he could not hear those particles speak; and therefore supposed they had no language.

"Besides, how should such imperceptible beings have the organs of speech? and what in the name of Jove can they say to one another? In order to speak, they must have something like thought, and if they think, they must surely have something equivalent to a soul. Now, to attribute anything like a soul to such an insect species appears a mere absurdity."

"But just now," replied the Sirian, "you believed they were engaged in devotional exercises; and do you think this could be done without thinking, without using some sort of language, or at least some way of making themselves understood? Or do you suppose it is more difficult to advance an argument than to engage in physical exercise? For my own part, I look upon all faculties as alike mysterious.

"I will no longer venture to believe or deny," answered the dwarf: "in short I have no opinion at all. Let us endeavor to examine these insects, and we will reason upon them afterward."

"With all my heart," said Micromegas, who, taking out a pair of scissors which he kept for paring his nails, cut off a paring from his thumb nail, of which he immediately formed a large kind of speaking trumpet, like a vast tunnel, and clapped the pipe to his ear: as the circumference of this machine included the ship and all the crew, the most feeble voice was conveyed along the circular fibres of the nail; so that, thanks to his industry, the philosopher could distinctly hear the buzzing of our insects that were below. In a few hours he distinguished articulate sounds, and at last plainly understood the French language. The dwarf heard the same, though with more difficulty.

The astonishment of our travelers increased every instant. They heard a nest of mites talk in a very sensible strain: and that *Lusus Naturœ* seemed to them inexplicable. You need not doubt but the Sirian and his dwarf glowed with impatience to enter into conversation with such atoms. Micromegas being afraid that his voice, like thunder, would deafen and confound the mites, without being understood by them, saw the necessity of diminishing the sound; each, therefore, put into his mouth a sort of small toothpick, the slender end of which reached to the vessel. The Sirian setting the dwarf upon his knees, and the ship and crew upon his nail, held down his head and spoke softly. In fine, having taken these and a great many more precautions, he addressed himself to them in these words:

"O ye invisible insects, whom the hand of the Creator hath deigned to produce in the abyss of infinite littleness! I give praise to his goodness, in that he hath been pleased to disclose unto me those secrets that seemed to be impenetrable."

If ever there was such a thing as astonishment, it seized upon the people who heard this address, and who could not conceive from whence it proceeded. The chaplain of the ship repeated exorcisms, the sailors swore, and the philosophers formed a system: but, notwithstanding all their systems, they could not divine who the person was that spoke to them. Then the dwarf of Saturn, whose voice was softer than that of Micromegas, gave them briefly to understand what species of beings they had to do with. He related the particulars of their voyage from Saturn, made them acquainted with the rank and quality of Monsieur Micromegas; and, after having pitied their smallness, asked if they had always been in that miserable state so near akin to annihilation; and what their business was upon that globe which seemed to be the property of whales. He also desired to know if they were happy in their situation? if they were inspired with souls? and put a hundred questions of the like nature.

A certain mathematician on board, braver than the rest, and shocked to hear his soul called in question, planted his quadrant, and having taken two observations of this interlocutor, said: "You believe then, Mr. what's your name, that because you measure from head to foot a thousand fathoms—"

"A thousand fathoms!" cried the dwarf, "good heavens! How should he know the height of my stature? A thousand fathoms! My very dimensions to a hair. What, measured by a mite! This atom, forsooth, is a geometrician and knows exactly how tall I am: while I, who can scarce perceive him through a microscope, am utterly ignorant of his extent!

"Yes, I have taken your measure," answered the philosopher, "and I will now do the same by your tall companion."

The proposal was embraced: his excellency reclined upon his side; for, had he stood upright, his head would have reached too far above the clouds. Our mathematicians planted a tall tree near him, and then, by a series of triangles joined together, they discovered that the object of their observation was a strapping youth, exactly one hundred and twenty thousand royal feet in length. In consequence of this calculation, Micromegas uttered these words:

"I am now more than ever convinced that we ought to judge of nothing by its external magnitude. O God! who hast bestowed understanding upon such seemingly contemptible substances, thou canst with equal ease produce that which is infinitely small, as that which is incredibly great: and if it be possible, that among thy works there are beings still more diminutive than these, they may nevertheless, be endued with understanding superior to the intelligence of those stupendous animals I have seen in heaven, a single foot of whom is larger than this whole globe on which I have alighted."

One of the philosophers assured him that there were intelligent beings much smaller than men, and recounted not only Virgil's whole fable of the bees; but also described all that Swammerdam hath discovered, and Reaumur dissected. In a word, he informed him that there are animals which bear the same proportion to bees, that bees bear to man; the same as the Sirian himself compared to those vast beings whom he had mentioned; and as those huge animals as to other substances, before whom they would appear like so many particles of dust. Here the conversation became very interesting, and Micromegas proceeded in these words:

"O ye intelligent atoms, in whom the Supreme Being hath been pleased to manifest his omniscience and power, without all doubt your joys on this earth must be pure and exquisite: for, being unencumbered with matter, and, to all appearances little else than soul, you must spend your lives in the delights of pleasure and reflection, which are the true enjoyments of a perfect spirit. True happiness I have no where found; but certainly here it dwells."

At this harangue all the philosophers shook their heads, and one among them, more candid than his brethren, frankly owned, that excepting a very small number of inhabitants who were very little esteemed by their fellows, all the rest were a parcel of knaves, fools, and miserable wretches.

"We have matter enough," said he, "to do abundance of mischief, if mischief comes from matter; and too much understanding, if evil flows from understanding. You must know, for example, that at this very moment, while I am speaking, there are one hundred thousand animals of our own species, covered with hats, slaying an equal number of their fellow-creatures, who wear turbans; at least they are either slaying or being slain; and this hath usually been the case all over the earth from time immemorial."

The Sirian, shuddering at this information, begged to know the cause of those horrible quarrels among such a puny race; and was given to understand that the subject of the dispute was a pitiful mole-hill [called Palestine,] no larger than his heel. Not that any one of those millions who cut one another's throats pretends to have the least claim to the smallest particle of that clod. The question is, whether it shall belong to a certain person who is known by the name of Sultan, or to another whom (for what reason I know not) they dignify with the appellation of King. Neither the one nor the other has seen or ever will see the pitiful corner in question; and probably none of these wretches, who so madly destroy each other, ever beheld the ruler on whose account they are so mercilessly sacrificed!

"Ah, miscreants!" cried the indignant Sirian, "such excess of desperate rage is beyond conception. I have a good mind to take two or three steps, and trample the whole nest of such ridiculous assassins under my feet."

"Don't give yourself the trouble," replied the philosopher, "they are industrious enough in procuring their own destruction. At the end of ten years the hundredth part of those wretches will not survive; for you must know that, though they should not draw a sword in the cause they have espoused, famine, fatigue, and intemperance, would sweep almost all of them from the face of the earth. Besides, the punishment should not be inflicted upon them, but upon those sedentary and slothful barbarians, who, from their palaces, give orders for murdering a million of men and then solemnly thank God for their success."

Our traveler was moved with compassion for the entire human race, in which he discovered such astonishing contrast. "Since you are of the small number of the wise," said he, "and in all likelihood do not engage yourselves in the trade of murder for hire, be so good as to tell me your occupation."

"We anatomize flies," replied the philosopher, "we measure lines, we make calculations, we agree upon two or three points which we understand, and dispute upon two or three thousand that are beyond our comprehension."

In a number of paragraphs omitted at this point, various earthlings impress the Sirian and "dwarf" by their knowledge of science. This leads to a discussion of the nature of the soul, in the course of which Voltaire mentions the views on this question of such authors as Aristotle, Descartes, Leibniz, and Locke. The conclusion follows.

[A man] in a square cap, who, taking the word from all his philosophical brethren, affirmed that he knew the whole secret. He surveyed the two celestial strangers from top to toe, and maintained to their faces that their persons, their fashions, their suns and their stars, were created solely for the use of man. At this wild assertion our two travelers were seized with a fit of that uncontrollable laughter, which (according to Homer) is the portion of the immortal gods: their bellies quivered, their shoulders rose and fell, and, during these convulsions, the vessel fell from the Sirian's nail into the Saturnian's pocket, where these worthy people searched for it a long time with great diligence. At length, having found the ship and set everything to rights again, the Sirian resumed the discourse with those diminutive mites, and promised to compose for them a choice book of philosophy which would demonstrate the very essence of things. Accordingly, before his departure, he made them a present of the book, which was brought to the Academy of Sciences at Paris, but when the old secretary came to open it he saw nothing but blank paper.

"Ay, ay," said he, "this is just what I suspected."

PIETY, POETRY, AND A PLURALITY OF WORLDS: "AN UNDEVOUT ASTRONOMER IS MAD"

As mentioned in the preface, numerous literary and religious figures embraced extraterrestrials and set them to work propagating messages of very varied nature. Included in what follows is a selection of such authors. The literary figures chosen—Pope, Gray, and Young—from the first half of the eighteenth century set the stage for later poets no less prominent than Coleridge and Wordsworth.

A certain playfulness (along with much that is serious) will be found in some of the poetry that follows, especially in Thomas Gray's "Luna Habitabilis" and in William Wordsworth's "Peter Bell." Such playfulness contrasts sharply with the somberness of tone that characterizes the writings of the religious authors included in this chapter. Fantastic as the claims made by Swedenborg may seem, he made them with full seriousness. The passages from Samuel Pye may appear playful, but such was not their author's intent—these passages were part of an attack Pye launched against the deistic writings of Lord Bolingbroke. In reading the selections from these authors, it is wise to keep in mind that these authors in their advocacy and employment of extraterrestrials were not exceptional. Dozens of other poets and religious writers writing in many languages could have been included, as is indicated by the evidence given in the preface that half or more of eighteenth-century intellectuals dealt with ideas of extraterrestrial life in their writings.

Alexander Pope (1688–1744)

One of the most famous English poets of the eighteenth century, Pope wrote in a philosophical manner, especially in his "Essay on Man," which is one of the last and among the most famous of his poetic creations. An idea prominent in that poem, including the passages cited next, is the idea of a Great Chain of Being, that all forms capable of existence are realized in the material world and that these forms can be ranked in a sort of linear scale from the lowest to the highest.

Alexander Pope, selections from *Essay on Man* (1733–1734), in his *Selected Works*, ed. Louis Kronenberger (New York: Random House, 1948), 97–137.

> Thro' worlds unnumber'd tho' the God be known,
> 'Tis ours to trace him only in our own.
> He, who thro' vast immensity can pierce,
> See worlds on worlds compose one universe,
> Observe how system into system runs,
> What other planets circle other suns,
> What vary'd Being people ev'ry star,
> May tell why Heav'n has made us as we are.
>
> Epistle I, lines 22–30

> Know then thyself, presume not God to scan;
> The proper study of Mankind is Man.
> Plac'd on this isthmus of a middle state,
> A Being darkly wise, and rudely great:
> With too much knowledge for the Sceptic side;
> With too much weakness for the Stoic's pride,
> He hangs between; . . . .
>
> Epistle II, lines 1–7

Superior beings, when of late they saw
A mortal man unfold all Nature's law,
Admired such wisdom in an earthly shape
And shewed a Newton as we shew an Ape.
Epistle II, lines 31–34

Thomas Gray (1716–1771)

Gray not only studied at Cambridge University but ended up as a don at that university. It was while he was a twenty-one year old student there that Gray wrote his "Luna Habitabilis" in Latin. Gray is above all remembered for his poem "Elegy Written in a Country Churchyard."

Thomas Gray, "Luna Habitabilis" (1737), trans. Sally Purcell, in *Poems of Science*, ed. J. Heath-Stubbs and P. Salman (Harmondsworth: Penguin, 1984), 146–48.

Surely these worlds, these jewels of high heaven,
Do not shine for us alone, a decorated ceiling,
Giant stage-flats or theatrical curtains?
. . . .
This world [the Moon] has clouds and dewfall of its own,
And icy cold, and showers for the grass;
Here too Iris' rainbow shines, rosy dawn, and twilight.
Can you believe such a world lacks men
To till the fields, and build their cities high,
To go to war, and celebrate
Triumphs for victory? Here too great deeds win honour;
Fear, love, and human passions touch their hearts.
And as it pleases us to let our eyes
Travel their shining plains and sound their sea,
They feel the same thrill when our earth appears,
Vaster than theirs, and golden, in clearer skies.
They surely study every sea and land,
Even the dwellers underneath our poles,

And stay up tireless through a summer night
To search the heaven and study its fires.
See, France appears, broad Germany behind,
And further rise the snowy Apennines;
Northward lies Britain, like a beauty patch,
Tiny, although far brighter than the rest.
. . . .
Still I shall sing out my praises, oracles
Of England's fame, and prophecies Fate long ago wrote down.
A time will come that sees great hastening crowds
Of colonists leaving for the moon, exchanging
Known for unknown homes, as old inhabitants
Watch in amazement flying fleets and novel 'birds' on high,
As when Columbus crossed the broad unknown
Sailing to seek new kingdoms, and the shores
And waters all around saw wondering
His mounted troops in armour clad, his ships
Like Trojan horses, and their man-made lightning—
Soon treaties are drawn up, and trade begins
Between the world, through now-familiar space.
Our England, that already rules the waves
And keeps the winds in awe, shall now extend
Her ancient triumphs over conquered air.

Edward Young (1683–1765)

An Anglican minister, Young was nearly sixty when he began to publish his book-length poem *Night Thoughts*. Although now rarely read, it attained during the eighteenth century and beyond extraordinary popularity. One of Young's biographers states that it was "for more than a hundred years . . . more frequently reprinted than any other book of the eighteenth century."[4] Nor was its popularity confined to Britain. It was translated into French, German, Italian, Magyar, Portuguese, Spanish, and Swedish.

4. Henry C. Shelley, *Life and Letters of Edward Young* (London: Pitman, 1914), 198.

Edward Young, *Night Thoughts* (1742–1745), "Night the Ninth," in his *Night Thoughts*, ed. George Gilfillan (Edinburgh: J. Nichol, 1853):

This gorgeous apparatus! this display!
This ostentation of creative power!
This theatre!—what eye can take it in?
By what divine enchantment was it raised,
For minds of the first magnitude to launch
In endless speculation, and adore?
One sun by day, by night ten thousand shine;
And light us deep into the Deity;
How boundless in magnificence and might!
(IX, 742–50)

Devotion! daughter of Astronomy!
An undevout astronomer is mad.
True, all things speak a God, but in the small
Men trace out Him; in great, he seizes man. . . .
(IX, 772–75)

. . . what swarms
Of worlds that laugh at earth; immensely great,
Immensely distant from each other's spheres
What, then, the wondrous space thro' which they roll?
At once it quite ingulfs all human thought:
"Tis comprehension's absolute defeat. . . .
(IX, 1102–7)

[Addressed to extraterrestrials]
What e'er your nature, this is past dispute,
Far other life you live, far other tongue
You talk, far other thought, perhaps, you think,
Than man. How various are the works of God?
But say, what thought? Is Reason here Inthroned
And Absolute? Or sense in arms against her?
Have you two lights? or need you no reveal'd? . . .
And had your Eden an abstemious Eve? . . .
Or if your mother fell, are you redeem'd? . . .

Is this your final residence? If not,
Change you your scene, translated? or by death?
And if by death: what death? . . .

(IX, 1766–81)

Samuel Pye

Whereas most of the authors treated in this chapter are so prominent that they scarcely need to be identified, Samuel Pye, M.D., is so obscure that efforts to determine his birth and death dates seem doomed to failure. His goal in his *Moses and Bolingbroke*, a curious if somewhat ingenious book, is to refute claims of the deistically inclined philosopher and statesman Henry St. John, Viscount Bolingbroke (1678–1751), by setting up a debate between Moses and Bolingbroke. In the process of doing this, Pye employed extraterrestrials on behalf of his cause, going even to the extent of composing a Jupiterian Genesis, a portion of which is provided.

Samuel Pye, *Moses and Bolingbroke* (London: L. Sandby, 1765), 60–63.

GENESIS

Chap. I

1. In the beginning God created the heaven and Jupiter.

. . .

16. And God made five great lights; the greater light to rule the day, and the lesser lights to rule the night: and stars also.

Chap. II

. . .

2. And on the fifteenth day God ended his work which he had made: and he rested on the fifteenth day from all his work, which he had made.

3. And God blessed the fifteenth day, and sanctified it.

Samuel Taylor Coleridge (1772–1834)

The five most prominent English Romantic poets were Byron, Coleridge, Shelley, Wordsworth, and Keats. At least the first four were involved in the extraterrestrial life debate. It was while he was a student at Cambridge University that Coleridge composed in Latin the poem given below in the translation made by his friend and fellow poet Robert Southey. Although expressing in this poem a desire to visit other inhabited or "vivified" worlds, Coleridge, as will be seen in the next chapter, eventually shifted to become an opponent of extraterrestrials.

Samuel Taylor Coleridge, "A Greek Ode on Astronomy" (1793), trans. Robert Southey, in *Coleridge's Poems*, ed. J. B. Beer (London: Dent, 1963), 15.

VI
Nor shalt thou escape my sight,
Who at the threshold of the sun-trod domes
Art trembling, . . . youngest Daughter of the Night!
And you, ye fiery-tressed strangers! you,
Comets who wander wide,
Will I along your pathless way pursue,
Whence bending I may view
The worlds whom elder Suns have vivified.

William Wordsworth (1770–1850)

Like Coleridge, Wordsworth was educated at Cambridge, but unlike him, incorporated extraterrestrials in both his early and late poetry. In fact, thirty-five years separate the composition of his "Peter Bell, A Tale" and his "To the Moon."

William Wordsworth, "Peter Bell, A Tale" (composed 1798; published 1819), in his *Poetical Works*, ed. Thomas Hutchinson (London: Oxford University Press, 1910), 236–37.

Away we go, my Boat and I—
Frail man ne'er sate in such another;
Whether among the winds we strive,
Or deep into the clouds we dive,
Each is contented with the other.

Away we go—and what care we
For treason, tumults, and for wars?
We are as calm in our delight
As is the crescent-moon so bright
Among the scattered stars.

Up goes my Boat among the stars
Through many a breathless field of light,
Through many a long blue field of ether,
Leaving ten thousand stars beneath her;
Up goes my little Boat so bright!

The Crab, the Scorpion, and the Bull—
We pry among them all; have shot
High o'er the red-haired race of Mars,
Covered from top to toe with scars:
Such company I like it not!

And towns in Saturn are decayed,
And melancholy Spectres throng among them;—
The Pleiads, that appear to kiss
Each other in this vast abyss,
With joy I sail among them.

Swift Mercury resounds with mirth,
Great Jove is full of stately bowers;
But these, and all that they contain,
What are they to that tiny grain,
That little Earth of ours?

Prologue, lines 21–50

William Wordsworth, "To the Moon" (composed 1835; published 1837), in his *Poetical Works*, 460–61.

> Then silent Monitress [the Moon]! let us—not blind
> To worlds unthought of till the searching mind
> Of Science laid them open to mankind—
> Told, also, how the voiceless heavens declare
> God's glory; and acknowledging thy share
> In that blest charge; let us—without offence
> To aught of highest, holiest, influence—
> Receive whatever good 'tis given thee to dispense
> May sage and simple, catching with one eye
> The moral intimations of the Sky. . . .
>
> lines 39–48

William Wordsworth, untitled poem, in *Poetical Works of William Wordsworth*, new and revised ed. (London: Edwin Moxon, 1849), 3:67.

> The stars are mansions built by Nature's hand,
> And, haply, there the spirits of the blest
> Dwell, clothed in radiance, their immortal vest; . . .
>
> Part II, xxv, lines 1–3

BENJAMIN FRANKLIN (1706–1790) AND JOHN ADAMS (1735–1826)

Benjamin Franklin

Ideas of extraterrestrial life have interacted with religion in numerous ways. Two especially interesting instances center on Benjamin Franklin and John Adams, the first playing a key role in the founding of the new nation, the latter serving as the second U.S. president. In 1728, shortly after spending eighteen months in England, Franklin drafted his "Articles of Belief," in which he not only expresses his deepest religious convictions but also discusses at some length his views regarding extraterrestrials.

Benjamin Franklin, "Articles of Belief and Acts of Religion," in *Works of Benjamin Franklin*, with notes and a life of the author by Jared Sparks (Boston: Hilliard, Gray, and Company, 1840), 2:1–3.

FIRST PRINCIPLES

I believe there is one Supreme most perfect Being, author and father of the gods themselves.

For I believe that man is not the most perfect Being but one, but rather that there are many degrees of beings superior to him.

Also, when I stretch my Imagination through and beyond our system of planets, beyond the visible fixed stars themselves, into that space that is every way infinite, and conceive it filled with suns like ours, each with a chorus of worlds for ever moving round him, then this little ball on which we move, seems, even in my narrow imagination, to be almost nothing, and my self less than nothing, and of no sort of consequence.

When I think thus, I imagine it great Vanity in me to suppose, that the *Supremely Perfect* does in the least regard such an inconsiderable nothing as man; more especially, since it is impossible for me to have any clear idea of that which is infinite and incomprehensible, I cannot conceive otherwise, than that he *the Infinite Father* expects or requires no worship or praise from us, but that he is even infinitely above it.

But, since there is in all men something like a natural principle, which inclines them to DEVOTION, or the worship of some unseen power.

And since men are endued with reason superior to all other animals, that we are in our world acquainted with;

Therefore, I think it seems required of me, and my duty as a man, to pay divine regards to SOMETHING.

I conceive, then, that the INFINITE has created many beings or gods, vastly superior to man, who can better conceive his perfections than we, and return him a more rational and glorious praise; as, among men, the praise of the ignorant or of children is not regarded by the ingenious painter or architect, who is rather honored and pleased with the approbation of wise men and artists.

It may be that these created gods are immortal, or it may be, that, after many ages, they are changed, and others supply their places.

Howbeit, I conceive that each of these is exceeding wise and good, and very powerful; and that Each has made for himself, one glorious sun, attended with a beautiful and admirable system of planets.

It is that particular wise and good God, who is the author and owner of our system, that I propose for the object of my praise and adoration.

For I conceive that he has in himself some of those passions he has planted in us; and that, since he has given us reason whereby we are capable of observing his wisdom in the creation, he is not above caring for us, being pleased with our praise, and offended when we slight him, or neglect his glory.

I conceive, for many reasons, that he is a *good Being*; and, as I should be happy to have so wise, good, and powerful a Being my friend, let me consider in what manner I shall make myself most acceptable to him.

Next to the praise resulting from and due to this wisdom, I believe he is pleased and delights in the happiness of those he has created; and, since without virtue a man can have no happiness in this world, I firmly believe he delights to see me virtuous, because he is pleased when he sees me happy.

And since he has created many things, which seem purely designed for the delight of man, I believe he is not offended, when he sees his children solace themselves in any manner of pleasant exercises and innocent delights; and I think no pleasure innocent, that is to man hurtful.

I *love* him therefore for his goodness, and I *adore* him for his wisdom.

Let me not fail, then to praise my God continually, for it is his due and it is all I can return for his many favors and great goodness to me; and let me resolve to be virtuous, that I may be happy, that I may please him, who is delighted to see me happy. Amen!

Scholars have investigated where Franklin may have derived this very non-traditional hierarchy of deities. The surprising but fully plausible answer is that his source was Isaac Newton. The story is that Newton on 7 March 1725 revealed to John Conduit his views (documented in chapter 6, above) concerning a plurality of gods presiding over separate solar systems, views quite similar to those advocated three years later by Franklin in his "Articles of Belief." Not only was Franklin in London at that time, but had come to know Henry Pemberton, who was among Newton's closest associates. This seems to have been the link through which this theological conception passed to Franklin.[5]

John Adams

As the case of Franklin suggests, the doctrine of a plurality of worlds could lead persons rather far from traditional Christianity. By the time John Adams graduated in 1755 from Harvard, that doctrine had been taught there, and in 1752 a disputation on the issue had been conducted at Yale.[6] Evidence from Adams's diary from 1756 indicates that during this period he was very impressed by the idea of extraterrestrial life and was, moreover, struggling to find his way between the Calvinism of his ancestry and the deism of his day.

John Adams, *Diary and Autobiography of John Adams*, by L. H. Butterfield (New York: Atheneum, 1964), 1:22.

Astronomers tell us, with good Reason, that not only all the Planets and Satellites in our Solar System, but all the

5. For details, see James Parton, *Life and Times of Benjamin Franklin* (New York: Mason Brothers, 1864), 1:174–75.

6. For this and other information on the history of American pluralism in this period, see Herbert Leventhal, *In the Shadow of the Enlightenment: Occultism and Renaissance Science in Eighteenth-Century America* (New York: New York University Press, 1976), 242–47.

unnumbered Worlds that revolve round the fixt Starrs are inhabited, as well as this Globe of Earth. If this is the Case all Mankind are no more in comparison of the whole rational Creation of God, than a point to the Orbit of Saturn. Perhaps all these different Ranks of Rational Beings have in a greater or less Degree, committed moral Wickedness. If so, I ask a Calvinist, whether he will subscribe to this Alternitive [*sic*], "either God almighty must assume the respective shapes of all these different Species, and suffer the Penalties of their Crimes, in their Stead, or else all these Being[s] must be consigned to everlasting Perdition?"

Many evidences indicate that deism won the day in Adams's case. Moreover, clear evidence reveals Adams's religious views in the next to last year of his life. On 22 January 1825, Adams wrote to his friend and former vice president Thomas Jefferson, who was then engaged in hiring faculty for the University of Virginia. On that process, Adams imparted the advice that he should avoid hiring European professors. The reason Adams offered was that "They all believe that great Principle which has produced this boundless universe, Newton's universe and Herschell's [*sic*] universe, came down to this little ball, to be spit upon by the Jews. And until this awful blasphemy is got rid of, there never will be any liberal science in the world."[7]

DAVID RITTENHOUSE (1732–1796)

In America around the time of the American Revolution, Philip Freneau, known as "the poet of the Revolution," discussed the idea of a plurality of worlds in various writings, citing at times as a source David Rittenhouse, whom Freneau labeled "the prince of astronomers." In fact, among eighteenth-century American physical scientists, Rittenhouse was rivaled only by Benjamin Franklin and certainly surpassed his fellow Philadelphian in astronomical expertise. Rittenhouse's reputation was such that in 1775 he was invited to address the American Philosophical Society.

7. Charles Francis Adams, *The Works of John Adams, Second President of the United States; with the Life of the Author* (Boston: Little, Brown and Company, 1856), 10:415.

The "Oration" he delivered on that occasion is a remarkable document, which is striking not only for the astronomical but also the religious claims made in it. One Rittenhouse biographer has described it as nothing less than Rittenhouse's "public confession of faith."[8] Although Rittenhouse cast his oration as a survey of the history of astronomy, he found that format capacious enough to allow him to make various suggestions about the reconcilability of religion and astronomy, to present a vision not only of the universe but also of its celestial inhabitants, and, on the eve of the American Revolution, to take a swipe at the "unfeeling British nabob" (20).

David Rittenhouse, "Oration," *The Scientific Writings of David Rittenhouse*, ed. Brooke Hindle (New York: Arno Press, 1980), 5, 7–8, 9, 13–14, 18–20, 22, 25–26, and 27.

GENTLEMEN,

IT was not without being sensible how very unequal I am to the undertaking, that I first consented to comply with the request of several gentlemen for whom I have the highest esteem, and to solicit your attention on a subject which an able hand might indeed render both entertaining and instructive, I mean ASTRONOMY. But the earnest desire I have to contribute something towards the improvement of *Science* in general, and particularly of *Astronomy*, in this my native country, joined with the fullest confidence that I shall be favoured with your most candid indulgence, however far I may fall short of doing justice to the noble subject, enables me chearfully to take my turn as a member of the society on this annual occasion.

The order I shall observe in the following discourse is this: In the first place I shall give a very short account of the rise and progress of Astronomy, then take notice of some of the most important discoveries that have been made in this science, and conclude with pointing out a few of its defects at the present time.

8. Brooke Hindle, *David Rittenhouse* (Princeton: Princeton University Press, 1964), 112.

Rittenhouse's comments about the "rise and progress of astronomy" have been omitted. In this section, he traces the history of astronomy from the ancient Egyptians, Babylonians, and Greeks, through the accomplishments of early modern Europe, highlighting at this point the work of Copernicus, Brahe, and Kepler.

Astronomy, like the Christian religion, if you will allow me the comparison, has a much greater influence on our knowledge in general, and perhaps on our manners too, than is commonly imagined. Though but a few men are its particular votaries, yet the light it affords is universally diffused amongst us; and it is difficult for us to divest ourselves of its influence so far, as to frame any competent idea of what would be our situation without it.

Rittenhouse maintains that the studies of the astronomers following Copernicus of the "celestial travelers"—the planets—support the notion of a plurality of worlds.

From a careful attendance to those newly discovered celestial travelers, and their various motions, direct and retrograde, the great discovery arose, that the sun is the center of their motions; and that by attributing a similar motion to the earth, and supposing the sun to be at rest, all the Phoenomena will be solved. Hence a hint was taken that opened a new and surprizing scene. The earth might be similar to them in other respects. The planets too might be habitable worlds.

In several paragraphs omitted at this point, Rittenhouse continues his history through Galileo and the telescopic discoveries made by its early users, including Christiaan Huygens, who was among the first to turn this new instrument to the Orion nebula.

Besides several other remarkable appearances, which Hugenius discovered amongst the fixed stars, there is one in Orion's Sword, which, I will venture to say, whoever shall attentively view, with a good telescope and experienced eye, will not find his curiosity disappointed. "Seven small stars, says he, of which three are very close together, seemed to shine through a cloud, so that a space round them appeared much brighter than any other part of heaven, which being very serene and black looked here as if there was an opening, through which one had a prospect into a much brighter region." Here some have supposed old night to be entirely dispossessed, and that perpetual daylight shines amongst numberless worlds without interruption.

Rittenhouse's survey then turns to Newton and his contemporaries. After discussing planetary distances and stellar parallax, he returns to the subject of the plurality of worlds.

The opinion of the earth's rotation on its axis was once violently opposed, from a notion of its dangerous tendency with respect to the interests of religion. But, as truth is always consistent with itself, so many new proofs were furnished from time to time by new discoveries, that a mistaken interpretation of some passages in the bible was compelled to give way to the force of astronomical evidence. The doctrine of a plurality of worlds, is inseparable from the principles of Astronomy; but the doctrine is still thought, by some pious persons, and by many more I fear, who do not deserve that title, to militate against the truths asserted by the Christian religion. If I may be allowed to give my opinion on a matter of such importance, I must confess that I think upon a proper examination the apparent inconsistency will vanish. Our religion teaches us what Philosophy could not have taught; and we ought to admire with reverence the great things it has pleased divine Providence to perform, beyond the ordinary course of Nature, for man, who is undoubtedly the most noble inhabitant of this globe. But neither Religion nor Philosophy forbids us to believe that infinite wisdom and power, prompted by infinite goodness, may throughout the vast extent of creation and duration, have frequently interposed in a manner quite incomprehensible to us, when it became necessary to the happiness of created beings of some other rank or degree.

How far indeed the inhabitants of the other planets may resemble man, we can not pretend to say. If like him they were created liable to fall, yet some, if not all of them, may still retain their original rectitude. We will hope they do. The thought is comfortable.—Cease Galileo to improve the optic tube. And thou great Newton, forbear thy ardent search into the distant mysteries of nature; lest ye make unwelcome discoveries. Deprive us not of the pleasure of believing that yonder radiant orbs, traversing in silent Majesty the etherial regions, are the peaceful seats of innocence and bliss. Where neither natural nor moral evil has ever yet intruded; where to enjoy with gratitude and adoration the creator's bounty, is the business of existence. If their inhabitants resemble man in their faculties and affections, let us suppose that they are wise enough to govern themselves according to the dictates of that reason their creator has given them, in such manner as to consult their own and each others true happiness on all occasions. But if on the contrary they have found it necessary to erect artificial fabrics of government, let us not suppose that they have done it with so little skill, and at such an enormous expense, as must render them a misfortune instead of a blessing. We will hope that their statesmen are patriots, and that their Kings, if that order of beings has found admittance there, have the feelings of humanity.—Happy people! and perhaps more happy still, that all communication with us is denied. We have neither corrupted you with our vices nor injured you by violence. None of your sons and daughters, degraded from their native dignity, have been doomed to endless slavery by us in America, merely because *their* bodies may be disposed to reflect or absorb the rays of light, in a way different from *ours*. Even you, inhabitants of the moon, situated in our very neighbourhood, are effectually secured, alike from the rapacious hand of the haughty Spaniard, and of the unfeeling British nabob. Even British thunder impelled by British thirst of gain, cannot reach you: And the utmost efforts of the mighty Frederick, that tyrant of the north and scourge of mankind, if aimed to disturb *your* peace, becomes inconceivably ridiculous and impotent.

In the paragraphs omitted at this point, Rittenhouse moves from the subject of the plurality of worlds to theology and European tyranny. After this, he suggests the need for an observatory in America. Rittenhouse's suggestion is a mixture of patriotism and practicality—not only would the observatory be good for America, it could be good for astronomy as well, providing a vantage point for phenomena that might not be viewed from the observatories of Europe. Rittenhouse then surveys the "defects" of astronomy, those aspects of the science worthiest of American effort.

The Astronomy of comets is still in its infancy; not that the attention of the learned and ingenious has at all been wanting for more than a century past; but because it will necessarily require many ages to bring it to perfection. I wish we were in a condition to promote it in some degree, by carefully observing such comets as may appear. As yet we scarce dare affirm that any one has or will return a second time. It has never that I know of been certainly proved by observation, that a comet has descended within a parabolic orbit, and until that is done we have only a coincidence of periods and orbits (none of which have been very precise) to depend on for their return. Far less are astronomers able to determine the changes that may, and probably do, happen in their orbits and velocities in every period, so as to predict their nearer or more remote approach to the earth or any planet. Whether their business be to repair or to destroy, whether they are worlds yet in formation or once habitable worlds in ruins; whether they are at present habitable and regular attendants of our Sun only, or whether they are the vast links that connect the distant parts of creation by surrounding more suns than one, we know not.

A few paragraphs have been omitted at this point, in which Rittenhouse discusses the rotational periods of the planets, their satellites, and the nature of sunspots.

But if all higher and more sublime discoveries are not reserved for us in a future and more perfect state; if Astronomy shall again break those limits that now seem to confine it, and expatiate freely in the superior celestial fields; what amazing discoveries may yet be made amongst the fixed stars! That grand phænomenon the Milkey-Way seems to be the clue that will one day guide us. Millions of small stars compose it, and many more bright ones lie in and near it, than in other parts of heaven. Is not this a strong indication that this astonishing system of worlds beyond worlds innumerable, is not alike extended every way, but confined between two parallel planes, of *immeasurable*, though not *infinite* extent? Or rather, is not the Milkey-Way a vein of a closer texture, running through this part of the material creation? Great things are sometimes best explained by small and small by great. Material substances, such as we daily handle, have been thought composed of impenetrable particles in actual contact: Then again it has seemed necessary to suppose them at a distance from each other, and kept in their relative situations by *attraction* and *repulsion*. Many appearances require that those distances should be very great in proportion to the size of the particles. Hence some, with no small reason, have concluded that matter consists of indivisible points endued with certain powers. Let us compare these small portions of it with that great aggregate of matter which is the object of Astronomy; *Light* will then appear to have as free passage through a piece of glass, as the comets have in the planetary regions; and several other new considerations will arise.

If instead of *descending* we *ascend* the scale. If we consider that infinite variety which obtains in those parts of Nature with which we are most intimate: How one order of most curiously organized bodies, infinitely diversified in other respects, all agree in being fixed to the earth, and receiving nourishment from thence: How another order have spontaneous motion, and seek their food on different parts of the earth, whilst by gravity they are confined to its surface, but in other respects diversified like the former. How a *Third* float in, and below the surface of, a dense fluid, of equal weight with their bodies, which would soon prove fatal to both the others. And a *Fourth* consisting of a vast variety too, have this property in common, that by a peculiar mechanism of their bodies, they can soar to great heights above the earth, and quickly transport themselves to distant regions in a fluid so rare as to be scarcely sensible to us. But not to pursue this boundless subject any further, I say, when we consider this great variety so obvious on *our* globe, and ever connected by some degree of uniformity, we shall find sufficient reason to conclude, that the visible creation, consisting of revolving worlds and central suns, even including all those that are beyond the reach of human eye and telescope, is but an inconsiderable part of the whole. Many other and very various orders of things unknown to, and inconceivable by us, may, and probably do exist, in the unlimited regions of space. And all yonder stars innumerable, with their dependencies, may perhaps compose but the leaf of a flower in the creator's garden, or a single pillar in the immense building of the divine architect.

Here is ample provision made for the all-grasping mind of man!

In the final paragraphs of his "Oration," he urges the immortality of the soul, suggesting that after our death, if we have lived wisely, we shall benefit.

[We shall] be promoted to a more exalted rank amongst the creatures of God, have our understandings greatly enlarged, be enabled to follow truth in all her labyrinths with a higher relish and more facility, and thus lay the foundation of an eternal improvement in knowledge and happiness.

EMANUEL SWEDENBORG (1688–1772)

Emanuel Swedenborg was a Swedish scientist turned religious prophet who, after serving many years on the Swedish board of mines, moved to London, where he had various remarkable visions and wrote an array of religious books. These books eventually formed the scriptures of the Swedenborgians, as the followers of Swedenborg came to be called. Their denomination is referred to more formally as the Church of the New Jerusalem. Swedenborg's *magnum opus*, his *Arcana Coelestia*, is an eight volume commentary on Genesis and Exodus, which he published from 1749 to 1756. In 1758, he published a shorter work drawn from the larger and which when translated from Latin into English carried the title: *Earths in Our Solar System Which Are Called Planets and Earths in the Starry Heavens; Their Inhabitants, and the Spirits and Angels There, From Things Heard and Seen.* The next selection is from that book.

Emanuel Swedenborg, *Earths in the Universe* (London: Swedenborg Society, 1970), 1–4.

ON THE EARTHS IN THE UNIVERSE

1. INASMUCH as, by the Divine mercy of the Lord, things interior are open to me, which appertain to my spirit, and thereby it has been granted me to discourse not only with spirits and angels who are near our earth, but also with those who are near other earths; and whereas I had a desire to know whether other earths exist, and of what sort they are, and what is the nature and quality of their inhabitants, therefore it has been granted me of the Lord to discourse and converse with spirits and angels who are from other earths, with some for a day, with some for a week, and with some for months; and to be instructed by them concerning the earths, from which and near which they were; and concerning the lives, customs, and worship of the inhabitants thereof, with various other things worthy to be noted: and whereas in this manner it has been granted me to become acquainted with such things, it is permitted to describe them according to what has been heard and seem. It is to be observed, that all spirits and angels are from the human race,[9] and that they are near their respective earths;[10] and that they are acquainted with things on those

9. There are no spirits and angels, but what were of the human race, no. 1880. [Notes in this selection are extracts from Swedenborg's *Arcana Coelestia*; the numbers refer to that text. MJC]

10. The spirits of the earth are near their own earth, because they are of the inhabitants of that earth, and of a similar genius; and they are meant to be of service to them, no. 9968.

earths; and that by them man may be instructed, if his interiors be so open as to be enabled to speak and converse with them; for man in his essence is a spirit[11] and together with spirits as to his interiors;[12] wherefore he whose interiors are opened by the Lord, may discourse with them, as man with man;[13] which privilege has been granted me now for twelve years daily.

2. That there are several earths, and men upon them, and thence spirits and angels, is a thing most perfectly well known in another life, for it is there granted to every one who desires it from a love of truth and consequent use, to discourse with the spirits of other earths, and thereby to be confirmed concerning a plurality of worlds, and to be informed, that the human race is not confined to one earth only, but extends to earths innumerable; and moreover to know, what is the particular genius, manner of life, and also divine worship, prevailing amongst the inhabitants of each particular earth.

3. I have occasionally discoursed on this subject with the spirits of our earth, and the result of our conversation was, that any man of an enlarged understanding may conclude, from various considerations, that there are several earths, and that they are inhabited by men; for it is a suggestion of reason, that so great material masses as the planets are, some of which far exceed this earth in magnitude, are not empty masses, and created only to be conveyed in their revolutions round the sun, and to shine with their scanty light for the benefit of one earth, but that their use must needs be more

11. The soul, which lives after death, is the spirit of man, which in a man is the man himself and also appears in the other life in a perfect human form, nos. 322, 1880, 1881, 3633, 4622, 4735, 6054, 6605, 6626, 7021, 10594.

12. Man, even during his abode in the world, is, as to his interiors, consequently as to his spirit or soul, in the midst of spirits and angels who are of such a character as he himself is, nos. 2378, 3645, 4067, 4073, 4077.

13. Man is capable of speaking with spirits and angels, and the ancients on our earth frequently spoke with them, nos. 67, 68, 69, 784, 1634, 1636, 7802. But at the present day it is dangerous for man to speak with them, unless he be in a true faith, and be led by the Lord, nos. 784, 9438, 10751.

enlarged and distinguished. He who believes, as every one ought to believe, that the Deity created the universe for no other end, than that mankind, and thereby heaven, might have existence, (for mankind is the seminary of heaven), must needs believe also, that wheresoever there is any earth, there are likewise men-inhabitants. That the planets, which are visible to our eyes, as being within the boundaries of this solar system, are earths, may appear manifest from this consideration, that they are bodies of earthy matter, because they reflect the light of the sun and when seen through optical glasses, they appear, not as stars glittering by reason of their flame, but as earths variegated by reason of their opaque spots: the same may further appear from this consideration, that they, in like manner as our earth, are conveyed by a progressive motion round the sun, in the way of the zodiac, whence they have their years, and seasons of the year, as spring, summer, autumn, and winter: and in like manner, revolve about their own axis, whence they have their days, and times of the day, as morning, mid-day, evening, and night; moreover some of them have moons, which are called satellites, and which perform their revolutions round their central globes, as the moon does round our earth; the planet Saturn has besides a large luminous belt, as being furthest distant from the sun, which belt supplies that earth with much light, although reflected. How is it possible for any reasonable person, acquainted with these circumstances, to assert, that such bodies are void, and without inhabitants?

4. Moreover, in my discourse with spirits, I have at such times suggested, that it is very credible that in the universe there are more earths than one, from this consideration, that the starry heaven is so immense, and the stars therein are so innumerable, each of which in its place, or in its world, is a sun, and like our sun, in various magnitude: every considerate person is led to conclude, that so immense a whole must needs be a means to some end, the ultimate of creation, which end is the kingdom of heaven, wherein the Divine [being or principle] may dwell with angels and men: for the visible universe, or the heaven resplendent with stars so innumerable,

which are so many suns, is only a means, or medium, for the existence of earths, and of men upon them, of whom may be formed a celestial kingdom. From these considerations every reasonable person must needs be led to conceive, that so immense a means, adapted to so great an end, was not constituted for a race of men from one earth only, and for a heaven thence derived; for what would this be to the Divine [being or principle] which is infinite, and to which thousands, yea ten thousands of earths, all full of inhabitants, are comparatively small, and scarce of any amount.

5. Moreover, the angelic heaven is so immense, that it corresponds with all and singular the things appertaining, to man, myriads corresponding to every member and organ, and to all the viscera, and the respective affections of each; and it has been given to know, that that heaven, as to all its correspondences, can by no means exist, except by the inhabitants of very many earths.[14]

6. There are spirits whose sole study is to acquire to themselves knowledges, because they are delighted only with knowledges; these spirits are permitted to wander at large, and even to pass out of this solar system into others, and to procure for themselves knowledges: these have declared, that there are not only earths inhabited by men, in this solar system, but also out of it in the starry heaven, to an immense number. These spirits are from the planet Mercury.

7. As to what in general concerns the divine worship of the inhabitants of other earths, such amongst them as are not idolators, all acknowledge the Lord to be the only God; for they adore the Divine [being or principle] not as

14. Heaven corresponds to the Lord, and man, as to all things general and particular, corresponds to heaven, and hence heaven, before the Lord, is a man in a large effigy, and may be called the Grand or Greatest Man, nos. 2996, 2998, 3624 to 3649, 3636 to 3643, 3741 to 3745, 4625. Concerning the correspondence of man, and of all things pertaining to him, with the Grand Man, which is heaven, in general, from experience, nos. 3021, 3624 to 3649, 3741 to 3751, 3883 to 3896, 4039 to 4055, 4218 to 4228, 4318 to 4331, 4403 to 4421, 4527 to 4533, 4622 to 4633, 4652 to 4660, 4791 to 4805, 4931 to 4953, 5050 to 5061, 5171 to 5189, 5377 to 5396, 5552 to 5573, 5711 to 5727, 10030.

invisible, but as visible, for this reason amongst others, because when the Divine [being or principle] appears to them, he appears in a human form, as he also formerly appeared to Abraham and others on this earth;[15] and they who adore the Divine [being or principle] under a Human Form are all accepted of the Lord.[16] They say also, that no one can rightly worship God, much less be joined to Him, unless he comprehends Him by some idea, and that God cannot be comprehended except in a Human Form; and if He be not so comprehended, the interior sight, which is of the thought, concerning God, is dissipated, as the sight of the eye is, when looking upon the boundless universe; and that in this case the thought must needs sink into nature, and worship nature instead of God.

8. When they were told that the Lord on our earth assumed the human [principle,] they mused awhile, and presently said, that it was done for the salvation of the human race.

Like many other religious authors who wrote about extraterrestrials, Swedenborg was concerned about the question of the place of Christ in a universe containing extraterrestrials beings. In that regard it is interesting to consider the following statement from his *Earths in the Universe:*

> 113. . . . it pleased the Lord to be born, and to assume the Human, on our Earth, and not in any other. THE PRINCIPLE REASON *was for the sake of the Word, that it might be written on our Earth, and when written might afterwards be published throughout the whole Earth; and that, once published, it might be preserved for all posterity; and thus it might be made manifest, even to all in the other life, that God did become Man.*

15. The inhabitants of all the earths adore the Divine Being under the Human form, consequently the Lord, nos. 8541–8547, 10159, 10736, 10737, 10738. And they rejoice when they hear that God actually became Man, no. 9361. It is impossible to think of God except in the Human Form, nos. 8705, 9359, 9972. Man is able to worship and love that of which he has some idea, but not that of which he has no idea, nos. 4733, 5110, 5663, 7211, 9167, 10067.

16. The Lord receives all who are in good, and who adore the Divine under the Human Form, nos. 9359, 7173.

As this passage indicates, Swedenborg believed that Christ came only to this Earth, not to any other planet. Moreover, it appears that the chief function of his coming was to communicate the scriptures rather than to perform a specific redemptive action. In addition, the message of Christ was conceived of as spreading from this Earth throughout the rest of the universe.

THOMAS PAINE (1737–1809)

The large number of eighteenth-century books, sermons, poems, and essays that employed the idea of a plurality of worlds to encourage piety might lead one to assume that by the end of that century, pluralism had been fully reconciled with Christianity. If ever that situation had prevailed, it rapidly changed in 1794, when Thomas Paine published Part I of his *Age of Reason*.

Born in England, Paine had employed his gift for polemics on behalf of the American revolutionaries in such writings as *Common Sense*. Turning then to France and its revolutionaries, Paine wrote his *Age of Reason*, Part I, in 1794, shortly before the French imprisoned him for ten months. One of the most vigorous attacks ever launched against Christianity, Paine's book contained a number of criticisms of that religion, including an extended argument that accepting extraterrestrial life entails rejecting, not theism, but rather Christianity with its central doctrines of the divine incarnation and redemption. Paine's *Age of Reason* created a sensation. Despite government efforts to suppress the book, 60,000 copies of it were snatched up by British and Irish readers. Across the Atlantic, one Philadelphia book store alone sold 15,000 copies. And throughout the nineteenth century, Paine's book continued to produce a powerful effect on those who read its polemical pages. Over fifty responses to it were published, some of which focused on Paine's arguments based on extraterrestrial life ideas against Christianity.

Thomas Paine, *The Age of Reason*, in *Representative Selections*, ed. Harry Hayden Clark (New York: Hill and Wang, 1961), 273–84.

Having thus shown from the internal evidence of things, the cause that produced a change in the state of learning, and the motive for substituting the study of dead languages, in the place of the sciences, I proceed, in addition to the several observations, already made in the former part of this work, to compare, or rather to confront the evidence that the structure of the universe affords, with the Christian system of religion; but, as I cannot begin this part better than by referring to the ideas that occurred to me at an early part of life, and which I doubt not have occurred in some degree to almost every other person at one time or other, I shall state what those ideas were, and add thereto such other matter as shall arise out of the subject, giving to the whole, by way of preface, a short introduction.

My father being of the Quaker profession, it was my good fortune to have an exceeding good moral education, and a tolerable stock of useful learning. Though I went to the grammar school,[17] I did not learn Latin, not only because I had no inclination to learn languages, but because of the objection the Quakers have against the books in which the language is taught. But this did not prevent me from being acquainted with the subjects of all the Latin books used in the school.

The natural bent of my mind was to science. I had some turn, and I believe some talent for poetry; but this I rather repressed than encouraged, as leading too much into the field of imagination. As soon as I was able, I purchased a pair of globes, and attended the philosophical lectures of Martin and Ferguson[18], and became afterwards acquainted with Dr. Bevis, of the society, called the Royal Society, then living in the Temple, and an excellent astronomer.

Two paragraphs are omitted at this point. In these paragraphs, Paine commented on his interest in politics and on the nature of the mind.

From the time I was capable of conceiving an idea, and acting upon it by reflection, I either doubted the truth of the Christian system, or thought it to be a strange affair; I scarcely knew which it was; but I well remember when about seven or

17. [A footnote in which Paine inserted some information on that school has been omitted. MJC]

18. [This is James Ferguson, whose pious advocacy of extraterrestrials was discussed in chapter 7. MJC]

eight years of age, hearing a sermon read by a relation of mine, who was a great devotee of the church, upon the subject of what is called redemption by the death of the Son of God. After the sermon was ended, I went into the garden, and as I was going down the garden steps (for I perfectly recollect the spot) I revolted at the recollection of what I had heard, and thought to myself that it was making God Almighty act like a passionate man, that killed his son, when he could not revenge himself any other way; and as I was sure a man would be hanged that did such a thing, I could not see for what purpose they preached such sermons. This was not one of that kind of thoughts that had anything in it of childish levity; it was to me a serious reflection, arising from the idea I had, that God was too good to do such an action, and also too almighty to be under any necessity of doing it. I believe in the same manner at this moment; and I moreover believe that any system of religion that has any thing in it that shocks the mind of a child, cannot be a true system.

It seems as if parents of the Christian profession were ashamed to tell their children any thing about the principles of their religion. They sometimes instruct them in morals, and talk to them of the goodness of what they call Providence; for the Christian Mythology has five deities — there is God the Father, God the Son, God the Holy Ghost, the God Providence, and the Goddess Nature. But the Christian story of God the Father putting his son to death, or employing people to do it, (for that is the plain language of the story), cannot be told by a parent to a child; and to tell him that it was done to make mankind happier and better, is making the story still worse; as if mankind could be improved by the example of murder; and to tell him that all this is a mystery, is only making an excuse for the incredibility of it.

How different is this to the pure and simple profession of Deism! The true Deist has but one Deity; and his religion consists in contemplating the power, wisdom and benignity of the Deity in his works, and in endeavoring to imitate him in every thing moral, scientifical and mechanical.

The religion that approaches the nearest of all others to true Deism, in the moral and benign part thereof, is that professed by the Quakers: but they have contracted themselves too much, by leaving the works of God out of their system. Though I reverence their philanthropy, I can not help smiling at the conceit, that if the taste of a Quaker could have been consulted at the creation what a silent and drab-colored creation it would have been! Not a flower would have blossomed its gaities, nor a bird been permitted to sing.

Quitting these reflections, I proceed to other matters. After I had made myself master of the use of the globes, and of the orrery,[19] and conceived an idea of the infinity of space, and the eternal divisibility of matter, and obtained, at least, a general knowledge of what was called natural philosophy, I began to compare, or, as I had before said to confront the eternal evidence those things afford with the Christian system of faith.

Though it is not a direct article of the Christian system, that this world that we inhabit is the whole of the habitable creation, yet it is so worked up therewith, from what is called the Mosaic account of the Creation, the story of Eve and the apple, and the counterpart of that story, the death of the Son of God, that to believe otherwise, that is to believe that God created a plurality of worlds, at least as numerous as what we called stars, renders the Christian system of faith at once little and ridiculous, and scatters it in the mind like feathers in the air. The two beliefs cannot be held together in the same mind; and he who thinks that he believes both, has thought but little of either.

Though the belief of a plurality of worlds was familiar to the ancients, it is only within the last three centuries that the extent and dimensions of this globe that we inhabit have been ascertained. Several vessels, following the tract of the ocean, have sailed entirely round the world, as a man may march in a circle, and come round by the contrary side of the circle to the spot he set out from. The circular dimensions of our world, in the widest part, as a man would measure the widest round

19. [At this point Paine inserted a long footnote explaining that an orrery is a clockwork device for representing planetary motions. MJC]

of an apple, or a ball, is only twenty-five thousand and twenty English miles, reckoning sixty-nine miles and a half to an equatorial degree, and may be sailed round in the space of about three years.[20]

A world of this extent may, at first thought, appear to us to be great; but if we compare it with the immensity of space in which it is suspended, like a bubble or balloon in the air, it is infinitely less, in proportion, than the smallest grain of sand is to the size of the world, or the finest particle of dew to the whole ocean, and is therefore but small; and, as will be hereafter shown, is only one of a system of worlds, of which the universal creation is composed.

It is not difficult to gain some faint idea of the immensity of space in which this and all the other worlds are suspended, if we follow a progression of ideas. When we think of the size or dimensions of a room, our ideas limit themselves to the walls, and there they stop; but when our eye or our imagination darts into space, that is, when it looks upwards into what we call the open air, we cannot conceive any walls or boundaries it can have; and if for the sake of resting our ideas, we suppose a boundary, the question immediately renews itself, and asks, what is beyond that boundary? and in the same manner, what beyond the next boundary? and so on till the fatigued imagination returns and says, *There is no end.* Certainly, then, the Creator was not pent for room, when he made this world no larger than it is; and we have to seek the reason in something else.

If we take a survey of our own world, or rather of this of which the Creator has given us the use, as our portion in the immense system of Creation, we find every part of it, the earth, the waters, and the air that surrounds it, filled, and as it were, crowded with life, down from the largest animals we know of to the smallest insects the naked eye can behold, and from thence to others still smaller, and totally invisible without the

20. Allowing a ship to sail, on an average, three miles in an hour, she would sail entirely round the world in less than one year, if she could sail in a direct circle; but she is obliged to follow the course of the ocean.

assistance of the microscope. Every tree, every plant, every leaf, serves not only as an habitation, but as a world to some numerous race, till animal existence becomes so exceedingly refined, that the effluvia of a blade of grass would be food for thousands.

Since, then, no part of our earth is left unoccupied, why is it to be supposed that the immensity of space is a naked void, lying in eternal waste? There is room for millions of worlds as large or larger than ours, and each of them millions of miles apart from each other.

Having now arrived at this point, if we carry our ideas only one thought further, we shall see, perhaps, the true reason, at least a very good reason, for our happiness, why the Creator, instead of making one immense world, extending over an immense quantity of space, has preferred dividing that quantity of matter into several distinct and separate worlds, which we called planets, of which our earth is one. But before I explain my ideas upon this subject, it is necessary (not for the sake of those who already know, but for those who do not) to show what the system of the universe is.

That part of the universe that is called the solar system (meaning the system of worlds to which our earth belongs, and of which Sol, or in English language, the Sun, is the center) consists, besides the Sun, of six distinct orbs, or planets, or worlds, besides the secondary bodies, called the satellites or moons of which our earth has one that attends her in her annual revolution round the Sun, in like manner as other satellites or moons, attend the planets or worlds to which they severally belong, as may be seen by the assistance of the telescope.

Six paragraphs, in which Paine described our solar system, are omitted at this point. The impressive detail included in this description suggest the depth of Paine's interest in astronomy—until one realizes that his list of planets is incomplete: Paine seems unaware that thirteen years before he published his *Age of Reason*, William Herschel had discovered a seventh planet, Uranus.

But this, immense as it is, is only one system of worlds. Beyond this, at a vast distance into space, far beyond all power of calculation, are the stars called the fixed stars. They are called fixed because they have no revolutionary motion, as the six worlds or planets have that I have been describing. Those fixed stars continue always at the same distance from each other, and always in the same place, as the Sun does, in the center of our system. The probability, therefore, is that each of those fixed stars is also a Sun, round which another system of worlds or planets, though too remote for us to discover, performs its revolutions, as our system of worlds does round our central Sun.

By this easy progression of ideas the immensity of space will appear to us to be filled with systems of worlds; and that no part of space lies at waste, any more than any part of the globe or earth and water is left unoccupied.

Having thus endeavored to convey, in a familiar and easy manner, some idea of the structure of the universe, I return to explain what I before alluded to, namely, the great benefits arising to man in consequence of the Creator having made a *plurality* of worlds, such as our system is, consisting of a central Sun and six worlds besides satellites, in preference to that of creating one world only of a vast extent.

It is an idea I have never lost sight of, that all our knowledge of science is derived from the revolutions (exhibited to our eye and from thence to our understanding) which those several planets or worlds, of which our system is composed, make in their circuit round the Sun.

Had then the quantity of matter which these six worlds contain been blended into one solitary globe, the consequence to us would have been, that either no revolutionary motion would have existed, or not a sufficiency of it to give us the idea and the knowledge of science we now have; and it is from the sciences that all the mechanical arts that contribute so much to our earthly felicity and comfort, are derived.

As, therefore, the Creator made nothing in vain, so also must it be believed that He organized the structure of the universe in the most advantageous manner for the benefit of man; and as we see, and from experience feel, the benefits we derive from the structure of the universe, formed as it is, which benefits we should not have had the opportunity of enjoying, if the structure, so far as relates to our system, had been a solitary globe—we can discover at least one reason why a plurality of worlds has been made, and that reason calls forth the devotional gratitude of man, as well as his admiration.

But it is not to us, the inhabitants of this globe, only, that the benefits arising from a plurality of worlds are limited. The inhabitants of each of the worlds of which our system is composed, enjoy the same opportunities of knowledge as we do. They behold the revolutionary motions of our earth, as we behold theirs. All the planets revolve in sight of each other; and, therefore, the same universal school of science presents itself to all.

Neither does the knowledge stop here. The system of worlds next to us exhibits, in its revolutions, the same principles and school of science, to the inhabitants of their system, as our system does to us, and in like manner throughout the immensity of space.

Our idea, not only of the almightiness of the Creator, but of his wisdom and his beneficence, become enlarged in proportion as we contemplate the extent and the structure of the universe. The solitary idea of a solitary world, rolling or at rest in the immense ocean of space, gives place to the cheerful idea of a society of worlds, so happily contrived as to administer, even by their motion, instruction to man. We see our own earth filled with abundance; but we forget to consider how much of that abundance is owing to the scientific knowledge the vast machinery of the universe has unfolded.

But, in the midst of those reflections, what are we to think of the Christian system of faith, that forms itself upon the idea of only one world, and that of no greater extent, as is before shown, than twenty-five thousand miles? An extent which a man, walking at the rate of three miles an hour, for twelve hours in the day, could he keep on in a circular direction, would walk entirely round in less than two years. Alas! what is this to the mighty ocean of space, and the almighty power of the Creator.

From whence then could arise the solitary and strange conceit, that the Almighty, who had millions of worlds equally dependent on his protection, should quit the care of all the rest, and come to die in our world, because they say one man and one woman had eaten an apple! And, on the other hand, are we to suppose that every world in the boundless creation, had an Eve, an apple, a serpent, and a redeemer? In this case, the person who is irreverently called the Son of God, and sometimes God himself, would have nothing else to do than to travel from world to world, in an endless succession of death, with scarcely a momentary interval of life.

It has been by rejecting the evidence, that the word or works of God in the creation afford to our senses, and the action of our reason upon that evidence, that so many wild and whimsical systems of faith, and of religion, have been fabricated and set up. There may be many systems of religion, that so far from being morally bad, are in many respects morally good: but there can be but one that is true; and that one necessarily must, as it ever will, be in all things consistent with the ever-existing world of God that we behold in his works. But such is the strange construction of the Christian system of faith, that every evidence the Heavens afford to man, either directly contradicts it, or renders it absurd.

PART THREE

FROM 1800 TO 1860

Fig. 8. Drawing by J. J. Grandville of a Juggler of Worlds, from *Un autre monde* (1844)

NINE

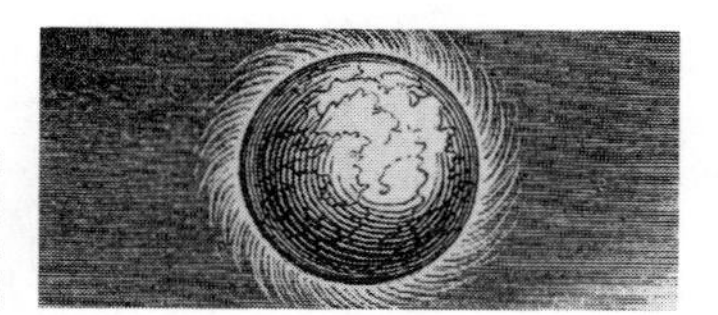

INTENSIFICATION OF THE DEBATE AFTER 1800

PERCY BYSSHE SHELLEY (1792–1822)

Thomas Paine was far from being the only author who saw the doctrine of a plurality of worlds as creating serious difficulties for Christianity. The Gothic novelist Horace Walpole, fourth lord of Orford (1717–1798), stated: "Fontenelle's Dialogues on a Plurality of Worlds, first rendered me an Infidel. Christianity, and a plurality of worlds, are, in my opinion, irreconcilable."[1] Lord Byron (1788–1824), who used pluralist themes in a number of his poems, admitted in an 1813 letter that he had come to question the immortality of the soul: "It was the comparative insignificance of ourselves and *our world,* when placed in competition with the mighty whole, of which it is an atom, that first led me to imagine that our pretensions to eternity might be over-rated."[2] A second leading romantic poet, Samuel Taylor Coleridge (1772–1834), reacted quite differently; although in his student days he had endorsed pluralism in his "Ode to Astronomy," in 1824 he remarked:

1. As quoted in "Walpoliana . . . Number IV," *Monthly Magazine* 6 (1798): 116.

2. *The Works of Lord Byron, Letters and Journals,* ed. R. E. Prothero (London: J. Murray, 1903), 2:221–22.

> I never could feel any force in the arguments for a plurality of worlds, in the common acceptation of that term. A lady once asked me—"What then could be the intention in creating so many great bodies, so apparently useless to us?" I said—I did not know, except perhaps to make dirt cheap. The vulgar inference is *in alio genere*. What in the eye of an intellectual and omnipotent Being is the whole sidereal system to the soul of one man for whom Christ died?[3]

A far more striking instance of the difficulties that some saw in reconciling Christianity with pluralism can be seen in the case of Percy Bysshe Shelley, whose interest in extraterrestrial life themes can be traced to his student days when he heard lectures from Adam Walker, a popular scientific lecturer, who endorsed that doctrine in his *System of Familiar Philosophy*. Shelley also encountered the doctrine through reading Lucretius and various writings of Erasmus Darwin, some of whose astronomical ideas were used in Shelley's *Prometheus Unbound* (1820). In response to the much discussed question of the cause of Shelley's break from Christianity, Ifor Evans has stated that "astronomical knowledge [led Shelley] to a rejection of the Christian faith."[4] At least two of Shelley's writings support this claim. The first consists of a note he added to his *Queen Mab*, privately printed in 1813. At one point in that work, Shelly referred to "Innumerable systems," adding the following note.

> The plurality of worlds,—the indefinite immensity of the universe, is a most awful subject of contemplation. He who rightly feels its mystery and grandeur is in no danger of seduction from the falsehoods of religious systems, or of deifying the principle of the universe. It is impossible to believe that the Spirit that pervades this infinite machine begat a son upon the body of a Jewish woman; or is angered at the consequences of that necessity, which is a synonym of itself. All this miserable tale of the Devil, and Eve, and an Intercessor, with the childish mummeries of the God of the Jews, is irreconcilable with the knowledge of the stars. The works of His fingers have born witness against Him.[5]

The second composition is his "On the Devil, and Devils," a satiric essay written shortly before his death but long held back from publication because of its shocking nature. A selection from that essay follows.

---

3. Samuel Taylor Coleridge, *Complete Works*, ed. Professor Shedd (New York: Harper and Brothers, 1853), 6:502–3.

4. Ifor Evans, *Literature and Science* (London: Allen and Urwin, 1954), 69.

5. Percy Bysshe Shelley, *The Complete Poetical Works*, ed. Neville Rogers (Oxford: Clarendon Press, 1972), 1:296.

Percy Shelley, "On the Devil, and Devils," in *The Works of Percy Bysshe Shelley*, ed. Harry Buxton Forman (London: Reeves and Turner, 1880), 6:390–91, 396–98, 400–402, 403, and 404.

After an opening section treating the history of ideas of the devil and especially the image of him created by John Milton, Shelley turns to what might be described as the cosmology of the devil.

I am afraid there is much laxity among the orthodox of the present day, respecting a belief in the Devil. I recommend to the Bishops to make a serious charge to their diocesans on this dangerous latitude. The Devil is the outwork of the Christian faith; he is the weakest point. You may observe that infidels, in their novitiate, always begin by tremulously doubting the existence of the Devil. Depend upon it, that when a person once begins to think that perhaps there is no Devil, he is in a dangerous way. There may be observed, in polite society, a great deal of coquetting about the Devil, especially among divines, which is singularly ominous. They qualify him as the evil spirit; they consider him as synonymous with the flesh. They seem to wish to divest him of all personality; to reduce him from his abstract to his concrete; to reverse the process by which he was created in the mind, which they will by no means bear with respect to God. It is popular, and well looked upon, if you deny the Devil "a local habitation and a name." Even the vulgar begin to scout him. Hell is popularly considered as metaphorical of the torments of an evil conscience, and by no means capable of being topographically ascertained. No one likes to mention the torments of everlasting fire and the poisonous gnawing of the worm that liveth forever and ever.

In a section omitted at this point, Shelley continues at some length to warn of the dangers of slackness in regard to belief in the devil and in hell, suggesting that it is an ominous sign that even the orthodox are soft-pedaling doctrines regarding the devil and hell. He also discusses various activities and powers of the devil.

The sphere of the operations of the Devil is difficult to determine. The late inventions and improvements in telescopes have considerably enlarged the notions of men respecting the bounds of the Universe. It is discovered that the Earth is a comparatively small globe, in a system consisting of a multitude of others, which roll round the sun; and there is no reason to suppose but that all these are inhabited by organized and intelligent beings. The fixed stars are supposed to be suns, each of them the centre of a system like ours. Those little whitish specks of light that are seen in a clear night, are discovered to consist of a prodigious multitude of suns, each probably the centre of a system of planets. The system of which our earth is a planet has been discovered to belong to one of those larger systems of suns, which, when seen at a distance, look like a whitish speck of light; and that lustrous streak called the milky way is found to be one of the extremities of the immense group of suns in which our system is placed. The heaven is covered with an incalculable number of these white specks, and the better the telescopes the more are discovered, and the more distinctly the confusion of white light is resolved into stars. All this was not known during the gradual invention of the Christian mythology, and was never even suspected by those barbarians on the obscure extremities of the Roman Empire, by whom it was first adopted. If these incalculable millions of suns, planets, satellites, and comets are inhabited, is it to be supposed that God formed their inhabitants better, or less liable to offend him than those primordial Spirits, those angels near his throne, those first and most admirable of his creatures, who rebelled and were damned? Or has he improved like a proficient in statuary or painting, proceeding from rude outlines and imperfect forms, to more perfect idealisms or imitations, so that his latter works are better than his first? Or has some fortunate chance, like that which, when the painter despaired of being able to depict the foam of a horse, directed the spunge so as to represent it accurately, interfered to confer stability and exactness upon one, or how many, among the numerous systems of animated nature? There is little reason to suppose that any considerable multitude of the planets were tenanted by beings better capable of resisting the temptations of the Devil than ours. But is the Devil, like God, omnipresent? If so, he interpenetrates God, and they both exist co-essentially; as metaphysicians have compared the omnipresence of God, pervading the infinity of space and being, to salt mixed with water. If not, he must send some inferior Angels, either to this or some other planet, first to tempt the Inhabitants to disobey God, and secondly to induce them to reject all terms of salvation; for which latter purpose, it seems equally requisite that he should take up his residence on the spot; nor do I see, how he or God, by whose providence he is permitted, that is to say, compelled to act, could commit a business of such high moment to an inferior Angel. It seems very questionable whether the Devil himself, or only some inferior Devil, tempted and betrayed the people of the Earth; or whether Jupiter, a planet capable of containing a hundred times more inhabitants than the earth,—to mention only the planets of our own system,—or the Sun, which would contain a million times more, were not entitled to the preference.

Any objection that might arise from the multitude of Devils, I think futile. You may suppose a million times as many devils as there are stars. In fact you may suppose anything you like on such a subject. That there are a great number of Devils, and that they go about in legions of six or seven, or more at a time, all mythologists are agreed. Christians, indeed, will not admit the actual substance and presence of the Devils upon Earth in modern times. Or, in proportion as any histories of them approach to the present epoch, or indeed any epoch in which there has been a considerable progress in historical criticism or natural science, they suppose their agency to be obscure and superstitious.

A section in which Shelley discusses the idea that devils may dwell in animals has been deleted at this point.

Some have supposed that the Devils live in the Sun, and that that glorious luminary is the actual Hell; perhaps that every fixed star is a distinct Hell appropriated to the use of its several systems of planets, so great a proportion of the inhabitants of which are probably devoted to everlasting damnation, if the belief of one particular creed is essential to their escape, and the testimony of its truth so very far remote and obscure as in the planet which we inhabit. I do not envy the theologians, who first invented this theory. The Magian worship of the Sun as the creator and Preserver of the world, is considerably more to the credit of the inventors. It is in fact a poetical exposition of the matter of fact, before modern science had so greatly enlarged the boundaries of the sensible world, and was, next to pure deism or a personification of all the powers whose agency we know or can conjecture, the religion of the fewest evil consequences.

If the sun is Hell, the Devil has a magnificent abode, being elevated as it were on the imperial throne of the visible world. If we assign to the Devil the greatest and most glorious habitation within the scope of our senses, where shall we conceive his mightier adversary to reside? Shall we suppose that the Devil occupies the centre and God the circumference of existence, and that one urges inwards with the centripetal, whilst the other is perpetually struggling outwards from the narrow focus with the centrifugal force, and that from their perpetual conflict results that mixture of good and evil, harmony and discord, beauty and deformity, production and decay, which are the general laws of the moral and material world? Alas! the poor theologian never troubled his fancy with nonsense of so philosophical a form. He contented himself with supposing that God was somewhere or other; that the Devil and all his angels together with the perpetually increasing multitude of the damned were burning above to all eternity in that prodigious orb of elemental light, which sustains and animates that multitude of inhabited globes, in whose company this earth revolves. Others have supposed Hell to be distributed among the comets, which constitute, according to this scheme, a number of floating prisons of intense and inextinguishable fire; a great modern poet adopts this idea when he calls a comet

"A wandering hell in the eternal space."[6]

A paragraph is omitted at this point in which Shelley suggests that poets find it easier to write about Hell than about Heaven, noting that many readers prefer Dante's *Inferno* to his *Paradiso.*

As to the Devil, and the imps, and the damned living in the Sun, why there is no great probability of it. The Comets are better fitted for this; except that some astronomer has suggested the possibility of their orbits gradually becoming ecliptical, until at last they might arrange themselves in orbits concentric with the planets, lose their heat and their substance, become subject to the same laws of animal and vegetable life as those according to which the substance of the surface of the others is arranged. The Devils and the damned, without some miraculous interposition would then be the inhabitants of a very agreeable world; and as they probably would have become very good friends from a community of misfortune and the experience which time gives those who live long enough of the folly of quarrelling—would probably administer the affairs of their Colony with great harmony and success.

Part of a long paragraph has been removed at this point. In it, Shelley discusses some of William Herschel's theories of the Sun, using these to critique the idea that Hell could be located in the Sun.

The idea of the sun being Hell, is an attempt at an improvement on the old-established idea of its occupying the centre of the earth. The Devils and the damned would be exceedingly crowded in process of ages, if they were confined within so inconsiderable a sphere.

6. [David Lee Clark in *Shelley's Prose* (Albuquerque: University of New Mexico Press, 1954) identifies this line as from Byron's *Manfred,* 1.1.46.]

THOMAS CHALMERS (1780–1847): EVANGELIZING THE EXTRATERRESTRIALS

It seemed hardly an auspicious event when at noon on Thursday, 21 November 1815, Thomas Chalmers, an almost unknown parson from the rural regions of Scotland, ascended the pulpit in Glasgow's Tron Church to begin a series of seven weekday sermons. Yet these sermons created a sensation, the book based on them became a best seller, and the preacher was himself launched on a career that culminated with Chalmers being recognized as the most important figure in Scottish religious life in the first half of the nineteenth century. In fact, Thomas Carlyle referred to Chalmers as "The Chief Scottish man of his time."[7] One of Chalmers's contemporaries gave an eye-witness account of the remarkable reaction to Chalmers's series of sermons:

> One or two of these "Discourses" . . . were heard by the writer . . . , then a boy. He had to wait nearly four hours before he could gain admission as one of a crowd in which he was nearly crushed to death. It was with no little effort that the great preacher could find his way to the pulpit. As soon as his fervid eloquence began to stream from it, the intense enthusiasm of the auditory became almost irrestrainable; and in that enthusiasm the writer, young as he was, fully participated. He has never since witnessed anything equal to the scene.[8]

When these sermons were published in early 1817 as *A Series of Discourses on the Christian Revelation Viewed in Connection with the Modern Astronomy*, the response was no less remarkable. Six thousand copies were sold within the first ten weeks and by year's end 20,000 copies in nine editions had been purchased. Writing in 1851, William Hanna, Chalmers's biographer, commented: "Never previously, nor ever since, has any volume of sermons met with such immediate and general acceptance. . . . It was, besides, the first volume of Sermons which fairly broke the lines which had separated too long the literary from the religious public."[9] The literary critic William Hazlitt described the book as running "like wild-fire through the country,"[10] and

7. Quoted from the Thomas Chalmers website: http://www.newble.co.uk/chalmers/ as accessed 14 July 2007.

8. [Samuel Warren], "Speculators among the Stars," *Blackwood's Edinburgh Magazine* 76 (1854), 373.

9. William Hanna, *Memoirs of the Life and Writings of Thomas Chalmers*, 4 vols. (Edinburgh: Thomas Constable, 1849), 2:89.

10. As quoted in Hanna, *Memoirs*, 2:89.

the enthusiasm at first evoked by it scarcely declined in subsequent decades. As late as 1851, by which time Chalmers's publications came to over twenty volumes, Hanna stated that Chalmers's discourses "to this day . . . command a larger sale than any other portion of Dr. Chalmers' writings."[11] Published in America already in 1817, they continued to be republished there as late as 1860, whereas British editions appeared in the 1870s. Moreover, in 1841, a German translation was published. All this gives support to Edward Hitchcock's claim, made in the 1850s, that "All the world is acquainted with Dr. Chalmers' splendid Astronomical Discourses."[12]

Chalmers came to his *Discourses* from a number of directions. Among these was his concern to reconcile science and religion, a concern intensified by the fact that Chalmers had been strongly tempted (as he viewed it) toward a career in science, for which he had substantial abilities. Another factor was that in the period immediately preceding the composition of his sermons he had undergone an intense religious experience, which turned him toward the evangelical movement in British Christianity. What brought these strands together was the idea of extraterrestrial life. Whereas Paine had urged that contemplation of the cosmos entailed rejection of Christianity, it was a mark of Chalmers's genius to see that the universe of pluralist astronomy provided ideas and images that would enable him to highlight themes prominent in evangelical Christianity. The success of his efforts, although testing the historical imagination of twentieth-first-century readers, is a matter of history.

Thomas Chalmers, *A Series of Discourses on the Christian Revelation Viewed in Connection with the Modern Astronomy* (New York: American Tract Society, 185?), 5–10, 12, 13–14, 16–17, 17–30, 35–37, 41–45, and 46–49.

PREFACE

The astronomical objection against the truth of the Gospel, does not occupy a very prominent place in any of our Treatises

11. Hanna, *Memoirs*, 2:92.

12. Edward Hitchcock, "Introductory Notice to the American Edition" of [William Whewell,] *The Plurality of Worlds* (Boston: Gould and Lincoln, 1854), x.

of Infidelity. It is often, however, met with in conversation—and we have known it to be the cause of serious perplexity and alarm in minds anxious for the solid establishment of their religious faith.

There is an imposing splendour in the science of astronomy; and it is not to be wondered at, if the light it throws, or appears to throw, over other tracks of speculation than those which are properly its own, should at times dazzle and mislead an inquirer. On this account, we think it were a service to what we deem a true and a righteous cause, could we succeed in dissipating this illusion; and in stripping Infidelity of those pretensions to enlargement and to a certain air of philosophical greatness by which it has often become so destructively alluring, to the young, and the ardent and the ambitious.

In my first Discourse I have attempted a sketch of the Modern Astronomy—nor have I wished to throw any disguise over that comparative littleness which belongs to our planet and which gives to the argument of Freethinkers all its plausibility

This argument involves in it an assertion and an inference. The assertion is that Christianity is a religion which professes to be designed for the single benefit of our world; and the inference is that God cannot be the author of this religion for he would not lavish on so insignificant a field such peculiar and such distinguishing attentions as are ascribed to him in the Old and New Testament.

Christianity makes no such profession. That it is designed for the single benefit of our world is altogether a presumption of the Infidel himself—and feeling that this is not the only example of temerity which can be charged on the enemies of our faith, I have allotted my second Discourse to the attempt of demonstrating the utter repugnance of such a spirit with the cautious and enlightened philosophy of modern times.

In the course of this Sermon I have offered a tribute of acknowledgment to the theology of Sir Isaac Newton; and in such terms as, if not farther explained, may be liable to misconstruction. The grand circumstance of applause in the character of this great man, is, that unseduced by all the magnificence of his own discoveries, he had a solidity of mind which could resist their fascination; and keep him in steady attachment to that book, whose general evidences stamped upon it the impress of a real communication from heaven. This was the sole attribute of his theology which I had in my eye when I presumed to eulogise it. I do not think that, amid the distraction and the engrossment of his other pursuits, he has at all times succeeded in his interpretation of the book; else he would never, in my apprehension, have abetted the leading doctrine of a sect or a system, which has now nearly dwindled away from public observation.

In my third Discourse I am silent as to the assertion, and attempt to combat the inference that is founded on it. I insist that upon all the analogies of nature and of providence, we can lay no limit on the condescension of God or on the multiplicity of his regards even to the very humblest departments of creation; and that it is not for us, who see the evidences of divine wisdom and care spread in such exhaustless profusion around us, to say that the Deity would not lavish all the wealth of his wondrous attributes on the salvation even of our solitary species.

At this point of the argument I trust that the intelligent reader may be enabled to perceive in the adversaries of the gospel, a twofold dereliction from the maxims of the Baconian philosophy: that, in the first instance the assertion which forms the groundwork of their argument, is gratuitously fetched out of an unknown region where they are utterly abandoned by the light of experience; and that, in the second instance, the inference they urge from it, is in the face of manifold and undeniable truths, all lying within the safe and accessible field of human observation.

In my subsequent Discourses, I proceed to the informations of the record. The Infidel objection, drawn from astronomy, may be considered as by this time disposed of; and if we have succeeded in clearing it away, so as to deliver the Christian testimony from all discredit upon this ground, then may we submit, on the strength of other evidences, to be guided by its information. We shall thus learn, that Christianity has a far more extensive bearing on the other orders of creation, than the Infidel is disposed to allow; and, whether he will own the authority of this information or not, he will, at least, be forced to admit, that the subject matter of the Bible itself is not chargeable with that objection which he has attempted to fasten upon it.

Thus, had my only object been the refutation of the Infidel argument, I might have spared the last Discourses of the Volume altogether. But the tracks of Scriptural information to which they directed me, I considered as worthy of prosecution on their own account—and I do think, that much may be gathered from these less observed portions of the field of revelation, to cheer, and to elevate, and to guide the believer.

But, in the management of such a discussion as this, though, for a great degree of this effect, it would require to be conducted in a far higher style than I am able to sustain, the taste of the human mind may be regaled, and its understanding put into a state of the most agreeable exercise. Now, this is quite distinct from the conscience being made to feel the force of a personal application; nor could I either bring this argument to its close in the pulpit, or offer it to the general notice of the world, without adverting, in the last Discourse, to a delusion, which, I fear, is carrying forward thousands, and tens of thousands, to an undone eternity.

I have closed the Volume with an Appendix of Scriptural Authorities. . . .

Chalmers's discussion of these scriptural authorities has been omitted at this point.

I shall only add, for the information of readers at a distance, that these Discourses were chiefly delivered on the occasion of the week-day sermon that is preached in rotation by the Ministers of Glasgow.

DISCOURSES ON THE CHRISTIAN REVELATION

DISCOURSE I.
A SKETCH OF THE MODERN ASTRONOMY.

"When I consider thy heavens, the work of thy fingers,
the moon and the stars, which thou hast ordained;
What is man, that thou art mindful of him?
and the son of man, that thou visitest him?"
—Psalm viii.3, 4.

In the reasonings of the Apostle Paul, we cannot fail to observe, how studiously he accommodates his arguments to the pursuits or principles or prejudices of the people whom he was addressing. He often made a favourite opinion of their own the starting point of his explanation; and educing a dexterous but irresistible train of argument from some principle upon which each of the parties had a common understanding, did he force them out of all their opposition, by a weapon of their own choosing—nor did he scruple to avail himself of a Jewish peculiarity, or a heathen superstition, or a quotation from Greek poetry, by which he might gain the attention of those whom he laboured to convince, and by the skillful application of which, he might "shut them up unto the faith."

Omitted at this point are nearly two paragraphs in which Chalmers discusses the merits of St. Paul's approach in accommodating his preaching to the special concerns of the congregation he was addressing.

It is under the impulse of these considerations, that I have, with some hesitation, prevailed upon myself to attempt an argument which I think fitted to soften and subdue those prejudices which lie at the bottom of what may be called the infidelity of natural science; if possible to bring over to the humility of the Gospel, those who expatiate with delight on the wonders and the sublimities of creation; and to convince them that a loftier wisdom still than that even of their high and honourable acquirements, is the wisdom of him who is resolved to know nothing but Jesus Christ and him crucified.

It is truly a most Christian exercise, to extract a sentiment of piety from the works and the appearances of nature. It has the authority of the Sacred Writers upon its side, and even our Saviour himself gives it the weight and the solemnity of his example. "Behold the lilies of the field; they toil not, neither do they spin, yet your heavenly Father careth for them." He expatiates on the beauty of a single flower, and draws from it the delightful argument of confidence in God. He gives us to see that taste may be combined with piety, and that the same heart may be occupied with all that is serious in the contemplations of religion, and be at the same time alive to the charms and the loveliness of nature.

The Psalmist takes a still loftier flight. He leaves the world, and lifts his imagination to that mighty expanse which spreads above it and around it. He wings his way through space, and wanders in thought over its immeasurable regions. Instead of a dark and unpeopled solitude, he sees it crowded with splendour, and filled with the energy of the Divine presence. Creation rises in its immensity before him, and the world, with all which it inherits, shrinks into littleness at a contemplation so vast and so overpowering. He wonders that he is not overlooked amid the grandeur and the variety which are on every side of him, and passing upward from the majesty of nature to the majesty of nature's Architects he exclaims, "What is man that thou art mindful of him, or the son of man that thou shouldest deign to visit him?"

It is not for us to say, whether inspiration revealed to the Psalmist the wonders of the modern astronomy. But even though the mind be a perfect stranger to the science of these enlightened times, the heavens present a great and an elevating spectacle,—an immense concave reposing upon the circular boundary of the world, and the innumerable lights which are suspended from on high, moving with solemn regularity along its surface. It seems to have been at night that the piety of the Psalmist was awakened by this contemplation, when the moon and the stars were visible; and not when the sun had risen in his strength, and thrown a splendour around him, which bore down and eclipsed all the lesser glories of the firmament. And there is much in the scenery of a nocturnal sky, to lift the soul to pious contemplation. That moon, and these stars, what are they? They are detached from the world, and they lift you above it. You feel withdrawn from the earth, and rise in lofty abstraction above this little theatre of human passions and human anxieties. The mind abandons itself to reverie, and is transferred in the ecstasy of its thoughts, to distant and unexplored regions. It sees nature in the simplicity of her great elements, and it sees the God of nature invested with the high attributes of wisdom and majesty.

But what can these lights be? The curiosity of the human mind is insatiable, and the mechanism of these wonderful heavens, has, in all ages, been its subject and its employment. It has been reserved for these latter times, to resolve this great and interesting question. The sublimest powers of philosophy have been called to the exercise, and astronomy may now be looked upon as the most certain and best established of the sciences.

Omitted at this point is a paragraph in which Chalmers informs his congregation of the vast distances of the heavenly bodies, noting also that in many cases, these bodies although appearing small are actually immense. For example, he notes that the Sun is many thousands of times larger than the Earth.

Now, what is the fair and obvious presumption? The world in which we live, is a round ball of a determined magnitude, and occupies its own place in the firmament. But, when we explore the unlimited tracks of that space which is every where around us, we meet with other balls of equal or superior magnitude; and from which our earth would either be invisible, or appear as small as any of those twinkling stars which are seen on the canopy of heaven. Why then suppose that this little spot, little at least in the immensity which surrounds it, should be the exclusive abode of life and of intelligence? What reason to think that those mightier globes which roll in other parts of creation, and which we have discovered to be worlds in magnitude, are not also worlds in use and in dignity? Why should we think that the great Architect of nature, supreme in wisdom as he is in power, would call these stately mansions into existence and leave them unoccupied? When we cast our eye over the broad sea, and look at the country on the other side, we see nothing but the blue land stretching obscurely over the distant horizon. We are too far away to perceive the richness of its scenery, or to hear the sound of its population. Why not extend this principle to the still more distant parts of the universe? What though, from this remote point of observation, we can see nothing but the naked roundness of yon planetary orbs? Are we therefore to say, that they are so many vast and unpeopled solitudes; that desolation reigns in every part of the universe but ours; that the whole energy of the divine attributes is expended on one insignificant corner of these mighty works; and that to this earth alone, belongs the bloom of vegetation, or the blessedness of life, or the dignity of rational and immortal existence?

But this is not all. We have something more than the mere magnitude of the planets to allege, in favour of the idea that they are inhabited. We know that this earth turns round upon itself; and we observe that all those celestial bodies, which are accessible to such an observation, have the same movement. We know that the earth performs a yearly revolution round the sun; and we can detect in all the planets which compose our system, a revolution of the same kind, and under the same circumstances. They have the same succession of day and night. They have the same agreeable vicissitude of the seasons. To them, light and darkness succeed each other; and the gaiety of summer is followed by the dreariness of winter. To each of them the heavens present as varied and magnificent a spectacle; and this earth, the encompassing of which would require the labour of years from one of its puny inhabitants, is but one of the lesser lights which sparkle in their firmament. To them, as well as to us, has God divided the light from the darkness, and he has called the light day, and the darkness he has called night. He has said, let there be lights in the firmament of their heaven, to divide the day from the night; and let them be for signs, and for seasons, and for days, and for years; and let them be for lights in the firmament of heaven, to give light upon their earth; and it was so. And God has also made to them great lights. To all of them he has given the sun to rule the day; and to many of them has he given moons to rule the night. To them he has made the stars also. And God has set them in the firmament of heaven, to give light upon their earth; and to rule over the day, and over the night, and to divide the light from the darkness; and God has seen that it was good.

In all these greater arrangements of divine wisdom, we can see that God has done the same things for the accommodation of the planets that he has done for the earth which we inhabit. And shall we say that the resemblance stops here, because we are not in a situation to observe it? Shall we say that this scene of magnificence has been called into being merely for the amusement of a few astronomers? Shall we measure the counsels of heaven by the narrow impotence of the human faculties; or conceive, that silence and solitude reign throughout the mighty empire of nature; that the greater part of creation is an empty parade; and that not a worshipper of the Divinity is to be found through the wide extent of yon vast and immeasurable regions?

It lends a delightful confirmation to the argument, when, from the growing perfection of our instruments, we can discover a new point of resemblance between our Earth and the other bodies of the planetary system. It is now ascertained, not merely that all of them have their day and night, and that all of them have their vicissitudes of seasons, and that some of them have their moons to rule their night and alleviate the darkness of it. We can see of one, that its surface rises into inequalities, that it swells into mountains and stretches into valleys; of another, that it is surrounded by an atmosphere which may support the respiration of animals; of a third, that clouds are formed and suspended over it, which may minister to it all the bloom and luxuriance of vegetation; and of a fourth, that a white colour spreads over its northern regions, as its winter advances, and that on the approach of summer this whiteness is dissipated—giving room to suppose, that the element of water abounds in it, that it rises by evaporation into its atmosphere, that it freezes upon the application of cold, that it is precipitated in the form of snow, that it covers the ground with a fleecy mantle, which melts away from the heat of a more vertical sun; and that other worlds bear a resemblance to our own, in the same yearly round of beneficent and interesting changes.

A paragraph that begins "Who shall assign a limit to the discoveries of future ages?" is omitted at this point. In it, Chalmers speculates (predicts) what future observational discoveries may provide evidence of life in other worlds. Suggesting that future astronomical discoveries "may lay open to us the unquestionable vestiges of art, and industry, and intelligence," he states that we can "assert with the highest probability, that yon planetary orbs are so many worlds."

Did the discoveries of science stop here, we have enough to justify the exclamation of the Psalmist, "What is man that thou art mindful of him, or the son of man that thou shouldest deign to visit him?" They widen the empire of creation far beyond the limits which were formerly assigned to it.

Chalmers then attempts to instill humility in his congregation by discussing the relatively small size of our planet. He concludes the paragraph with an admonishment.

We should learn not to look on our earth as the universe of God, but as one paltry and insignificant portion of it; that it is only one of the many mansions which the Supreme Being has created for the accommodation of his worshippers, and only one of the many worlds rolling in that flood of light which the sun pours around him to the outer limits of the planetary system.

But is there nothing beyond these limits? The planetary system has its boundary, but space has none; and if we wing our fancy there, do we only travel through dark and unoccupied regions? There are only five, or at most six, of the planetary orbs visible to the naked eye. What, then, is that multitude of other lights which sparkle in our firmament, and fill the whole concave of heaven with innumerable splendours? The planets are all attached to the sun; and, in circling around him, they do homage to that influence which binds them to perpetual attendance on this great luminary. But the other stars do not own his dominion. They do not circle around him. To all common observation, they remain immovable; and each, like the independent sovereign of his own territory, appears to occupy the same inflexible position in the regions of immensity. What can we make of them? Shall we take our adventurous flight to explore these dark and untravelled dominions? What mean these innumerable fires lighted up in distant parts of the universe? Are they only made to shed a feeble glimmering over this little spot in the kingdom of nature? or do they serve a purpose worthier of themselves, to light up other worlds, and give animation to other systems?

Three paragraphs have been omitted at this point in which Chalmers discusses stars, showing how greatly distant they are from us and stressing their similarities to the Sun, the point of this being to show that the stars are suns. As he states, the stars "are so many suns, each throned in the centre of his own dominions, and pouring a flood of light over his own portion of these unlimitable regions."

Shall we say, then, of these vast luminaries, that they were created in vain? Were they called into existence for no other purpose than to throw a tide of useless splendour over the solitudes of immensity? Our sun is only one of these luminaries, and we know that he has worlds in his train. Why should we strip the rest of this princely attendance? Why may not each of them be the centre of his own system, and give light to his own worlds? It is true that we see them not; but could the eye of man take its flight into those distant regions, it should lose sight of our little world, before it reached the outer limits of our system — the greater planets should disappear in their turn — before it had described a small portion of that abyss which separates us from the fixed stars, the sun should decline into a little spot, and all its splendid retinue of worlds be lost in the obscurity of distance — he should, at last, shrink into a small indivisible atom, and all that could be seen of this magnificent system, should be reduced to the glimmering of a little star. Why resist any longer the grand and interesting conclusion? Each of these stars may be the token of a system as vast and as splendid as the one which we inhabit. Worlds roll in these distant regions; and these worlds must be the mansions of life and of intelligence. In yon gilded canopy of heaven, we see the broad aspect of the universe, where each shining point presents us with a sun, and each sun with a system of worlds—where the Divinity reigns in all the grandeur of his high attributes—where he peoples immensity with his wonders; and travels in the greatness of his strength through the dominions of one vast and unlimited monarchy.

The contemplation has no limits. If we ask the number of suns and of systems—the unassisted eye of man can take in a thousand, and the best telescope which the genius of man has constructed can take in eighty millions. But why subject the dominions of the universe to the eye of man, or to the powers of his genius? Fancy may take its flight far beyond the ken of eye or of telescope. It may expatiate on the outer regions of all that is visible—and shall we have the boldness to say, that there is nothing there? that the wonders of the Almighty are at end, because we can no longer trace his footsteps? that his omnipotence is exhausted, because human art can no longer follow him? that the creative energy of God has sunk into repose, because the imagination is enfeebled by the magnitude of its efforts, and can keep no longer on the wing through those mighty tracks, which shoot far beyond what eye hath seen, or the heart of man hath conceived—which sweep endlessly along, and merge into an awful and mysterious infinity?

A section has been omitted at this point in which Chalmers discusses astronomical investigations of whether the Sun has a motion through space. The point he draws from this is that this idea reduces "our planetary seasons, and our planetary movements, to a very humble and fractionary rank in the scale of a higher astronomy. There is room for all this in immensity; and there is even argument for all this, in the records of actual observation; and, from the whole of this speculation, do we gather a new emphasis to the lesson, how minute is the place, and how secondary is the importance of our world, amid the glories of such a surrounding magnificence" (40–41).

But, there is still another very interesting track of speculation, which has been opened up to us by the more recent observations of astronomy. What we allude to, is the discovery of the *nebulae*. We allow that it is but a dim and indistinct light which this discovery has thrown upon the structure of the universe; but still it has spread before the eye of the mind a field of very wide and lofty contemplation. Anterior to this discovery, the universe might appear to have been composed of an indefinite number of suns, about equidistant from each other, uniformly scattered over space, and each encompassed by such a planetary attendance as takes place in our own system. But, we have now reason to think, that, instead of lying uniformly, and in a state of equidistance from each other, they are arranged into distinct clusters—that, in the same manner as the distance of the nearest fixed stars so inconceivably superior to that of our planets from each other, marks the separation of the solar systems; so the distance of two contiguous clusters may be so inconceivably superior to the reciprocal distance of those fixed stars which belong to the same cluster, as to mark an equally distinct separation of the clusters, and to constitute each of them an individual member of some higher and more extended arrangement. This carries us upwards through another ascending step in the scale of magnificence, and there leaves us wildering in the uncertainty, whether even here the wonderful progression is ended; and, at all events, fixes the assured conclusion in our minds, that, to an eye which could spread itself over the whole, the mansion which accommodates our species might be so very small as to lie wrapped in microscopical concealment; and, in reference to the only Being who possesses this universal eye, well might we say, "What is man that thou art mindful of him, or the son of man that thou shouldest deign to visit him?"

And, after all, though it be a mighty and difficult conception, yet who can question it? What is seen may be nothing to what is unseen; for what is seen is limited by the range of our instruments. What is unseen has no limit; and, though all which the eye of man can take in, or his fancy can grasp at, were swept away, there might still remain as ample a field, over which the Divinity may expatiate, and which he may have peopled with innumerable worlds. If the whole visible creation were to disappear, it would leave a solitude behind it—but to the infinite Mind, that can take in the whole system of nature, this solitude might be nothing; a small unoccupied point in that immensity which surrounds it, and which he may have filled with the wonders of his omnipotence. Though this earth were to be burned up, though the trumpet of its dissolution were sounded, though yon sky were to pass away as a scroll, and every visible glory, which the finger of the Divinity has inscribed on it, were to be put out for ever—an event, so awful to us, and to every world in our vicinity, by which so many suns would be extinguished, and so many varied scenes of life and of population would rush into forgetfulness—what is it in the high scale of the Almighty's workmanship? a mere shred, which, though scattered into nothing, would leave the universe of God one entire scene of greatness and of majesty. Though this earth, and these heavens, were to disappear, there are other worlds which roll afar; the light of other suns shines upon them; and the sky which mantles them, is garnished with other stars. Is it presumption to say, that the moral world extends to these distant and unknown regions? that they are occupied with people? that the charities of home and of neighbourhood flourish there? that the praises of God are there lifted up, and his goodness rejoiced in? that piety has its temples and its offerings? and 'the richness' of the divine attributes is there felt and admired by intelligent worshippers?

And what is this world in the immensity which teems with them—and what are they who occupy it? The universe at large would suffer as little, in its splendour and variety, by the destruction of our planet, as the verdure and sublime magnitude of a forest would suffer by the fall of a single leaf. The leaf quivers on the branch which supports it. It lies at the mercy of the slightest accident. A breath of wind tears it from its stem, and it lights on the stream of water which passes underneath. In a moment of time, the life, which we know, by the microscope, it teems with, is extinguished; and, an occurrence so insignificant in the eye of man, and on the scale of his observation, carries in it, to the myriads which people this little leaf, an event as terrible and as decisive as the destruction of a world. Now, on the grand scale of the universe, we, the occupiers of this ball, which performs its little round among the suns and the systems that astronomy has unfolded—we may feel the same littleness, and the same insecurity. We differ from the leaf only in this circumstance, that it would require the operation of greater elements to destroy us. But these elements exist.

In the remainder of this long apocalyptic paragraph, Chalmers employs powerful prose to conjure up images of such possible cosmic catastrophes as volcanic eruptions, worldwide floods, collision with a comet, darkening and death of the Sun or the Earth falling into it.

Now, it is this littleness, and this insecurity, which make the protection of the Almighty so dear to us, and bring, with such emphasis, to every pious bosom, the holy lessons of humility and gratitude. The God who sitteth above, and presides in high authority over all worlds, is mindful of man; and, though at this moment his energy is felt in the remotest provinces of creation, we may feel the same security in his providence, as if we were the objects of his undivided care. It is not for us to bring our minds up to this mysterious agency. But, such is the incomprehensible fact, that the same Being, whose eye is abroad over the whole universe, gives vegetation to every blade of grass, and motion to every particle of blood which circulates through the veins of the minutest animal; that, though his mind takes into its comprehensive grasp, immensity and all its wonders, I am as much known to him as if I were the single object of his attention; that he marks all my thoughts; that he gives birth to every feeling and every movement within me; and that, with an exercise of power which I can neither describe nor comprehend, the same God who sits in the highest heaven, and reigns over the glories of the firmament, is at my right hand, to give me every breath which I draw, and every comfort which I enjoy.

But this very reflection has been appropriated to the use of Infidelity, and the very language of the text has been made to bear an application of hostility to the faith. "What is man, that God should be mindful of him, or the son of man, that he should deign to visit him?" Is it likely, says the Infidel, that God would send his eternal Son, to die for the puny occupiers of so insignificant a province in the mighty field of his creation? Are we the befitting objects of so great and so signal an interposition? Does not the largeness of that field which astronomy lays open to the view of modern science, throw a suspicion over the truth of the gospel history; and how shall we reconcile the greatness of that wonderful movement which was made in heaven for the redemption of fallen man, with the comparative meanness and obscurity of our species?

This is a popular argument against Christianity, not much dwelt upon in books; but we believe, a good deal insinuated in conversation, and having no small influence on the amateurs of a superficial philosophy. At all events, it is right that every such argument should be met, and manfully confronted; nor do we know a more discreditable surrender of our religion, than to act as if she had any thing to fear from the ingenuity of her most accomplished adversaries.

The author of the following Discourses engages in his present undertaking, under the full impression, that a something may be found with which to combat Infidelity in all its forms; that the truth of God and of his message, admits of a noble and decisive manifestation, through every mist which the pride, or the prejudice, or the sophistry of man may throw around it; and elevated as the wisdom of him may be, who has ascended the heights of science, and poured the light of demonstration over the most wondrous of nature's mysteries, that even out of his own principles, it may be proved how much more elevated is the wisdom of him who sits with the docility of a little child, to his Bible, and casts down to its authority, all his lofty imaginations.

Summary of the Remainder of Chalmers's *Discourses*

Humility is the main virtue encouraged in Chalmers's second sermon, entitled "On the Modesty of True Science." Herein he praises Newton as an empiricist whose modest conception of the powers of intellect would have led him to avoid the inconsistency of the infidel who, while speculating on the botany of distant planets, denies the possibility of Christianity being communicated to those locales. Newton, contrasted with Voltaire, is portrayed as approaching both nature and scripture with a childlike modesty.

In the third discourse, "On the Extent of the Divine Condescension," Chalmers attempts to resolve the earlier question as to how God can care for this Earth, which modern astronomy has revealed to be but a minute part of a vast cosmos. To this he suggests that the microscope, invented nearly simultaneously with the telescope, provides much of the answer: "The one [the telescope] led me to see a system in every star. The other leads me to see a world in every atom."[13] After asserting that the microscope shows God's solicitude extending even to the smallest of animals, Chalmers advocates the acceptance of a God who "while magnitude does not overpower him, minuteness cannot escape him, and variety cannot bewilder him."[14] Similarly, Chalmers asks how can the person who knows that God oversees every thought doubt that God's beneficence extends also to minds on other orbs. Concerning Christ and the redemption, Chalmers at one point raises a question that seems to imply that sin occurred only on this Earth:

> tell me, O tell me, would it not throw the softening of a most exquisite tenderness over the character of God, should we see him putting forth his every expedient to reclaim to himself those children who had wandered away from him . . . that, rather than lose the single world which had turned to its own way, he should . . . lay upon his own Son the burden of its atonement, that he might again smile upon the world, and hold out the sceptre of invitation to all its families?[15]

Nonetheless, shortly after this, he asks his auditors to contemplate the truth that God who created this vast universe "came to this humblest of its provinces, in the disguise of a servant, and took upon him the form of our degraded species, and let himself down to sorrows, and to sufferings, and to death, for us."[16] Then Chalmers

13. Thomas Chalmers, *A Series of Discourses on the Christian Revelation Viewed in Connection with the Modern Astronomy* (New York: American Tract Society, 185?), 103.

14. Chalmers, *Discourses*, 95.

15. Chalmers, *Discourses*, 94.

16. Chalmers, *Discourses*, 111.

addresses the questions whether Christ's actions on Earth extend to extraterrestrials, phrasing his position on this question only in the negative sense:

> for any thing we can know by reason, the plan of redemption may have its influences and its bearings on those creatures of God who people other regions, and occupy other fields in the immensity of his dominions; that to argue, therefore, on this plan being instituted for the single benefit of . . . the species to which we belong, is a mere presumption of the Infidel himself.[17]

When we compare this statement with the extended quotation cited earlier in this paragraph, the implication seems to be that Chalmers is saying that we cannot tell with certainty which of these alternatives is correct. Sin may have occurred only on Earth, making it unnecessary for Christ to redeem extraterrestrials, or at least some extraterrestrials sinned but they can be redeemed by Christ's redemptive actions on Earth.

Entitled "On the Knowledge of Man's Moral History in the Distant Places of Creation," Chalmers's fourth sermon suggests that just as the effects of Christ's redemptive action have not diminished through millennia of time upon this Earth, so also may they reach to remote planets. Learning what Christ has done for this puny planet, how could minds elsewhere doubt God's generosity? The darkness of our minds and the ambiguity of scripture prevent us, Chalmers notes, from knowing the redemptive plan in its fullness, but he quotes some passages from scripture suggesting that they can be reinterpreted in terms of the enlarged cosmos of modern astronomy.

An example of this is the passage from Luke 15:7 that heads the fifth discourse: "I say unto you, that likewise joy shall be in heaven over one sinner that repenteth, more than over ninety and nine just persons who need no repentance." Taking this as his text, he sketches its significance on a cosmic scale. Poetic imagery helps him evoke an image of the entire universe rejoicing at the repentance of one earthling, yet without asserting such acclamation occurs. Introducing angels, extraterrestrials with scriptural warrant, he portrays them as ringing "throughout all their mansions the hosannas of joy, over every one individual of [earth's] repentant population."[18] He suggests other ways in which the tension postulated by the infidel may be relieved: "For anything I know, the very planet that rolls in the immensity around me,

17. Chalmers, *Discourses*, 113–14.
18. Chalmers, *Discourses*, 159.

may be a land of righteousness."[19] This leads to the awesome thought that the universe may be "one secure and rejoicing family [wherein our] alienated world is the only strayed, or only captive member."[20] Such thoughts should do more than silence the infidel; they should call us to repentance.

Satan enters the sixth discourse and with him the "Contest for an Ascendency over Man, amongst the Higher Orders of Intelligence," as this sermon is entitled. Chalmers presents the Earth as "the actual theatre of a keen and ambitious contest amongst the upper orders of creation,"[21] and speculates that the Earth "has become the theatre of such a competition, as may have all the desires and all the energies of a divided universe embarked upon it. It involves in it other objects than the single recovery of our species. It decides higher questions."[22] Although he admits that "I know not if our rebellious world be the only strong-hold which Satan is possessed of, or if it be but the single post of an extended warfare, that is now going on between the powers of light and of darkness,"[23] he later uses this idea to explain "why on the salvation of our solitary species so much attention appears to have been concentrated."[24] After reading passages such as these, one suspects that the burgers of Glasgow returned to their work with an elevated sense of the significance of their every action. Donne's "No man is an island" pales in comparison to Chalmers's imagery in loftiness and grandeur of scope.

Chalmers's seventh and final discourse stresses that a sensitivity to majestic sights and luxurious imagery is different from a readiness to accept Jesus as personal Lord. Such sensitivity is commended, but this skillful preacher, who could not have been unaware of the impact of his eloquence, presents his sermons as above all a call to a humble life of Christian commitment. It is the Christian's will, not the fineness of the person's sensitivities, that brings redemption. This serves as another reminder that Chalmers spoke and wrote, not as an astronomer, not as a philosopher or theologian, not even as a poet, but as an evangelical preacher striving to reach the hearts and souls of those who came to him. We cannot know whether he succeeded in this, but it is clear from the reception of his book that many left it convinced pluralists, certain that Christianity not only could be reconciled with that doctrine, but also attained thereby a new grandeur.

---

19. Chalmers, *Discourses*, 170.
20. Chalmers, *Discourses*, 173.
21. Chalmers, *Discourses*, 156.
22. Chalmers, *Discourses*, 188–89.
23. Chalmers, *Discourses*, 189.
24. Chalmers, *Discourses*, 197.

THOMAS DICK (1774–1857)

Although the doctrine of a plurality of worlds concerned Chalmers for a only a matter of months, it became a core component of the career of Chalmers's fellow Scotsman Thomas Dick. From Dundee where he had erected an observatory, Dick deluged both sides of the Atlantic with volumes in which ideas of extraterrestrial life appeared with great frequency. This was true not only of his first book, his *Christian Philosopher* (1823), which established his reputation, but also of such subsequent volumes as his *Philosophy of Religion* (1826), *Philosophy of a Future State* (1828), *Celestial Scenery* (1837), and *Sidereal Heavens* (1840).

In his *Celestial Scenery*, having discussed pluralist ideas in nearly every section, Dick devoted his last chapter to five arguments for the plurality of worlds:

I. *[T]here are bodies in the planetary system of such* MAGNITUDES *as to afford ample scope for the abodes of myriads of inhabitants.*
II. *There is a* GENERAL SIMILARITY *among all the bodies of the Planetary System, which tends to prove that they are intended to subserve the same ultimate designs in the arrangements of the Creator.*
III. *In the bodies which constitute the solar system, there are* SPECIAL ARRANGEMENTS *which indicate their* ADAPTION *to the enjoyments of sensitive and intelligent beings; and which prove that this was the ultimate design of their creation.*
IV. The scenery of the heavens, as viewed from the surfaces of the larger planets and their satellites, forms a presumptive proof that both the planets and their moons are inhabited by *intellectual* beings.
V. [I]n the world we inhabit, *every part of nature is destined to the support of animated beings.*[25]

In his *Sidereal Heavens*, he added yet three more arguments:

I. *That the doctrine of a plurality of worlds is more worthy of the perfections of the Infinite Creator, and gives us a more glorious and magnificent idea of his character and operations than to suppose his benevolent regards confined to the globe on which we dwell.*

---

25. Thomas Dick, *Celestial Scenery, or the Wonders of the Planetary System Displayed; Illustrating the Perfections of Deity and a Plurality of Worlds* (Hartford: Sumner & Goodman, 1848), 161, 163, 166, 171–72, and 172.

II. *[W]herever one perfection of Deity is exerted, there also* ALL *his attributes are in operation,* and must be displayed, in a greater or less degree, to certain orders of intelligences.

III. *There is an absurdity involved in the contrary supposition*—namely, that the distant regions of creation are devoid of inhabitants.[26]

Dick, like Chalmers, attained a wide audience. Persons as prominent as Ralph Waldo Emerson, William Lloyd Garrison, and Harriet Beecher Stowe came to Dundee to meet the "Christian Philosopher," while Emily Brontë seems to have borrowed from his *Sidereal Heavens* to enrich her poetry. Nonetheless, important differences separated the two Scottish pluralists; for example, Dick went far beyond Chalmers in the boldness of the claims that he made, as the following selections from his *Christian Philosopher* and his *Celestial Scenery* illustrate.

Thomas Dick, *The Christian Philosopher, or the Connexion of Science and Philosophy with Religion* (Hartford: A. C. Goodman & Co., 1850), 83–84 and 150–53.

The Moon.—Next to the sun, the moon is to us the most interesting of all the celestial orbs. She is the constant attendant of the earth, and revolves around it in 27 days, 8 hours; but the period from one new or full moon to another is about 29 days, 12 hours. She is the nearest of all the heavenly bodies; being only about two hundred and forty thousand miles distant from the earth. She is much smaller than the earth; being only 2,180 miles in diameter. Her surface, when viewed with a telescope, presents an interesting and a variegated aspect; being diversified with mountains, valleys, rocks and plains, in every variety of form and position. Some of these mountains form long and elevated ridges, resembling the chains of the Alps and the Andes; while others, of a conical

26. Thomas Dick, *Sidereal Heavens and Other Subjects Connected with Astronomy, as Illustrative of the Character of the Deity, and of an Infinity of Worlds* (Hartford: Sumner & Goodman, 1848), 118, 122, and 123.

form, rise to a great height, from the middle of level plains, somewhat resembling the Peak of Teneriffe. But the most singular feature of the moon, is, those circular ridges and cavities which diversify every portion of her surface. A range of mountains of a circular form, rising three or four miles above the level of the adjacent districts, surrounds, like a mighty rampart, an extensive plain; and, in the middle of this plain or cavity, an insulated conical hill rises to a considerable elevation. Several hundreds of these circular plains, most of which are considerably below the level of the surrounding country, may be perceived, with a good telescope, on every region of the lunar surface. They are of all dimensions, from two or three miles to forty miles in diameter; and, if they be adorned with verdure, they must present to the view of a spectator, placed among them, a more variegated, romantic, and sublime scenery than is to be found on the surface of our globe. An idea of some of these scenes may be acquired by conceiving a plain of about a hundred miles in circumference, encircled with a range of mountains, of various forms, three miles in perpendicular height, and having a mountain near the centre, whose top reaches a mile and a half above the level of the plain. From the top of this central mountain, the whole plain, with all its variety of objects, would be distinctly visible; and the view would appear to be bounded on all sides by a lofty amphitheatre of mountains, in every diversity of shape, rearing their summits to the sky. From the summit of the circular ridge, the conical hill in the centre, the opposite circular range, the plain below, and some of the adjacent plains, which encompass the exterior ridge of the mountains, would form another variety of view; and a third variety would be obtained from the various aspects of the central mountain, and the surrounding scenery, as viewed from the plains below.

The lunar mountains are of all sizes, from a furlong to five miles in perpendicular elevation. Certain luminous spots, which have been occasionally seen on the dark side of the moon, seem to demonstrate that fire exists in this planet. Dr. Herschel and several other astronomers suppose, that they are volcanoes in a state of eruption. It would be a more pleasing idea, and perhaps as nearly corresponding to fact, to suppose, that these phenomena are owing to some occasional splendid illuminations, produced by the lunar inhabitants, during their long nights. Such a scene as the burning of Moscow, the conflagration of an extensive forest, or the splendid illumination of a large city with gas-lights, might present similar appearances to a spectator in the moon. The bright spots of the moon are the mountainous regions; the dark spots are the plains, or more level parts of her surface. There may probably be rivers or small lakes on this planet; but there are no seas or large collection of water. It appears highly probable, from the observations of Schroeter, that the moon is encompassed with an atmosphere; but no clouds, rain, nor snow seem to exist in it. The illuminating power of the light derived from the moon, according to the experiments made by Professor Leslie, is about the *one hundred and fifty thousandth part* of the illuminating power of the sun. According to the experiments of M. Bo[u]guer, it is only as 1 to 300,000.

The Moon always presents the same face to us; which proves, that she revolves round her axis in the same time that she revolves round the earth. As this orb derives its light from the sun, and reflects a portion of it upon the earth, so the earth performs the same office to the moon. A spectator on the lunar surface would behold the earth, like a luminous orb, suspended in the vault of heaven, presenting a surface about 13 times larger than the moon does to us, and appearing sometimes gibbous, sometimes horned, and at other times with a round full face. The light which the earth reflects upon the dark side of the moon may be distinctly perceived by a common telescope, from three to six or eight days after the change. The lunar surface contains about 15 millions of square miles, and is, therefore, capable of containing a population equal to that of our globe, allowing only about 53 inhabitants to every square mile. That this planet is inhabited by sensitive and intelligent beings, there is every reason to conclude, from a consideration of the sublime scenery with which its surface is adorned, and of the general beneficence of the Creator, who appears to have left no large portion of his material creation without animated existences; and it is highly probable that *direct proofs* of the moon's being inhabited may hereafter be obtained, when all the varieties on her surface shall have been more minutely explored.

At this point, Dick inserts a reference to the first of three appendices in which he reports on various matters relevant to extraterrestrial life. He titles this appendix "—*On the means by which it may probably be ascertained whether the Moon be a Habitable World.*" It appears on pp. 150–51 of the 1850 edition of his *Christian Philosopher.*

About six years ago, the author published, in the Monthly Magazine, a few observations on the surface of the moon, in which a few remarks were offered on this subject. The following is an extract from that communication:—

"If we be ever to obtain an ocular demonstration of the habitability of any of the celestial orbs, the moon is the only one, where we can expect to trace, by our telescopes, indications of the agency of sentient or intelligent beings; and I am pretty much convinced, that a long continued series of observations on this planet, by a number of individuals in different places, might completely set at rest the question, 'Whether the moon be a habitable world?' Were a vast number of persons, in different parts of the world, to devote themselves to a particular survey of the moon—were different portions of her surface allotted to different individuals, as the object of their particular research—were every mountain, hill, cavern, cliff, and plain accurately inspected—and every change and modification in the appearance of particular spots carefully marked and represented in a series of delineations, it might lead to some certain conclusions, both as to her physical constitution, and her ultimate destination. It can be demonstrated, that a telescope which magnifies 100 times will show a spot on the moon's surface, whose diameter is 1223 yards; and one which magnifies a thousand times, will, of course, enable us to perceive a portion of her surface, whose size is only 122 yards: and consequently, an object, whether natural or artificial, of no greater extent than one of our large edifices, (for example, St. Paul's church, London,) may by such an instrument, be easily distinguished. Now, if every minute point on the lunar surface were accurately marked by numerous observers, it might be ascertained whether any changes are taking place, either from physical causes, or from the operations of intelligent agents. If a large forest were cutting down — if a city were building in an open plain, or extending its former boundaries — if a barren waste were changing into a scene of vegetation — or, if an immense concourse of animated beings were occasionally assembled on a particular spot, or shifting from one place to another — such changes would be indicated by certain modifications of shade, colour, or motion; and, consequently, would furnish a direct proof of the agency of intelligent beings analogous to man, and of the moon being a habitable globe. For although we may never be able to distinguish the *inhabitants* of the moon, (if any exist,) yet if we can trace those *effects* which can flow only from the operations of intelligent agents, it would form a complete demonstration of their existence, on the same ground on which a navigator concludes an unknown island to be inhabited, when he perceives human habitations, and cultivated fields.

"That changes occasionally happen on the lunar hemisphere next the earth, appears from the observations of Herschel and Schroeter, particularly from those of the latter. In the transactions of the 'Society of Natural Philosophy,' at Berlin, Schroeter relates, that on the 30th December, 1791; at five o'clock, P. M. with a seven feet reflector, magnifying 161 times, he perceived the commencement of a small crater on the south-west declivity of the volcanic mountain in the *Mare Crisium*, having a shadow of at least 2"5. On the 11th January, at twenty minutes past five, on looking at this place again, he could see neither the new crater nor its shadow. Again, on the 4th January, 1792, he perceived, in the eastern crater of Helicon, a central mountain, of a clear gray colour, 3" in diameter, of which, during many years' observations, he had perceived no trace. 'This appearance,' he adds, 'is remarkable, as probably from the time of Hevelius, the western part of Helicon has been forming into its present shape, and nature seems, in that district, to be particularly active.'—In making such minute observations as those to which I allude, it would be proper, along with an inspection of the moon's luminous disk, to mark the appearances of different portions of her dark hemisphere, when it is partially enlightened by the reflected light from the earth, soon after the appearance of new moon. These researches would require a *long-continued* series of the most minute observations, by numerous observers in different regions of the globe, which could be effected only by exciting, among the bulk of mankind, a general attention to such investigations. But were this object accomplished, and were numerous observations made from the tops of mountains, and in the serene sky of southern climes, where the powers of the telescope are not counteracted by dense vapours, there can be little doubt that direct proofs would be obtained that the moon is a habitable world; or, at least, that the question in relation to this point would be completely set at rest."

Dick's next appendix is titled *"Remarks on the late pretended discovery of a Lunar Fortification."* It appears on pp. 151–52 of the 1850 edition of his *Christian Philosopher.*

The British public was lately amused by the announcement of a discovery said to have been made by Professor Frauenhofer, of Munich. This gentleman[27] was said to have discovered a *fortification* in the moon, and to have distinguished several lines of road, supposed to be the work of the lunar inhabitants. It is scarcely necessary to say, that such announcements are obviously premature. To perceive distinctly the shape of an object in the moon, which resembles

27. [Dick is incorrect in attributing these "discoveries" to Josef Frauenhofer. They were rather claims made by Franz von Paula Gruithuisen, who subsequently became the professor of astronomy at the University of Munich. MJC]

a fortification, it is requisite, that that object be of a much larger size than our terrestrial ramparts. Besides, although an object resembling one of our fortifications were perceived on the surface of the moon, there would be no reason to conclude, that it served the same purpose as fortifications do among us. We are so much accustomed to war in our terrestrial system, and reflect so little on its diabolical nature, that we are apt to imagine that it must form a necessary employment even in other worlds. To be assured that a fortification existed in the moon for the same purpose as with us, would indeed be dismal tidings from another world; for it would be a necessary conclusion, from such intelligence, that the inhabitants of that globe are actuated by the same principles of depravity, ambition, and revenge, which have infected the moral atmosphere of our sublunary world. With regard to the pretended discovery of the lunar roads, it may not be improper to remark, that such roads behooved to be at least 400 feet broad, or ten times the breadth of ours, in order to be perceived as faint lines through a telescope which magnifies a thousand times; which is a higher power, I presume, than Frauenhofer can apply with distinctness to any of his telescopes. It is not at all likely that the lunar inhabitants are of such a gigantic size, or employ carriages of such an enormous bulk, as to require roads of such dimensions, since the whole surface of the moon is only the thirteenth part of the area of our globe.

Schroeter conjectures the existence of a great city to the north of *Marius*, (a spot in the moon,) and of an extensive canal towards *Hygena*, (another spot,) and he represents part of the spot named *Mare Imbrium*, to be as fertile as the Campania. See *Edin. Phil. Jour.* No. 21, *for July*, 1824. Similar remarks to those now stated will apply to these conjectures of Schroeter. We are too apt to imagine, that the objects we perceive in the moon must bear a certain resemblance to those with which we are acquainted on the earth; whereas, there is every reason to believe, from the variety we perceive in nature, that no one world resembles another, except in some of its more prominent and *general* arrangements. The moon bears a general resemblance to the earth, in its being diversified with mountains and valleys; but the positions and arrangement of these objects in the moon, and the scenery they exhibit, are materially different from what appears on the surface of the terraqueous globe.

Dick's next appendix on extraterrestrial life issues is titled *"On a Plurality of Worlds."* It appears on pp. 152–53 of the 1850 edition of his *Christian Philosopher.*

In the first paragraph of this appendix, Dick discourses on the need for a good presentation of the plurality of worlds doctrine, especially from the point of view of religion. From this perspective he remarks that "Fontenelle's 'Plurality of Worlds' contains a number of ingenious reasonings; but he treats the subject in too light and flippant a manner, and without the least reference to a Supreme Intelligence."

That the scriptures are silent on this head, has been assumed by some as a presumptive argument that this doctrine is without a solid foundation. I have already endeavoured to show that this assumption is unfounded; (see page 90.) A plurality of worlds is more than once asserted in scripture, and in numerous passages is evidently taken for granted. Celestial intelligences are represented as ascribing "glory, honour, wisdom, and power" to the king of heaven, "because he hath created all things," and because they perceive his works to be "great and marvellous." But if all the great globes in the firmament were only so many frightful deserts, destitute of inhabitants, such a universe could never inspire superior intelligences with admiration of the *wisdom* of the Creator. For wisdom consists in proportioning means to ends; but, in the case supposed, there would be no proportion between the means and the end. The means are indeed great and astonishing; but no end appears to justify such a display of creating energy. The psalmist, when he contemplated the heavens, was so affected with the idea of the immense population of the universe, that he seems to have been almost afraid lest he should be overlooked amidst the immensity of beings that are under the superintendence of God. "When I consider thy heavens—what is man that thou art mindful of him!" There would be no propriety nor emphasis in this exclamation, if the heavenly orbs were devoid of inhabitants; for if no intelligent beings exist beside man, and a colony of angels, it would not appear wonderful that the Creator should exercise a particular care over the one-half of his intelligent offspring. But, if we conceive the universe as composed of ten thousand times ten thousand worlds, people with myriads of intellectual beings of various orders, the sentiment of admiration implied in the passage is extremely natural and emphatic, and conveys to us an impressive idea of the intelligence, the beneficence, and the condescension of the Founder and Governor of all worlds.

From Thomas Dick, *Celestial Scenery, or the Wonders of the Planetary System Displayed; Illustrating the Perfections of Deity and a Plurality of Worlds* (Hartford: Sumner & Goodman, 1848), 135–36.

CHAPTER VI.

Summary View of the Magnitude of the Planetary System.

Having, in the preceding pages, given a brief description of the principal facts and phenomena connected with the solar system, and offered a few reflections suggested by the subject, it may not be inexpedient to place before the reader a summary view of the magnitude of the bodies belonging to this system, as compared with the population and magnitude of the globe on which we live. In this summary statement I shall chiefly attend to the area or superficial contents of the different planets, which is the only accurate view we can take of their magnitudes, when we compare them with each other as habitable worlds. The population of the different globes is estimated, as in the preceding description, at the rate of 280 inhabitants to a square mile, which is the rate of population in England, and yet this country is by no means overstocked with inhabitants, but could contain, perhaps, double its present population.

Table 6. Thomas Dick's population figures for the solar system

	Square Miles	*Population*	*Solid Contents*
Mercury	32,000,000	8,960,000,000	17,157,324,800
Venus	191,131,911	53,500,000,000	248,475,427,200
Mars	55,417,824	15,500,000,000	38,792,000,000
Vesta	229,000	64,000,000	10,035,000
Juno	6,380,000	1,786,000,000	1,515,250,000
Ceres	8,285,580	2,319,962,400	2,242,630,320
Pallas	14,000,000	4,000,000,000	4,900,000,000
Jupiter	24,884,000,000	6,967,520,000,000	368,283,200,000,000
Saturn	19,600,000,000	5,488,000,000,000	261,326,800,000,000
Saturn's outer ring	9,058,803,600	⇩⇩⇩⇩⇩⇩⇩⇩⇩⇩⇩	
Inner ring	19,791,561,636	8,141,963,826,080	1,442,518,261,800
Edges of the rings	228,077,000	⇧⇧⇧⇧⇧⇧⇧⇧⇧⇧⇧	
Uranus	3,848,460,000	1,077,568,800,000	22,437,804,620,000
The Moon	15,000,000	4,200,000,000	5,455,000,000
Jupiter's satellites	95,000,000	26,673,000,000	45,693,970,126
Saturn's satellites	197,920,800	55,417,824,000	98,960,400,000
Uranus's satellites	169,646,400	47,500,992,000	84,823,200,000
Amount	**78,196,916,784**	**21,894,974,404,480**	**654,038,348,119,246**

From the above statement, the real magnitude of all the moving bodies connected with the solar system may at once be perceived. If we wish to ascertain what proportion these magnitudes bear to the amplitude of our own globe, we have only to divide the different amounts stated at the bottom of the table by the area, solidity, or population of the earth. The amount of area, or the superficial contents of all the planets, primary and secondary, is 78,195,916,784; or above *seventy-eight thousand millions* of square miles. If this sum be divided by 197,000,000, the number of square miles on the surface of our globe, the quotient will be 397; showing that the surfaces of these globes are 397 times more expansive than the whole surface of the terraqueous globe; or, in other words, that they contain an amplitude of space for animated beings equal to nearly *four hundred worlds* such as ours. If we divide the same amount by 49,000,000, the number of square miles in the habitable parts of the earth, the quotient will be 1595; showing that the surface of all the planets contains a space equal to one thousand five hundred and ninety-five times the area of all the continents and islands of our globe. If the amount of population which the planets might contain, namely, 21,894,974,404,480, or nearly *twenty-two billions*, be divided by 800,000,000, the population of the earth, the quotient will be 27,368; which shows that the planetary globes could contain a population more than twenty-seven thousand times the population of our globe; in other words, if peopled in the proportion of England, they are equivalent to twenty-seven thousand worlds such as ours in its present state of population. The amount of the third column expresses the number of solid miles comprised in all the planets, which is 654,038,348,119,246, or more than *six hundred and fifty-four billions*. If this number be divided by 263,000,000,000, the number of cubical miles in the earth, the quotient will be 2483; which shows that the solid bulk of the other planets is two thousand four hundred and eighty-three times the bulk of our globe. Such is the immense magnitude of our planetary system, without taking into account either the sun or the hundreds of comets which have been observed to traverse the planetary regions.

In the remaining portion of this section, Dick discourses on the sizes of the Sun, planets, satellites, and asteroids compared to the Earth. His most striking conclusion is that "the area of the surface of the sun is *thirty-one* times greater than the area of all the primary planets, with their rings and satellites" (136), the implication being that the Sun's population is proportionally immense. He also presents diagrams illustrating how much larger Jupiter, Saturn, and Uranus are than the Earth.

"THE GREAT MOON HOAX"

The 25 August 1835 issue of the *New York Sun* carried the first installment of what may be the most remarkable report ever published. The series of articles purported to be an announcement by the famous astronomer Sir John Herschel that he had telescopically discovered a civilization on the Moon. So sensational was the first installment of the report that over nineteen thousand copies of the 26 August issue of the *Sun* were sold, giving the *Sun* the largest circulation of any newspaper on our planet. Moreover, the installment of 29 August noted that the entire report (including the final installment, which appeared on 31 August) was being made available in pamphlet form. Soon 60,000 copies of this booklet were bought up. Moreover, reprintings have appeared periodically ever since. Lithographs (see figure 9) of the lunarians were also soon made available. Many translations of this remarkable document also appeared. Already by the end of 1836, French translations had been published in Bordeaux, Lausanne, Lyon, Paris, and Strasbourg; Italian editions in Florence, Livorno, Naples, and Ravenna; Spanish translations in Cuba and Mexico; and a German translation published at Hamburg.

The 1 September issue of the *Sun* recounted reactions from other newspapers, stating that the *Mercantile Advertiser* had begun reprinting the series, noting that "It appears to carry intrinsic evidence of being an authentic document."[28] The *Daily Advertiser*, it was reported, expressed its enthusiasm by stating: "No article, we believe, has appeared for years, that will command so general a perusal and publication. Sir John has added a stock of knowledge to the present age that will immortalize his name, and place it high on the page of science."[29] The *Albany Daily Advertiser*, calling it a "Stupendous Discovery," told of having read the story "with unspeakable emotions of pleasure and astonishment."[30] The *New York Times*, according to the *Sun* report, pronounced the discoveries "probable and plausible,"[31] while the *New Yorker* described them as creating "a new era in astronomy and science generally."[32]

---

28. As quoted in Richard Adams Locke, *The Moon Hoax, or A Discovery that the Moon Has a Vast Population of Human Beings* (New York: William Gowans, 1859; facsimile ed. pub. Boston: Gregg Press, 1975), 61.

29. As quoted in Locke, *Moon Hoax*, 61.

30. As quoted in Locke, *Moon Hoax*, 61.

31. As quoted in Locke, *Moon Hoax*, 62.

32. As quoted in Locke, *Moon Hoax*, 62.

Fig. 9. Lithograph of the Lunarians from 1835

According to other contemporary reports, "Some of the grave religious journals made the great discovery a subject of pointed homilies"[33] and an American clergyman warned his congregation that he might have to solicit them for funds for Bibles for the inhabitants of the Moon. It has even been claimed that "the philanthropists of England had frequent and crowded meetings at Exeter Hall, and appointed committees to inquire . . . in regard to the condition of the people of the moon, for purposes of relieving their wants, . . . and, above all, abolishing slavery if it should be found to exist among the lunar inhabitants."[34] A person who had been in New Haven in 1835 described the situation there: "Yale was alive with staunch supporters. The literati—students and professors, doctors in divinity and law—and all the rest of the reading community, looked daily for the arrival of the New York mail with unexampled avidity and implicit faith."[35] Professors Loomis and Olmsted of Yale, it is claimed, went to New York to examine the deleted mathematical sections, but were sent on a wild goose chase.[36] Edgar Allan Poe later reported:

> Not one person in ten discredited it, and (strangest point of all!) the doubters were chiefly those who doubted without being able to say why—the ignorant, those uninformed in astronomy, people who would not believe because the thing was so novel, so entirely "out of the usual way." A grave professor of mathematics in a Virginia college told me seriously that he had no doubt of the truth of the whole affair![37]

Finally the bubble burst. The *Journal of Commerce* sent a reporter to the *Sun* to secure a copy so that *Journal* might republish it. The reporter met with a *Sun* reporter named Richard Adams Locke, who told him: "Don't print it right away.

33. As quoted in Sidney Moss, *Poe's Literary Battles* (Durham, N.C.: Duke University Press, 1963), 87.

34. This claim from "Locke among the Moonlings" (*Southern Quarterly Review* 24 (1853): 501–14) should perhaps not be taken seriously; the tone of the article suggests that satire may have been mixed with the factual information in it.

35. "Locke among the Moonlings," 502.

36. Willy Ley, *Watchers of the Skies* (London: Sidgwick & Jackson, 1963), 273. David F. Musto disputes this in his "Yale Astronomy in the Nineteenth Century," *Ventures in Astronomy* 8 (1968): 9.

37. Edgar Allan Poe, "Richard Adams Locke," *Complete Works of Edgar Allan Poe*, ed. James A. Harrison (New York: G. D. Sproul, 1902), 15:134.

I wrote it myself."[38] The *Journal of Commerce* then denounced the articles as a hoax and the New York *Herald* specified Locke as its perpetrator.

The author of what has repeatedly been called "The Great Moon Hoax" was, in fact, Richard Adams Locke (1800–1871), who, according to some reports, studied at the University of Cambridge (curiously Cambridge has no evidence that Locke ever enrolled). His chief claim to fame remains the articles he composed for the *Sun*, which show that he was fairly widely read in the scientific literature of the period, especially that regarding extraterrestrial life.

In reading the selection that follows, it may be useful to keep in mind that later in this chapter, the claim will be made that there is solid historical evidence indicating that the "Great Moon Hoax" was not in fact a hoax.

The Moon Hoax, or A Discovery that the Moon Has a Vast Population of Human Beings. By Richard Adams Locke (New York: William Gowans, 1859; facsimile ed. pub. Boston: Gregg Press, 1975), 7–9, 21–39, 42–46, 48–50.

GREAT
ASTRONOMICAL DISCOVERIES
LATELY MADE
BY SIR JOHN HERSCHEL, L.L., D.F.R.S, &C.,
AT THE
CAPE OF GOOD HOPE.

FIRST PUBLISHED IN THE NEW YORK SUN
IN AUGUST AND SEPTEMBER, 1835,
FROM THE SUPPLEMENT TO
THE EDINBURGH JOURNAL OF SCIENCE

In this unusual addition to our Journal, we have the happiness of making known to the British public, and thence to the whole civilized world, recent discoveries in Astronomy which will build an imperishable monument to the age in

38. Frank M. O'Brien, *The Story of the Sun*, new ed. (New York: Appleton, 1928), 34–35.

which we live, and confer upon the present generation of the human race a proud distinction through all future time. It has been poetically said, that the stars of heaven are the hereditary regalia of man, as the intellectual sovereign of the animal creation. He may now fold the Zodiack around him with a loftier consciousness of his mental supremacy.

It is impossible to contemplate any great Astronomical discovery without feelings closely allied to a sensation of awe, and nearly akin to those with which a departed spirit may be supposed to discover the realities of a future state. Bound by the irrevocable laws of nature to the globe on which we live, creatures "close shut up in infinite expanse," it seems like acquiring a fearful supernatural power when any remote mysterious works of the Creator yield tribute to our curiosity. It seems almost a presumptuous usurpation of powers denied us by the divine will, when man, in the pride and confidence of his skill, steps forth, far beyond the apparently natural boundary of his privileges, and demands the secrets and familiar fellowship of other worlds. We are assured that when the immortal philosopher to whom mankind is indebted for the thrilling wonders now first made known, had at length adjusted his new and stupendous apparatus with a certainty of success, he solemnly paused several hours before he commenced his observations, that he might prepare his own mind for discoveries which he knew would fill the minds of myriads of his fellow-men with astonishment, and secure his name a bright, if not transcendant [*sic*] conjunction with that of his venerable father to all posterity. And well might he pause! From the hour the first human pair opened their eyes to the glories of the blue firmament above them, there has been no accession to human knowledge at all comparable in sublime interest to that which he has been the honored agent in supplying; and we are taught to believe that, when a work, already preparing for the press, in which his discoveries are embodied in detail, shall be laid before the public, they will be found of incomparable importance to some of the grandest operations of civilized life. Well might he pause! He was about to become the sole depository of wondrous secrets which had been hid from the eyes of all men that had lived since the birth of time. He was about to crown himself with a diadem of knowledge which would give him a conscious pre-eminence above every individual of his species who then lived, or who had lived in the generations that are passed away. He paused ere he broke the seal of the casket which contained it.

To render our enthusiasm intelligible, we will state at once that by means of a telescope of vast dimensions and an entirely new principle, the younger Herschel, at his observatory in the Southern Hemisphere, has already made the most extraordinary discoveries in every planet of our solar system; has discovered planets in other solar systems; has obtained a distinct view of objects in the moon, fully equal to that which the unaided eye commands of terrestrial objects at the distance of a hundred yards; has affirmatively settled the question whether this satellite be inhabited, and by what order of beings; has firmly established a new theory of cometary phenomena; and has solved or corrected nearly every leading problem of mathematical astronomy.

For our early and almost exclusive information concerning these facts, we are indebted to the devoted friendship of Dr. Andrew Grant, the pupil of the elder, and for several years past the inseperable [*sic*] coadjutor of the younger Herschel. The amanuensis of the latter at the Cape of Good Hope, and the indefatigable superintendent of his telescope during the whole period of its construction and operation, Dr. Grant has been enabled to supply us with intelligence equal, in general interest at least, to that which Dr. Herschel himself has transmitted to the Royal Society. Indeed our correspondent assures us that the voluminous documents now before a committee of that institution contain little more than details and mathematical illustrations of the facts communicated to us in his own ample correspondence. For permission to indulge his friendship in communicating this invaluable information to us, Dr. Grant and ourselves are indebted to the magnanimity of Dr. Herschel, who, far above all mercenary considerations, has thus signally honored and rewarded his fellow-laborer in the field of science. The engravings of lunar animals and other objects and of the phases of the several planets, are accurate copies of drawings taken in the observatory by Herbert Home,

Esq., who accompanied the last powerful series of reflectors from London to the Cape, and superintended their erection; and he has thus recorded the proofs of their triumphant success. The engravings of the belts of Jupiter is a reduced copy of an imperial folio drawing by Dr. Herschel himself, and contains the results of his latest observation of that planet. The segment of the inner ring of Saturn is from a large drawing by Dr. Grant.

We first avail ourselves of the documents which contain a description and history of the instrument by which these stupendous discoveries have been made. A knowledge of the one is essential to the credibility of the other.

Omitted at this point is a section titled "The Younger Herschel's Telescope," which tells of John Herschel's successes in telescopic construction. Readers are informed that around 1820 John Herschel constructed a telescope that attained magnifications of 6000. This instrument enabled Herschel "to confirm some discoveries of former observers, and to confute those of others. The existence of volcanoes discovered by his father and by Schroeter of Berlin, and the changes by the latter in the volcano in the *Mare Crisium* or Lucid Lake, were corroborated and illustrated, as was also the prevalence of far more extensive volcanic phenomena. The disproportionate height attributed to the lunar mountains was corrected from careful admeasurement; whilst the celebrated conical hills, encircling valleys of vast diameter, and surrounding the lofty central hills, were distinctly perceived. The formation which Professor Frauenhofer uncharitably conjectured to be a lunar fortification, he ascertained to be a tabular buttress of a remarkably pyramidical mountain; lines which had been whimsically pronounced roads and canals, he found to be keen ridges of singularly regular rows of hills; and that which Schroeter imagined to be a great city in the neighborhood of *Marius*, he determined to be a valley of disjointed rocks scattered in fragments, which averaged at least a thousand yards in diameter. Thus the general geography of the planet, in its grand outlines of cape, continent, mountain, ocean, and island, was surveyed with greater particularity and accuracy than by any previous observer; and the striking dissimilarity of many of its local features to any existing on our own globe, was clearly demonstrated." Despite this progress, Herschel had been unable to determine whether life exists on the Moon.

Early in the installment for the *Sun* of Wednesday, 26 August 1835, the report informs readers that Herschel, drawing on some suggestions of Sir David Brewster, devised a telescope that was new not only in design but also in size, having an "object-glass of twenty-four feet in diameter" and permitting magnification of 42,000. So pleased was Herschel that he "expressed confidence in his ultimate ability to study even the entomology of the moon." Readers are informed that with this powerful instrument John Herschel set sail for the Cape of Good Hope, where he erected an observatory, which is duly described, and commenced observations, some of which are next described.

New Lunar Discoveries.

Until the 10th of January, the observations were chiefly directed to the stars in the southern signs, in which, without the aid of the hydro-oxygen reflectors, a countless number of new stars and nebulae were discovered. But we shall defer our correspondent's account of these to future pages, for the purpose of no longer withholding from our readers the more generally and highly interesting discoveries which were made in the lunar world. And for this purpose, too, we shall defer Dr. Grant's elaborate mathematical details of the corrections which Sir John Herschel has made in the best tables of the moon's tropical, sidercal [*sic*], and synodic revolutions, and of those phenomena of syzygies on which a great part of the established lunar theory depends.

It was about half-past nine o'clock on the night of the 10th, the moon having then advanced within four days of her mean libration, that the astronomer adjusted his instruments for the inspection of her eastern limb. The whole immense power of his telescope was applied, and to its focal image about one half of the power of his microscope. On removing the screen of the latter, the field of view was covered throughout its entire area with a beautifully distinct, and even vivid representation of *basaltic rock*. Its color was a greenish brown, and the width of the columns, as defined by their interstices on the canvass, was invariably twenty-eight inches. No fracture whatever appeared in the mass first presented, but in a few seconds a shelving pile appeared of five or six columns width, which showed their figure to be hexagonal, and their articulations similar to those of the basaltic formation at Staffa. This precipitous shelf was profusely covered with a dark red flower, "precisely similar," says Dr. Grant, "to the Papaver Rhoeas, or rose-poppy of our sublunary cornfields; and this was the first organic production of nature, in a foreign world, ever revealed to the eyes of men."

The rapidity of the moon's ascension, or rather of the earth's diurnal rotation, being nearly equal to five hundred yards in a second, would have effectually prevented the inspection, or even the discovery of objects so minute as these, but for the admirable mechanism which constantly regulates, under the guidance of the sextant, the required altitude of the lens. But its operation was found to be so consummately perfect, that the observers could detain the object upon the field of view for any period they might desire. The specimen of lunar vegetation, however, which they had already seen, had decided a question of too exciting an interest to induce them to retard its exit. It had demonstrated that the moon has an atmosphere constituted similarly to our own, and capable of sustaining organized, and therefore, most probably, animal life. The basaltic rocks continued to pass over the inclined canvass plane, through three successive diameters, when a verdant declivity of great beauty appeared, which occupied two more. This was preceded by another mass of nearly the former height, at the base of which they were at length delighted to perceive that novelty, a lunar forest. "The trees," says Dr. Grant, "for a period of ten minutes, were of one unvaried kind, and unlike any I have seen, except the largest kind of yews in the churchyards, which they in some aspects resemble. These were followed by a level green plain, which, as measured by the painted circle on our canvass of forty-nine feet, must have been more than half a mile in breadth; and then appeared as a fine forest of firs, unequivocal firs, as I have ever seen cherished in the bosom of my native mountains. Wearied with the long continuance of these, we greatly reduced the magnifying power of the microscope, without eclipsing either of the reflectors, and immediately perceived that we had been insensibly descending, as it were, a mountainous district of a highly diversified and romantic character, and that we were on the verge of a lake, or inland sea; but of what relative locality or extent, we were yet too greatly magnified to determine. On that the water, whose boundary we had just discovered, answered in general outline to the Mare Nubium of Ricc[i]oli, by which we detected that, instead of commencing, as we supposed, on the eastern longitude of the planet, some delay in the elevation of the great lens had thrown us nearly upon the axis of her equator. However, as she was a free country, and we not, as yet, attached to any particular province, and moreover, since we could at any moment occupy our intended position, we again slid in our magic lenses to survey the shores of the Mare Nubium. Why Ricc[i]oli so termed it, unless in ridicule of Cleomedes, I know not; for fairer shores never angels coasted on a tour of pleasure. A beach of brilliant white sand, girt with wild castellated rocks, apparently of green marble, varied at chasms, occurring every two or three hundred feet, with grotesque blocks of chalk or gypsum, and feathered and festooned at the summit with the clustering foliage of unknown trees, moved along the bright wall of our apartment until we were speechless with admiration. The water, wherever we obtained a view of it, was nearly as blue as that of the deep ocean, and broke in large white billows upon the strand. The action of very high tides was quite manifest upon the face of the cliffs for more than a hundred miles; yet diversified as the scenery was during this and a much greater distance, we perceived no trace of animal existence, notwithstanding we could command at will a perspective or a foreground view of the whole. Mr. Holmes, indeed, pronounced some distance in the interior of a cavern, to be bona fide specimens of a large cornu ammonis; but to me they appeared merely large pebbles, which had been chafed and rolled there by the tides. Our chase of animal life was not yet to be rewarded.

The first part of a long paragraph is omitted at this point. In it, Grant reports on visual surveys made at reduced magnification of various lunar regions, noting such matters as geological features and forests. At one point, they sight certain geometrically shaped structures, believing at first "we had assuredly fallen on productions of art." Herschel, however, concluded that rather than being buildings, they were "giant amethysts." After reporting magnificent waterfalls, they reward readers' attention by recounting observations of forms of animal life.

At the foot of this boundary of hills was a perfect zone of woods surrounding the whole valley, which was about eighteen or twenty miles wide, at its greatest breadth, and about thirty in length. Small collections of trees, of every imaginable kind, were scattered about the whole of the luxuriant area; and here our magnifiers blest our panting hopes with specimens of conscious existence. In the shade of the woods on the south-eastern side, we beheld continuous herds of brown quadrupeds, having all the external characteristics of the bison, but more diminutive than any species of the bos genus in our natural history. Its tail is like that of our grunniens; but in its semicircular horns, the hump on its shoulders, and the depth of its dewlap, and the length of its shaggy hair, it closely resembled the species to which I first compared it. It had, however, one widely distinctive feature, which we afterwards found common to nearly every lunar quadruped we have discovered; namely, a remarkable fleshy appendage over the eyes, crossing the whole breadth of forehead and united to the ears. We could most distinctly perceive this hairy veil, which was shaped like the upper front outline of a cap known to the ladies as Mary Queen of Scots' cap, lifted and lowered by means of the ears. It immediately occurred to the acute mind of Dr. Herschel, that this was a providential contrivance to protect the eyes of the animal from the great extremes of light and darkness to which all the inhabitants of our side of the moon are periodically subjected.

The next animal perceived would be classed on earth as a monster. It was of a bluish lead color, about the size of a goat, with a head and beard like him, and a *single horn*, slightly inclined forward from the perpendicular. The female was destitute of the horn and beard, but had a much longer tail. It was gregarious, and chiefly abounded on the acclivitous glades of the woods. In elegance of symmetry it rivalled the antelope, and like him it seemed an agile sprightly creature, running with great speed, and springing from the green turf with all the unaccountable antics of a young lamb or kitten. This beautiful creature afforded us the most exquisite amusement. The mimicry of its movements upon our white painted canvass was as faithful and luminous as that of animals within a few yards of the camera obscura, when seen pictured upon its tympan. Frequently when attempting to put our fingers upon its beard, it would suddenly bound away into oblivion, as if conscious of our earthly impertinence; but then others would appear, whom we could not prevent nibbling the herbage, say or do what we would to them.

A paragraph omitted at this point is devoted chiefly to discussing in considerable detail various water birds that they see, including pelicans and cranes. Their hopes of spotting fish were, however, disappointed. Nonetheless, they report sighting what previous lunar observers had failed to see: lunar clouds, thereby establishing that the Moon has an atmosphere.

The moon being now low on her descent, Dr. Herschel inferred that the increasing refrangibility of her rays would prevent any satisfactory protraction of our labors, and our minds being actually fatigued with the excitement of the high enjoyments we had partaken, we mutually agreed to call in the assistants at the lens, and reward their vigilant attention with congratulatory bumpers of the best "East Indian Particular." It was not, however, without regret that we left the splendid valley of the red mountains, which, in compliment to the arms of our royal patron, we denominated "the Valley of the Unicorn;" and it may be found in Blunt's map, about midway between the Mare Foecunditatis and the Mare Nectaris.

The nights of the 11th and 12th being cloudy, were unfavorable to observation; but on those of the 13th and 14th further animal discoveries were made of the most exciting interest to every human being. We give them in the graphic language of our accomplished correspondent:—

[The installment of Thursday, 27 August 1835 began here.]

Omitted at this point is a long paragraph devoted chiefly to lunar geology, especially lunar mountains. These are reported to be mainly of volcanic origin, speculations being offered as to the time of their formation. They also come upon "twelve luxuriant forests, divided by open plains, which waved in an ocean of verdure." They also sight "numerous herds of quadrupeds similar to our friends the bison in the Valley of the Unicorn" and "flocks of white and red birds upon the wing."

"At length we carefully explored the Endymion. We found each of the three ovals volcanic and sterile within; but, without, most rich, throughout the level regions around them, in every imaginable production of a bounteous soil. Dr. Herschel has classified not less than thirty-eight species of forest trees, and nearly twice this number of plants, found in this tract alone, which are widely different to those found in more equatorial latitudes. Of animals, he classified nine species of mammalia, and five of ovipara. Among the former is a small kind of rein-deer, the elk, the moose, the horned bear, and the biped beaver. The last resembles the beaver of the earth in every other respect than in its destitution of a tail, and its invariable habit of walking upon only two feet. It carries its young in its arms like a human being, and moves with an easy gliding motion. Its huts are constructed better and higher than those of many tribes of human savages, and from the appearance of smoke in nearly all of them, there is no doubt of its being acquainted with the use of fire. Still its head and body differ only in the points stated from that of the beaver, and it was never seen except on the borders of lakes and rivers, in which it has been observed to immerse for a period of several seconds.

The next long paragraph in this installment continues the frequently very detailed descriptions of the geology, fauna, and flora of the Moon. Many of these descriptions would do credit to a tourist booklet. Among the chief items sighted are a crescent-shaped inland sea, a miniature zebra-like animal, and palm trees. Nonetheless, as yet they have nothing to report on intelligent beings, but the installment ends with a hint that striking results will be forthcoming.

"We were admonished by this to lose no time in seeking the next proposed object of our search, the Langrenus, or No. 26, which is almost within the verge of the libration in longitude, and of which, for this reason, Dr. Herschel entertained some singular expectations.

[The installment of Friday, 28 August 1835 began here.]

Omitted at this point is a paragraph reporting on Herschel's sighting a seventy-mile-long lake and the surrounding scenery including "a truly magnificent amphitheatre," remarkable lunar hills including one with a huge perpendicular face, "long lines of some yellow metal" that the scientists judge to be "virgin gold," and various animals, some essentially identical to terrestrial sheep, which appear in flocks. These call forth from Grant the remark: "I need not say how desirous we were of finding shepherds to these flocks, and even a man with blue apron and rolled up sleeves would have been a welcome sight to us, if not to the sheep; but they fed in peace, lords of their own pastures, without either protector or destroyer in human shape" (36).

"We at length approached the level opening to the lake, where the valley narrows to a mile in width, and displays scenery on both sides picturesque and romantic beyond the powers of a prose description. Imagination, borne on the wings of poetry, could alone gather similes to portray the wild sublimity of this landscape, where dark behemoth crags stood over the brows of lofty precipices, as if a rampart in the sky; and forests seemed suspended in mid air. On the eastern side there was one soaring crag, crested with trees, which hung over in a curve like three-fourths of a Gothic arch, and being of a rich crimson color, its effect was most strange upon minds unaccustomed to the association of such grandeur with such beauty.

"But whilst gazing upon them in a perspective of about half a mile, we were thrilled with astonishment to perceive four successive flocks of large winged creatures, wholly unlike any kind of birds, descend with a slow even motion from the cliffs on the western side, and alight upon the plain. They were first noticed by Dr. Herschel, who exclaimed, 'Now, gentlemen, my theories against your proofs, which you have often found a pretty even bet, we have here something worth looking at: I was confident that if ever we found beings in human shape, it would be in this longitude, and that they would be provided by their Creator with some extraordinary powers of locomotion: first exchange for my number D.' This lens being soon introduced, gave us a fine half-mile distance, and we counted three parties of these creatures, of twelve, nine, and fifteen in each, walking erect towards a small wood near the base of the eastern precipices. Certainly they *were* like human beings, for their wings had now disappeared, and their attitude in walking was both erect and dignified. Having observed them at this distance for some minutes, we introduced lens H *z* which brought them to the apparent proximity of eighty yards; the highest clear magnitude we possessed until the latter end of March, when we effected an improvement in the gas-burners. About half of the first party had passed beyond our canvass; but of all the others we had a perfectly distinct and deliberate view. They averaged four feet in height, were covered, except on the face, with short and glossy copper-colored hair, and had wings composed of a thin membrane, without hair, lying snugly upon their backs, from the top of the shoulders to the calves of the legs. The face, which was of a yellowish flesh color, was a slight improvement upon that of the large orang outang, being more open and intelligent in its expression, and having a much greater expansion of forehead. The mouth, however, was very prominent, though somewhat relieved by a thick beard upon the lower jaw, and by lips far more human than those of any species of the simia genus. In general symmetry of body and limbs they were infinitely superior to the orang outang; so much so, that, but for their long wings, Lieut. Drummond said they would look as well on a parade ground as some of the old cockney militia! The hair on the head was a darker color than that of the body, closely curled, but apparently not woolly, and arranged in two curious semicircles over the temples of the forehead. Their feet could only be seen as they were alternately lifted in walking; but, from what we could see of them in so transient a view, they appeared thin, and very protuberant at the heel.

"Whilst passing across the canvass, and whenever we afterwards saw them, these creatures were evidently engaged in conversation; their gesticulation, more particularly the varied action of their hands and arms, appeared impassioned and emphatic. We hence inferred that they were rational beings, and although not perhaps of so high an order as others which we discovered the next month on the shores of the Bay of Rainbows, that they were capable of producing works of art and contrivance. The next view we obtained of them was still more favorable. It was on the borders of a little lake, or expanded stream, which we then for the first time perceived running down the valley to a large lake and having on its eastern margin a small wood.

"Some of these creatures had crossed this water and were lying like spread eagles on the skirts of the wood. We could then perceive that they possessed wings of great expansion, and were similar in structure to those of the bat, being a semitransparent membrane expanded in curvilineal divisions by means of straight radii, united at the back by the dorsal integuments. But what astonished us very much was the circumstance of this membrane being continued from the shoulders to the legs, united all the way down, though gradually decreasing in width. The wings seemed completely under the command of volition, for those of the creatures whom we saw bathing in the water, spread them instantly to their full width, waved them as ducks do theirs to shake off the water, and then as instantly closed them again in a compact form. Our further observation of the habits of these creatures, who were of both sexes, led to results so very remarkable, that I prefer they should first be laid before the public in Dr. Herschel's own work, where I have reason to know they are fully and faithfully stated, however incredulously they may be received.—**** The three families then almost simultaneously spread their wings, and were lost in the dark confines of the canvass before we had time to breathe from our paralyzing astonishment. We scientifically denominated them the Vespertilio-homo, or man-bat; and they are doubtless innocent and happy creatures, notwithstanding that some of their amusements would but ill comport with our terrestrial notions of decorum. The valley itself we called the Ruby Colosseum, in compliment to its stupendous southern boundary, the six mile sweep of precipices two thousand feet high. And the night, or rather morning, being far advanced, we postponed our tour to Petavius (No. 20), until another opportunity." We have, of course, faithfully obeyed Dr. Grant's private injunction to omit those highly curious passages in his correspondence which he wished us to suppress, although we do not perceive the force of the reason assigned for it. It is true, the omitted paragraphs contain facts which would be wholly incredible to readers who do not carefully examine the principles and capacity of the instrument with which these marvellous discoveries have been made; but so will nearly all of those which he has kindly permitted us to publish; and it was for this reason that we considered the explicit description which we have given of the telescope so important a preliminary. From these, however, and other prohibited passages, which will be published by Dr. Herschel, with the certificates of the civil and military authorities of the colony, and of several Episcopal, Wesleyan, and other ministers, who, in the month of March last, were permitted, under stipulation of temporary secrecy, to visit the observatory, and become eye-witnesses of the wonders which they were requested to attest, we are confident his forthcoming volumes will be at once the most sublime in science, and the most intense in general interest, that ever issued from the press.

The night of the 14th displayed the moon in her mean libration, or full; but the somewhat humid state of the atmosphere being for several hours less favorable to a minute inspection than to a general survey of her surface, they were chiefly devoted to the latter purpose. But shortly after midnight the last veil of mist was dissipated, and the sky being as lucid as on the former evenings, the attention of the astronomers was arrested by the remarkable outlines of the spot marked No. 18, in Blunt's lunar chart; and in this region they added treasures to human knowledge which angels might well desire to win. Many parts of the following extract will remain forever in the chronicles of time:—

[The installment of Saturday, 29 August 1835 began here.]

Omitted at this point are some paragraphs reporting on three oceans and numerous seas as well as countless lakes and associated geological features sighted on the Moon. The most amazing formation discussed is a quartz crystal 340 miles long.

"The dark expanse of waters to the south of the first great ocean has often been considered a fourth; but we found it to be merely a sea of the first class, entirely surrounded by land, and much more encumbered with promontories and islands than it has been exhibited in any lunar chart. One of its promontories runs from the vicinity of Pitatus (No. 19), in a slightly curved and very narrow line, to Bullialdus (No. 29), which is merely a circular head to it, 264 miles from its starting place. This is another mountainous ring, a marine volcano, nearly burnt out, and slumbering upon its cinders. But Pitatus, stranding upon a bold cape of the southern shore, is apparently exulting in the might and majesty of its fires. The atmosphere being now quite free from vapor, we introduced the magnifiers to examine a large bright circle of hills which sweep close beside the western abutments of this flaming mountain. The hills were either of snow-white marble or semi-transparent crystal, we could not distinguish which, and they bounded another of those lovely green valleys, which, however monotonous in my descriptions, are of paradisaical beauty and fertility, and like primitive Eden in the bliss of their inhabitants. Dr. Herschel here again predicated another of his sagacious theories. He said the proximity of the flaming mountain, Bullialdus, must be so great a local convenience to dwellers in this valley during the long periodical absence of solar light, as to render it a place of populous resort for the inhabitants of all the adjacent regions, more especially as its bulwark of hills afforded an infallible security against any volcanic eruption that could occur. We therefore applied our full power to explore it, and rich indeed was our reward.

"The very first object in this valley that appeared upon our canvass was a magnificent work of art. It was a temple—a fane of devotion, or of science, which, when consecrated to the Creator, *is* devotion of the loftiest order; for it exhibits his attributes purely free from the masquerade, attire, and blasphemous caricature of controversial creeds, and has the seal and signature of his own band to sanction its aspirations. It was an equitriangular temple, built of polished sapphire, or of some resplendent blue stone, which, like it, displayed a myriad points of golden light twinkling and scintillating in the sunbeams. Our canvass, though fifty feet in diameter, was too limited to receive more than a sixth part of it at one view, and the first part that appeared was near the centre of one of its sides, being three square columns, six feet in diameter at its base, and gently tapering to a hight [*sic*] of seventy feet.

The remainder of this paragraph describes this temple in lavish detail and pious tone, devoting less attention to two similar temples that they had also sighted.

[The installment of Monday, 31 August 1835 began here.]

"But we had not far to seek for inhabitants of this 'Vale of the Triads.' Immediately on the outer border of the wood which surrounded, at the distance of half a mile, the eminence on which the first of these temples stood, we saw several detached assemblies of beings whom we instantly recognized to be of the same species as our winged friends of the Ruby Colosseum near the lake Langrenus. Having adjusted the instrument for a minute examination, we found that nearly all the individuals in these groups were of a larger stature than the former specimens, less dark in color, and *in every respect* an improved variety of the race. They were chiefly engaged in eating a large yellow fruit like a gourd, sections of which they divided with their fingers, and ate with rather uncouth voracity, throwing away the rind. A smaller red fruit, shaped like a cucumber, which we had often seen pendant from trees having a broad dark leaf, was also lying in heaps in the centre of several of the festive groups; but the only use they appeared to make of it was sucking its juice, after rolling it between the palms of their hands and nibbling off an end. They seemed eminently happy, and even polite, for we saw, in many instances, individuals sitting nearest these piles of fruit, select the largest and brightest specimens and throw them archwise across the circle to some opposite friend or associate who had extracted the nutriment from those scattered around him, and which were frequently not a few. While thus engaged in their rural banquets, or in social converse, they were always seated with their knees flat upon the turf, and their feet brought evenly together in the form of a triangle. And for some mysterious reason or other this figure seemed to be an especial favorite among them; for we found that every group or social circle arranged itself in this shape before it dispersed, which was generally done at the signal of an individual who stepped into the centre and brought his hands over his head in an acute angle. At this signal each member of the company extended his arms forward so as to form an acute horizontal angle with the extremity of the fingers. But this was not the only proof we had that they were creatures of order and subordination. **** We had no opportunity of seeing them actually engaged in any work of industry or art; and so far as we could judge, they spent their happy hours in collecting various fruits in the woods, in eating, flying, bathing, and loitering about upon the summits of precipices. **** But although evidently the highest order of animals in this rich valley, they were not its only occupants. Most of the other animals which we had discovered elsewhere in very distant regions, were collected here; and also at least eight or nine new species of quadrupeds. The most attractive of these was a tall white stag with lofty spreading antlers, black as ebony. We several times saw this elegant creature trot up to the seated parties of the semi-human beings I have described, and browse the herbage close beside them, without the least manifestation of fear on its part or notice on theirs. The universal state of amity among all classes of lunar creatures, and the apparent absence of every carnivorous or ferocious species, gave us the most refined pleasure, and doubly endeared to us this lovely nocturnal companion of our larger, but less favored world. Ever again when I 'eye the blue vault and bless the *useful* light,' shall I recall the scenes of beauty, grandeur, and felicity, I have beheld upon her surface, not 'as *through* a glass darkly, but face to face;' and never shall I think of that line of our thrice noble poet,

—'Meek Diana's crest
Sails through the azure air, an island of the blest,'[39]

without exulting in my knowledge of its truth."

A few paragraphs are omitted at this point informing readers of a fire that nearly destroyed Herschel's observatory and of some amazing observations of Saturn made after the telescope was put back into operation.

**** "It was not until the new moon of the month of March, that the weather proved favorable to any continued series of lunar observations; and Dr. Herschel had been too enthusiastically absorbed in demonstrating his brilliant discoveries in the southern constellations, and in constructing tables and catalogues of his new stars, to avail himself of the few clear nights which intervened.

On one of these, however, Mr. Drummond, myself, and Holmes, made those discoveries near the Bay of Rainbows, to which I have somewhere briefly alluded. The bay thus fancifully denominated is a part of the northern boundary of the first great ocean which I have lately described, and is marked in the chart with the letter O. The tract of country which we explored on this occasion is numbered 6, 5, 8, 7, in the catalogue, and the chief mountains to which these numbers are attached are severally named Atlas, Hercules, Heraclides Verus, and Heraclides Falsus. Still farther to the north of these is the island circle called Pythagoras, and numbered 1; and yet nearer the meridian line is the mountainous district

39. [This is from Lord Byron's poem "Childe Harold's Pilgrimage" as found in *Living Thoughts in Words that Burn, from Poet, Sage, and Humorist*, ed. by Charles F. Beezley (Elliott & Beezley, 1891), 129, where it appears as "Where the day joins the past eternity; While on the other hand, meek Diana's crest / Floats through the azure air—an island of the blest." MJC]

marked R, and called the Land of Drought, and Q, the Land of Hoar Frost; and certainly the name of the latter, however theoretically bestowed, was not altogether inapplicable, for the tops of its very lofty mountains were evidently covered with snow, though the valleys surrounding them were teeming with the luxuriant fertility of midsummer. But the region which we first particularly inspected was that of Heraclides Falsus (No. 7), in which we found several new specimens of animals, all of which were horned and of a white or grey color; and the remains of three ancient triangular temples which had long been in ruins. We thence traversed the country southeastward, until we arrived at Atlas (No. 6), and it was in one of the noble valleys at the foot of this mountain that we found the very superior species of the Vespertilio-homo. In stature they did not exceed those last described, but they were of infinitely greater personal beauty, and appeared in our eyes scarcely less lovely than the general representations of angels by the more imaginative schools of painters. Their social economy seemed to be regulated by laws or ceremonies exactly like those prevailing in the Vale of the Triads, but their works of art were more numerous, and displayed a proficiency of skill quite incredible to all except actual observers. I shall, therefore, let the first detailed account of them appear in Dr. Herschel's authenticated natural history of this planet."

[This concludes the Supplement, with the exception of forty pages of illustrative and mathematical notes, which would greatly enhance the size and price of this work, without commensurably adding to its general interest.—*Ed Sun.*]

Commentary: Evidence that the Moon Hoax Was Not a Hoax

Research that I have done on the "Great Moon Hoax" has led me to the conclusions that despite the numerous accounts that have centered on the claim that Richard Locke perpetrated one of the most remarkable hoaxes of all time, in fact, this event was not a hoax nor was Locke himself a hoaxster. The focus of this commentary is to present some of the evidence that has led me to this conclusion. I have presented much of this evidence in two earlier publications,[40] but recently important new evidence has been discovered.

To begin with the key point: Locke himself explicitly stated that what he had written was *satire*, a satire that alas failed to find an appropriate audience. In editing an 1852 edition of Locke's articles, William Griggs stressed this point:

> we have the assurance of the author, in a letter published some years since, in the *New World*, that it was written expressly to satirize the unwarranted and extravagant anticipations upon this subject, that had been first excited by a prurient coterie of German astronomers, and then aggravated almost to the point of lunacy itself . . . by the religio-scientific rhapsodies of Dr. Dick. At that time the astronomical works of this author enjoyed a degree of popularity . . . almost unexampled in the history of scientific literature.[41]

Griggs added in regard to Dick that

> it would be difficult to name a writer who, with sincere piety, much information, . . . and the best intentions, has done greater injury, at once, to the cause of rational religion and inductive science, by the fanatical, fanciful, and illegitimate manner in which he has attempted to force each into the service of the other.[42]

That Locke intended to write satire is, moreover, clear from the content of his articles, at least if they are read by persons possessing a knowledge of the very satirizable writings of Thomas Dick.

One objection that has been made to this claim is that it appears that no scholar has located the letter written by Locke and referred to but not specifically referenced by Griggs in which Locke explicitly stated this claim. This situation changed in early 2004 when James Secord, a well-known historian of science at Cambridge University, located the letter. He generously shared this discovery with me and sent a copy of Locke's letter. It is dated 6 May 1840 and appears in the 16 May 1840 issue of the

---

40. M. J. Crowe, "New Light on the 'Moon Hoax,'" *Sky and Telescope* 62 (1981): 428–29, and M. J. Crowe, *The Extraterrestrial Life Debate, 1750–1900: The Idea of a Plurality of Worlds from Kant to Lowell* (New York: Cambridge University Press, 1986; new ed., Mineola, N.Y.: Dover Publications, 1999), 210–15.

41. William Griggs, *The Celebrated "Moon Story": Its Origin and Incidents with a Memoir of Its Author* (New York: Bunnell and Price, 1852), 8.

42. Griggs, *Moon Story*, 8–9.

New World. Although of letter form, it is of article length, running six columns and about 9000 words. Twice in this letter Locke explicitly states that what he was writing was satire. Early in the letter Locke recounts that "He had become convinced that the *imaginative school of philosophy* . . . was emasculating the minds of our studious youth" and making them unfit for regular science. Moreover, he was disturbed by the "*theological and devotional encroachments* upon the traditional province of science." He then adds: "One of the most conspicuous of the jingling heads of this school, is the famous Dr. Dick of Dundee, who pastes together so many books about the moon and stars, and devoutly helps out the music of the spheres with the nasal twang of the conventicle. It was this cyphering sage's 'Christian Philosopher,' that suggested the moon-story." Locke notes that Dick, in his *Christian Philosopher,* had promised that all sorts of observational discoveries could be made regarding the Moon and indicated that German astronomers had already attained some of these. In his *Celestial Scenery* of 1837, Dick had taken Locke to task for his Moon essays and labeled him "in the class of liars and deceivers."[43] Locke then describes Dick's discussion of the idea of erecting a huge structure on the Earth as a signal to the Lunarians that intelligent beings exist on Earth. Locke even speculates how well this idea might fare with the U.S. government, with Wall Street, and with the newspapers. Locke responds to the attack on him that Dick in his *Celestial Scenery* had launched by making a statement that reveals precisely what Locke was writing:

> Dr. Dick preaches a hypocritical homily against my moon-hoax, which was intended merely as a good-natured *satire;* telling me that God requires "truth in the inward parts;" that upon the universal observance of this law of truth "depend the happiness of the whole intelligent system and the foundations of the throne of the Eternal." (emphasis added)

Locke then suggests that it was Dick who has perpetrated hoaxes. Late in the letter Locke again explicitly states that what he had been aiming to do was to "satirize."

> So far from feeling that I deserve the coarse reproaches of Dr. Dick, I think it is quite laudable in any man to *satirize,* as I did, that school of crude speculation and cant of which he is so eminent a professor. My hasty speculation succeeded beyond all expectation or idea, and thus proved how deplorably the public mind had been trained to gullibility on matters of science by those who had preceded me in the field. (emphasis added)

43. Dick, *Celestial Scenery,* 121.

Efforts to explain the acceptance of the incredible claims in these articles as a result of Locke's skill in writing a semitechnical account miss the point. The main reason why thousands of Americans believed these fantastic fictions is that for a number of decades they had been prepared for them by the preachings and proclamations of such authors as Paine, Chalmers, Emerson, and Dick. As one author put it in 1852: "The soil had been thoroughly ploughed, harrowed and manured in the mental fields of our wiser people, and the seed of farmer Locke bore fruit a hundred fold."[44]

44. "Locke among the Moonlings," 502.

TEN

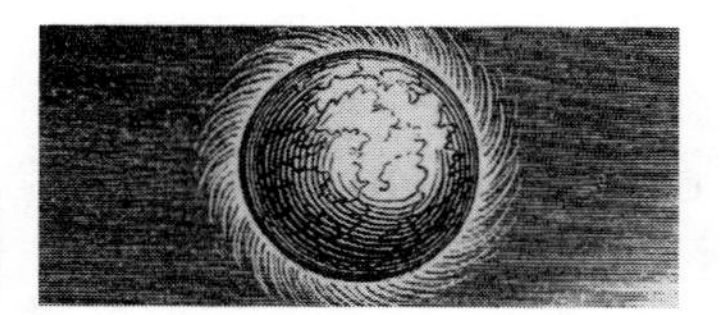

BEFORE THE WHEWELL DEBATE

SIR JOHN HERSCHEL (1792–1871)

John Herschel, the only son of William Herschel, attained, like his father, great eminence in astronomy. Moreover, John's accomplishments in mathematics, physics, photography, philosophy of science, and other areas led to his being seen by his contemporaries as the leading British scientist of the mid-nineteenth century. In the field of astronomy, the younger Herschel wrote a text that was authoritative for much of the nineteenth century. This was his *Treatise on Astronomy* (1833), which he expanded into his *Outlines of Astronomy* (1849). Also like his father, John Herschel used a giant telescope to observe nebulous objects; in fact, in the 1830s, he spent four years at the Cape of Good Hope observing the heavens of the Southern Hemisphere. John Herschel also endorsed the doctrine of a plurality of worlds and incorporated it into a number of his writings.

John Herschel, *Treatise on Astronomy* (London: Longman, Rees, Orme, Brown, Green, & Longman, 1833), 1–2.

INTRODUCTION.

(1.) In entering upon any scientific pursuit, one of the student's first endeavours ought to be, to prepare his mind for the reception of truth, by dismissing, or at least loosening his hold on, all such crude and hastily adopted notions respecting the objects and relations he is about to examine as may tend to embarrass or mislead him; and to strengthen himself, by something of an effort and a resolve, for the unprejudiced admission of any conclusion which shall appear to be supported by careful observation and logical argument, even should it prove of a nature adverse to notions he may have previously formed for himself, or taken up, without examination, on the credit of others. Such an effort is, in fact, a commencement of that intellectual discipline which forms one of the most important ends of all science. It is the first movement of approach towards that state of mental purity which alone can fit us for a full and steady perception of moral beauty as well as physical adaptation. It is the "euphrasy and rue"[1] with which we must "purge our sight" before we can receive and contemplate as they are the lineaments of truth and nature.

(2.) There is no science which, more than astronomy, stands in need of such a preparation, or draws more largely on that intellectual liberality which is ready to adopt whatever is demonstrated, or concede whatever is rendered highly probable, however new and uncommon the points of view may be in which objects the most familiar may thereby become placed. Almost all its conclusions stand in open and striking contradiction with those of superficial and vulgar observation, and with what appears to every one, until he has understood and weighed the proofs to the contrary, the most positive evidence of his senses. Thus, the earth on which he stands, and which has served for ages as the unshaken foundation of the firmest structures, either of art or nature, is divested by the astronomer of its attribute of fixity, and conceived by him as turning swiftly on its centre, and at

1. [This is a phrase from John Milton's famous epic poem *Paradise Lost,* where the Archangel Michael uses the herbs "euphrasy and rue" to cleanse Adam's eyes. See Milton's *Paradise Lost,* book 11, line 414. MJC]

> the same time moving onwards through space with great rapidity. The sun and the moon, which appear to untaught eyes round bodies of no very considerable size, become enlarged in his imagination into vast globes,—the one approaching in magnitude to the earth itself, the other immensely surpassing it. The planets, which appear only as stars somewhat brighter than the rest, are to him spacious, elaborate, and habitable worlds; several of them vastly greater and far more curiously furnished than the earth he inhabits, as there are also others less so, and the stars themselves properly so called, which to ordinary apprehension present only lucid sparks or brilliant atoms, are to him suns of various and transcendent glory—effulgent centres of life and light to myriads of unseen worlds: so that when, after dilating his thoughts to comprehend the grandeur of those ideas his calculations have called up, and exhausting his imagination and the powers of his language to devise similes and metaphors illustrative of the immensity of the scale on which his universe is constructed, he shrinks back to his native sphere; he finds it, in comparison, a mere point; so lost—even in the minute system to which it belongs—as to be invisible and unsuspected from some of its principal and remoter members.

What is most striking within the present context in the juxtaposition of these two paragraphs is that whereas the first paragraph impresses on the reader the need for rigorous scientific method and for "intellectual discipline," especially the readiness to attend to "conclusion which shall appear to be supported by careful observation and logical argument," the very next paragraph illustrates the conclusions that result from following this demanding method by maintaining that the planets are "spacious, elaborate, and habitable worlds" and that the stars are "suns of various and transcendent glory—effulgent centres of life and light to myriads of unseen worlds." Despite Herschel's claim that such notions were seen as "adverse," these ideas were among the most popular of the period. Moreover, no observation of the nineteenth century, nor of later periods, had revealed such results.

WILLIAM WHEWELL (1794–1866)

In 1829, a rather eccentric Englishman living in Paris, Francis Henry Egerton, the Eighth Earl of Bridgewater, died, leaving £8,000 to go to some person or persons to be selected by the president of England's Royal Society who would "write, print and publish, one thousand copies of a work on the Power, Wisdom and Goodness of God, as manifested in the Creation." Davies Gilbert, the president of the Royal Society, after consulting the Archbishop of Canterbury, the Bishop of London, and others, selected eight authors to write treatises, which are known as the Bridgewater Treatises. The various areas of science were more or less divided among the authors. The authors chosen and the titles of their treatises are:

Rev. William Buckland, *On Geology and Mineralogy*

Sir Charles Bell, *The Hand: Its Mechanism and Vital Endowments as Evincing Design*

Rev. Thomas Chalmers, *On the Power, Wisdom, and Goodness of God as Manifested in the Adaptation of External Nature to the Moral and Intellectual Constitution of Man*

John Kidd, *On the Adaptation of External Nature to the Physical Conditions of Man*

William Kirby, *On the History, Habits, and Instincts of Animals*

William Prout, *On Chemistry, Meteorology, and the Function of Digestion*

Peter Frank Roget, *Treatise on Animal and Vegetable Physiology*

William Whewell, *Astronomy and General Physics Considered with Reference to Natural Theology*

The treatise of present concern is that listed last. Rev. William Whewell, its author, was at that time a teacher at the University of Cambridge, where he had taken his undergraduate degree. A man of broad and deep learning, Whewell, who was ordained in 1826, had already authored some mathematical texts for Cambridge students, one of whom was Alfred Tennyson, for whom Whewell had served as tutor. In 1828, Whewell attained the professorship of mineralogy, but went on to make his main contribution in the form of his multivolume *History of the Inductive Sciences* (1837) and *Philosophy of the Inductive Sciences* (1840).

Whewell's involvement with extraterrestrial life ideas extended throughout much of his life; for example, in 1827 he preached a series of four sermons in the University Church in Cambridge in which he made such statements as the following:

> [T]he earth . . . is one among a multitude of worlds . . . with resemblances and subordinations among them suggesting . . . that [they may be inhabited by] crowds of sentient . . . beings—these are not the reveries of idle dreamers or busy connivers;—but for the most part truths collected by wise and patient men on evidence indisputable, from unwearied observation and thought; and for the rest, founded upon analogies which most will allow to reach at least some degree of probability.[2]

As shown in the next selection, extraterrestrial life themes also entered Whewell's Bridgewater treatise. His volume, which appeared in 1833, proved quite popular and served to bring Whewell to the attention of the English public. It went through seven English editions, various American editions, and was translated into German.

In writing his treatise, Whewell faced at least four problems:

1) David Hume, in his *Dialogues Concerning Natural Religion* (1779), had published a penetrating critique of the argument for God's existence from the design seen in nature, and thereby forcefully challenged the validity of the design argument itself. Whewell, the most philosophically sensitive of the eight authors, felt this problem with special severity.
2) A second problem was that even if the design argument and natural theology were valid, the tension between natural and revealed religion would remain. Natural theology, after all, was a favorite subject of deists, who rejected revealed religion.
3) In writing a tract in natural theology centered on astronomy, Whewell encountered both the problem that astronomy, as William Paley had suggested in his *Natural Theology*, "is not the best medium through which to prove the agency of an intelligent Creator [because we] are destitute of the means of examining the constitution of the heavenly bodies,"[3] and also the problem that astronomy as a branch of physical science seemed barred, as Francis Bacon had urged, from the use of final causality. The severity of this problem is indicated by a remark made in a notebook of that author whose writings did most to frustrate the cause so fully financed by the Earl of Bridgewater. Charles Darwin is now

2. As quoted in Isaac Todhunter, *William Whewell, D.D.* (London: Macmillan, 1876), 1:326.

3. William Paley, *Natural Theology*, 15th edition (London: F. C. and J. Rivington, 1815), 378–79.

known to have written in one of his 1838 notebooks: "Mayo (Philosophy of Living) quotes Whewell as profound because he says length of day adapted to duration of sleep in man!! whole universe so adapted!!! & not man to Planets.—instance of arrogance!!"[4] In other words, Darwin was opposed to Whewell's claim that God in creating humans had made them well adapted to the length of terrestrial days. Darwin's own conviction was that humans had evolved in such a way as to accommodate the length of day on Earth. Whewell dealt with this problem partly by advocating what may be called the nomological form of the design argument, rather than one centered on particular adaptations. What he did was to stress that God operates by means of general laws, which laws result in the specific features of the natural world.

(4) A final problem was that such authors as Buffon, Kant, William Herschel, and Laplace had introduced evolutionary modes of explanation into astronomy, especially the nebular hypothesis. The nebular hypothesis has existed in a number of forms, the most common of which consists in the idea that the solar system formed from the rotation and gradual condensation of a nebular mass. Various authors saw this hypothesis as in conflict with religious views of creation. Whewell in fact coined the phrase "nebular hypothesis" and presented it as a legitimate speculation in his treatise. Although aware of all these problems, Whewell did not *at that time* view ideas of extraterrestrial life as religiously or scientifically problematic.

William Whewell, *Astronomy and General Physics Considered with Reference to Natural Theology* (Philadelphia: Carey, Lea & Blanchard, 1833), 205–12.

CHAPTER II.

On the Vastness of the Universe.

1. The aspect of the world, even without any of the peculiar lights which science throws upon it, is fitted to give us an idea of the greatness of the power by which it is directed and

4. Charles Darwin, "D Notebook: Transmutation of Species," in *Darwin on Man*, Howard Gruber and Paul H. Barrett (New York: E. P. Dutton, 1974), 455.

governed, far exceeding any notions of power and greatness which are suggested by any other contemplation. The number of human beings who surround us—the various conditions requisite for their life, nutrition, well-being, all fulfilled;—the way in which these conditions are modified, as we pass in thought to other countries, by climate, temperament, habit;—the vast amount of the human population of the globe thus made up; yet man himself but one among almost endless tribes of animals;—the forest, the field, the desert, the air, the ocean, all teeming with creatures whose bodily wants are as carefully provided for as his;—the sun, the clouds, the winds, all attending, as it were, on these organized beings;—a host of beneficent energies, unwearied by time and succession, pervading every corner of the earth;—this spectacle cannot but give the contemplator a lofty and magnificent conception of the Author of so vast a work, of the Ruler of so wide and rich an empire, of the Provider for so many and varied wants, the Director and Adjuster of such complex and jarring interests.

But when we take a more exact view of this spectacle, and aid our vision by the discoveries which have been made of the structure and extent of the universe, the impression is incalculably increased.

The number and variety of animals, the exquisite skill displayed in their structure, the comprehensive and profound relations by which they are connected, far exceed any thing which we could in any degree have imagined. But the view of the universe expands also on another side. The earth, the globular body thus covered with life, is not the only globe in the universe. There are, circling about our own sun, six others, so far as we can judge, perfectly analogous in their nature: besides our moon and other bodies analogous to it. No one can resist the temptation to conjecture, that these globes, some of them much larger than our own, are not dead and barren;—that they are, like ours, occupied with organization, life, intelligence. To conjecture is all that we can do, yet even by the perception of such a possibility, our view of the kingdom of nature is enlarged and elevated. The outermost of the planetary globes of which we have spoken is so far from the sun, that the central luminary must appear to the inhabitants of that planet, if any there are, no larger than Venus does to us; and the length of their year will be eighty-two of ours.

But astronomy carries us still onwards. It teaches us that, with the exception of the planets already mentioned, the stars which we see have no immediate relation to our system. The obvious supposition is that they are of the nature and order of our sun: the minuteness of their apparent magnitude agrees, on this supposition, with the enormous and almost inconceivable distance which, from all the measurements of astronomers, we are led to attribute to them. If then these are suns, they may, like our sun, have planets revolving round them; and these may, like our planet, be the seats of vegetable and animal and rational life;—we may thus have in the universe worlds, no one knows how many, no one can guess how varied:—but however many, however varied, they are still but so many provinces in the same empire, subject to common rules, governed by a common power.

But the stars which we see with the naked eye are but a very small portion of those which the telescope unveils to us. The most imperfect telescope will discover some that are invisible without it; the very best instrument perhaps does not show us the most remote. The number which crowds some parts of the heavens is truly marvellous. Dr. Herschel calculated that a portion of the milky way, about ten degrees long and two and a half broad, contained two hundred and fifty-eight thousand. In a sky so occupied, the moon would eclipse two thousand of such stars at once.

We learn too from the telescope that even in this province the variety of nature is not exhausted. Not only do the stars differ in colour and appearance, but some of them grow periodically fainter and brighter, as if they were dark on one side, and revolved on their axes. On other cases two stars appear close to each other, and in some of these cases it has been clearly established, that the two have a motion of revolution about each other; thus exhibiting an arrangement before unguessed, and giving rise, possibly, to new conditions of worlds. In other instances again, the telescope shows, not luminous points, but extended masses of dilute light, like bright clouds, hence called nebulae. Some have supposed (as we have noticed in the last book) that such nebulae by further condensation might become suns; but for such opinions we have nothing but conjecture. Some stars again have undergone permanent changes, or have absolutely disappeared, as the celebrated star of 1572, in the constellation Cassiopea.

If we take the whole range of created objects in our own system, from the sun down to the smallest animalcule, and suppose such a system, or something in some way analogous to it, to be repeated for each of the millions of stars thus revealed to us, we have a representation of the material part of the universe, according to a view which many minds receive as a probable one; and referring this aggregate of systems to the Author of the universe, as in our own system we have found ourselves led to do, we have thus an estimate of the extent to which his creative energy would thus appear to have been exercised in the material world.

If we consider further the endless and admirable contrivances and adaptations which philosophers and observers have discovered in every portion of our own system, every new step of our knowledge showing us something new in this respect; and if we combine this consideration with the thought how small a portion of the universe our knowledge includes, we shall, without being able at all to discern the extent of the skill and wisdom thus displayed, see something of the character of the design, and of the copiousness and ampleness of the means which the scheme of the world exhibits. And when we see that the tendency of all the arrangements which we can comprehend is to support the existence, to develope the faculties, to promote the well-being of these countless species of creatures; we shall have some impression of the beneficence and love of the Creator, as manifested in the physical government of his creation.

2. It is extremely difficult to devise any means of bringing before a common apprehension the scale on which the universe is constructed, the enormous proportion which the larger dimensions bear to the smaller, and the amazing number of steps from large to smaller, or from small to larger,

which the consideration of it offers. The following comparative representations may serve to give the reader to whom the subject is new some idea of these steps.

If we suppose the earth to be represented by a globe a foot in diameter, the distance of the sun from the earth will be about two miles; the diameter of the sun, on the same supposition, will be something above one hundred feet, and consequently his bulk such as might be made up of two hemispheres, each about the size of the dome of St. Paul's. The moon will be thirty feet from us, and her diameter three inches, about that of a cricket ball. Thus the sun would much more than occupy all the space within the moon's orbit. On the same scale, Jupiter would be above ten miles from the sun, and Uranus forty. We see then how thinly scattered through space are the heavenly bodies. The fixed stars would be at an unknown distance, but, probably, if all distances were thus diminished, no star would be nearer to such a one-foot earth, than the moon now is to us.

On such a terrestrial globe the highest mountains would be about an eightieth of an inch high, and consequently only just distinguishable. We may imagine therefore how imperceptible would be the largest animals. The whole organized covering of such a globe would be quite undiscoverable by the eye, except perhaps by colour, like the bloom on a plum.

In order to restore this earth and its inhabitants to their true dimensions, we must magnify them forty millions of times; and to preserve the proportions, we must increase equally the distances of the sun and of the stars from us. They seem thus to pass off into infinity; yet each of them thus removed, has its system of mechanical and perhaps of organic processes going on upon its surface.

But the arrangements of organic life which we can see with the naked eye are few, compared with those which the microscope detects. We know that we may magnify objects thousands of times, and still discover fresh complexities of structure; if we suppose, therefore, that we increase every particle of matter in our universe in such a proportion, in length, breadth, and thickness, we may conceive that we tend thus to bring before our apprehension a true estimate of the quantity of organized adaptations which are ready to testify the extent of the Creator's power.

Two paragraphs have been deleted at this point in which Whewell continues to discuss various astronomical quantities, such as the speed at which the Earth rotates on its axis and revolves around the Sun and also the velocity of light. The point is not only to impress the reader by the vast quantities involved in astronomy, but also to remind the reader of God's remarkable powers.

4. The above statements are vast in amount, and almost oppressive to our faculties. They belong to the measurement of the powers which are exerted in the universe, and of the spaces through which their efficacy reaches (for the most distant bodies are probably connected both by gravity and light.) But these estimates cannot be said so much to give us any notion of the powers of the Deity, as to correct the errors we should fall into by supposing his powers at all to resemble ours:—by supposing that numbers, and spaces, and forces, and combinations, which would overwhelm us, are any obstacle to the arrangements which his plan requires.

In the remainder of this paragraph, Whewell discusses further the significance of such immense numbers and spaces in "tracing the Divine Perfections," suggesting that we should not be overwhelmed in contemplating such magnitudes.

On the contrary, those magnitudes and proportions which leave our powers of conception far behind;—that ever-expanding view which is brought before us, of the scale and mechanism, the riches and magnificence, the population and activity of the universe;—may reasonably serve, not to disturb, but to enlarge and elevate our conceptions of the Maker and Master of all; to feed an ever-growing admiration of His wonderful nature; and to excite a desire to be able to contemplate more steadily and conceive less inadequately the scheme of his government and the operation of his power.

ALFRED LORD TENNYSON (1809–1892) AND OTHER LITERARY FIGURES

Ranked by most scholars as the preeminent Victorian poet, Alfred Lord Tennyson was deeply interested in ideas of extraterrestrial life and in astronomy in general extending from his youth, including his student days at Cambridge University, to his old age. In fact, passages from his writings appear at two additional locations in this collection. If read together and compared, these selections indicate the diverse ways in which extraterrestrials entered his writings. They also reveal that, as the century advanced, his views on extraterrestrials evolved in interesting ways.

Some commentary will illuminate the first quotation from Tennyson. One of William Herschel's main achievements was his discovery and observation of about twenty-five hundred nebulous-appearing celestial objects, which objects were until well into the twentieth century called nebulae. These are dim patches of light of cloudy appearance, sometimes irregularly shaped, sometimes elliptical in form. They are seen scattered throughout the sky. Herschel furthermore speculated that these objects are distant Milky Ways or universes more or less comparable in size and structure to our Milky Way. This idea, which provided yet more locales for extraterrestrial beings, was no doubt in Tennyson's mind when he remarked to his brother, who suffered from shyness: "Fred, think of Herschel's great star-patches, and you will soon get over all that."[5]

The following selection is from "Timbuctoo," a poem that in 1829, when Tennyson was a student at Cambridge University, won him the Chancellor's medal for poetry.[6]

I saw
The smallest grain that dappled the dark earth,
The indistinctest atom in deep air,
The Moon's white cities, and the opal width
Of her small glowing lakes, her silver heights
Unvisited with dew of vagrant cloud,
And the unsounded, undescended depth
Of her black hollows. The clear galaxy

5. Hallam Lord Tennyson, *Alfred Lord Tennyson, A Memoir* (New York: Macmillan, 1897), 1:20.

6. Alfred Lord Tennyson, "Timbuctoo," in *Poems*, ed. Hallam Lord Tennyson (London: Macmillan, 1908), 1:320.

Shorn of its hoary lustre, wonderful,
Distinct and vivid with sharp points of light,
Blaze within blaze, and unimagin'd depth
And harmony of planet-girded suns
And moon-enriched planets, wheel in wheel,
Arch'd the wan sapphire. Nay—the hum of men,
Or other thing talking in unknown tongues,
And notes of busy life in distant worlds
Beat like a far wave on my anxious ear.

In a later poem, "The Two Voices" (1833),[7] Tennyson writes:

I said: 'When first the world began,
Young Nature thro' five cycles ran,
And in the sixth she moulded man.

"She gave him mind, the lordliest
Proportion, and, above the rest,
Dominion in the head and breast.'

Thereto the silent voice replied:
'Self-blinded are you by your pride;
Look up thro' night; the world is wide.

"This truth within thy mind rehearse
That in a boundless universe,
Is boundless better, boundless worse.

'Think you this mould of hopes and fears
Could find no statelier than his peers
In yonder hundred million spheres?'
lines 16–30

And still later, in "Ode on the Death of the Duke of Wellington" (1852),[8] Tennyson writes:

7. "The Two Voices," in *Poems and Plays of Alfred Lord Tennyson* (New York: Random House Modern Library, 1938), 59–72.

8. "Ode on the Death of the Duke of Wellington," in *Poems and Plays*, 424–30.

And Victor he must always be
For tho' the Giant Ages heave the hill
And break the shore, and evermore
Make and break, and work their will,
Tho' world on world in myriad myriads roll
Round us, each with different powers,
And other forms of life than ours,
What know we greater than the soul?
lines 258–65

Philip James Bailey (1816–1902)

Trained in the law, Bailey decided to attempt a Christianized version of the Faust myth. This was his *Festus,* a book-length poem published in 1839, which went through a number of editions. Bailey is known as the father of the 'Spasmodic School' of poetry. In this selection, Christ conveys a message to an angel assigned to minister to earthlings.[9]

Think not I lived and died for thine alone,
And that no other sphere hath hailed me Christ.
My life is ever suffering for love.
In judging and redeeming worlds is spent
Mine everlasting being.

James Anthony Froude (1818–1894)

Trained at Oxford, Froude wrote an autobiographical novel, *The Nemesis of Faith* (1849), which led to his forced resignation from his Oxford fellowship. Later he wrote a well known *History of England* and served as editor of *Fraser's Magazine.* In 1892, he was appointed Regius Professor of Modern History at Oxford.

My eyes were opening slowly to see for myself the strangeness of this being of ours. I had flung myself off into space, and seen this little earth ball caree[n]-ing through its depths; this miserable ball, not a sand-grain in the huge uni-

---

9. Philip James Bailey, *Festus: A Poem,* 1st American ed. (Boston: B. B. Mussey, 1845), 25, but see also 326.

verse of suns, and yet to which such a strangely mysterious destiny was said to have been attached. I had said to myself, Can it be that God, Almighty God, He, the Creator himself, went down and took the form of one of these miserable insects crawling on its surface, and died Himself to save their souls? I have asked the question. Did ever any man ask it honestly, and answer *yes?*[10]

AUGUSTE COMTE (1798–1857)

Auguste Comte is primarily known as the founder of the philosophical position known as positivism, which, described overly briefly, maintains that proper methodology requires that claims be based on careful observation of phenomena rather than on metaphysical or theological ideas or doctrines. As part of this empiricist philosophy, Comte warned against speculation. Comte developed this position in a number of publications, most notably his multivolume *Cours de philosophie positive* (1830–1842). Comte viewed astronomy as the paradigm science and as the earliest to detach itself from philosophical and theological encumbrances. A major portion of his discussion of the methodology of astronomy in his *Course of Positive Philosophy* is provided in the next selection.

Auguste Comte, *Cours de philosophie positive* (Paris: Ancienne libraire Schleicher, 1924), 2:1–7, 10. This selection has been translated by Keith R. Lafortune.

NINETEENTH LECTURE

PHILOSOPHICAL CONSIDERATIONS ON ASTRONOMICAL SCIENCE

Astronomy is, to date, the only branch of natural philosophy in which the human intellect has finally been rigorously freed from all theological and metaphysical influence, direct or indirect. This makes it particularly easy to present astronomy's

10. James Anthony Froude, *The Nemesis of Faith* (London: John Chapman, 1849), 162.

true philosophical character. But, in order to develop a general idea of the nature and composition of this science, it is indispensable to discard its vague definitions; to this end, we begin by defining limits to the positive knowledge that we acquire from the stars.

Of the three senses that reveal the existence of distant bodies, vision is clearly the only one of use regarding celestial bodies. As such, astronomy could not exist for a sightless species, however intelligent one imagines it to be. And for ourselves, the unseen stars—more numerous than the visible ones—preclude study. Their existence can be surmised only through induction. All research that is not finally reducible to our simple visual observations is necessarily forbidden from discussion of the stars—the same may be said of all natural things that we can know. We can imagine the possibility of determining their forms, distances, sizes, and movements, while we could never by any means study their chemical composition, mineralogical structure, or—by stronger reason—the nature of the organic forms that live on their surface, etc. To put it scientifically, our positive knowledge of the stars is limited to their geometric and mechanical phenomena; our knowledge falls short of illuminating other physical, chemical, physiological, and social phenomena that would otherwise be accessible to our other means of observation.

It would certainly be rash to fix the boundaries of our knowledge for all of natural philosophy too precisely; by immersing ourselves in detail, we would inevitably be too strict or too lenient. Also relevant to this is the state of our intellectual development: those who are strangers to mathematics would not understand that we can estimate the distances and dimensions of the celestial bodies with certainty, since (they would object) the celestial bodies are inaccessible; likewise, the half-enlightened would easily admit the possibility of such measurements, but would deny that one could indirectly weigh the sun and the planets. Despite these obvious remarks, general limits in the matter are indispensable, so the human intellect neither loses itself in vague or inaccessible goals nor forbids those goals that are accessible by more or less indirect procedures, whatever difficulty one encounters in reconciling these two equally fundamental conditions. In astronomy, this delicate reconciliation appears, to me, to be established by virtue of the philosophical maxim stated above that limits inquiry to geometric and mechanical phenomena. Such a rule is not at all arbitrary, since it results from a general comparison between the objects of study and our means of exploring them. Only its application might present some difficulty. A special, deeper examination (with adherence to the same fundamental principle) will almost always make this difficulty disappear in every particular case. To clarify these ideas, we turn to the traditional question of the atmospheres of the celestial bodies. We can certainly imagine—even before the discovery of ingenious methods for their investigation—that we could devise some accessible cause for the luminous phenomena that would be more or less appreciable from the standpoint of the luminous phenomena these atmospheres most evidently produce. But it is also sensible, by the same consideration, that our knowledge in regard to these gaseous wrappings is necessarily confined to our knowledge of their existence, their greater or lesser extent, and their true refractive power. We can by no means determine their chemical composition or their density. For that matter, it would be a grave leap to suppose that the atmosphere of Venus is as dense as our atmosphere, as some have often supposed. The supposition derives from the half-degree difference in the refractive indices of Venus and Earth. This example is a leap because the refractive index of a gas is as much a function of its chemical nature as it is a function of its density.

Generally speaking, in each sort of question that we can imagine about the stars: either we clearly perceive that the question does not depend on the latest state of visual observation—in which case we do not hesitate to declare the answer sooner or later accessible; or better, we acknowledge that the nature of the question demands some other kind of exploration—and in this case we must admit that the answer is utterly inaccessible; or finally, we see neither clearly, in which case we must suspend our judgment until the progress of our knowledge provides us with some decisive direction. This is, unfortunately, extremely rare, but quite necessary.

Here is an easy rule to follow: scientific observation has never employed—nor could it ever employ—any other means but simple observation in these circumstances, though the observations may improve or find extended use.

The determination of temperatures is probably the only matter in which the stated limit may appear too strict. But despite the expectations raised by our immortal Fourier's creation of mathematical thermology—especially his beautiful calculation of the temperature of space—I still regard the search for stellar temperatures to be forever confused. Even if all of the relative thermological influences on the exchanges of heat between the many celestial bodies were known, then the internal state of each star and, in many cases, the manner of heat absorption of their respective atmospheres both remain in question. Thus, for example, Newton's attempt to calculate the temperature of the comet of 1680 by its perihelion was illusory; this is because such a calculation, even if it were remade with the present means, would teach only what the temperature of the Earth would be if, without changing its constitution, it were transported to this position. In light of the physical and chemical differences, this can stray extremely far from the temperature of the comet.

According to the preceding considerations, I believe that it is reasonable to confine the goal of astronomy to the discovery of the laws of geometric and mechanical phenomena present in the celestial bodies.

It is necessary to add to this limitation on the nature of observable phenomena another limitation, relative to the bodies subject to exploration. This second restriction is not absolute like the first (this is important to remark), but, in the present state of our knowledge, it is almost as rigorous.

Philosophers to whom the profound study of astronomy is unknown—and even some astronomers—have not yet sufficiently separated the *solar* point of view from the *universal.* This distinction appears indispensable in order to distinguish the part of science that permits total perfection from the part that is by nature confined to speculative infancy. Our solar system is a subject that is well defined, worthy of consideration, capable of complete exploration, and that guides us to the most satisfactory knowledge. By contrast, the *universe* is indefinite in itself, in the sense that our real knowledge of it brings us no closer to a true conception of all of the stars. The difference is quite striking today, since, despite the lofty perfection acquired in the past two centuries by solar astronomy, we do not possess the same level of knowledge in sidereal astronomy. Absent is the first and simplest element of all positive research in sidereal astronomy, the determination of stellar distances. Without doubt, we have reason to believe that these distances will be determined soon for some stars (at least between certain bounds) and that we will know for these same stars several other important elements, such as their masses, etc. But, the important distinction established above will remain intact. When we have finally completed our study of the relative movements of some multitude of stars—invaluable though this study may be, especially if these stars relate to the group to which our sun likely belongs—we would still be little closer to a true knowledge of the universe, which knowledge will forever escape us.

In all classes of research, there exists a constant and necessary harmony between the extent of our true intellectual needs and the range of our real knowledge. This harmony is not, as common philosophers assume, the result of indication of the final cause. It simply derives from this evident necessity: we need to know only what can act on us more or less directly; and by virtue of the fact that this influence exists, it must eventually serve as a certain means to knowledge. The present example verifies this in a remarkable way. Study of the laws of our solar system interests us, so we have brought this study to an admirable precision. By contrast, if the exact notion of the universe is necessarily forbidden, then it is unimportant to everything but our insatiable curiosity. Routine astronomical observations demonstrate that the internal phenomena of each solar system—the only phenomena that can affect their inhabitants—are essentially independent from the most general phenomena relative to the mutual action of the suns. This is similar to the differences we witness between meteorological and planetary phenomena. Our tables of celestial events, long established and concerned only with our world, correspond to direct observations with minute precision. This manifest independence of our solar system finds explanation in the immense disproportion that we know exists between the great distances between suns and the tiny distances between planets. If—as is highly probable—the planets provided with an atmosphere, such as Mercury, Venus, Jupiter, etc. are in fact inhabited, we can regard those inhabitants as being in some fashion our fellow citizens, since, hailing from what is a sort of common fatherland, there ought of necessity result a certain community of thought and even of interests, whereas the inhabitants of other solar systems will be complete strangers. Therefore, it is necessary to distinguish more clearly than in the past between the solar and universal points of view, between the idea of the world and that of the universe: the first provides the most advanced knowledge that we can expect, and it is the only one with which we are truly interested.

Thus, without abandoning all hope of obtaining some sidereal knowledge, it is necessary to imagine positive astronomy as consisting of the geometric and mechanical study of the small number of celestial bodies that comprise the *world* of which we are a part. It is only between such limits that astronomy earns its perfection and the supreme rank that it currently occupies among the natural sciences. As for those innumerable stars distributed in the heavens, they are little more than convenient landmarks for our astronomical observations, their positions regarded as fixed relative to the interior movements of our system, which is the only essential object of our study.

• • •

By scrupulously examining the actual philosophical state of the many fundamental sciences, we will have cause to acknowledge that astronomy is today the only science that has finally been truly purged of all theological and metaphysical considerations. Such is its first title to supremacy. In this, the philosophers can study what truly constitutes a science; and it is according to this model that we must reconstitute all of the other fundamental sciences, always regarding the more

> or less profound differences that necessarily result from the growing complexity of phenomena.

Comte extended his position regarding astronomy somewhat further in his *Système de politique positive,* the first volume of which appeared in 1851, five years after astronomers had discovered the planet Neptune. Having earlier warned astronomers to avoid giving attention to the stars because of their remoteness from the Earth, he issued a comparable lament against the "mad infatuation . . . which some years ago possessed, not only the public, but even the whole group of Western astronomers, on the subject of the alleged discovery [of Neptune], which, even if real, would not be of interest to anyone except the inhabitants of Uranus."[11]

In reading Comte's views on the proper methodology of astronomy, one may be struck by the fact that three decades before the development of spectrum analysis, which made it possible to determine the chemical composition of various celestial bodies or of their atmospheres, Comte warned astronomers against even asking questions about the chemistry of the heavens; that at a time when stellar astronomy was beginning to emerge as the astronomy of the future, Comte excoriated those astronomers who pursued this area of research, and the French philosopher similarly disparaged the discovery of Neptune, which astronomers rightly regarded as one of their greatest achievements. Comte's failures in methodology in these areas are perhaps all the more striking when one finds him, immediately after enunciating his methodology for astronomy, accepting without question one of the most speculative claims in astronomy, the idea of extraterrestrial intelligent life. One may suspect that the source of this blindness on Comte's part may have been that he wanted to employ his extraterrestrials to attack theology. In this context, Comte complains about those who see astronomy as allied with religion, "as if the famous verse 'The Heavens declare the glory of God' had preserved its meaning. It is however certain that all true science is in radical and necessary opposition to all theology." Comte asserts that for those who follow the true philosophy of astronomy, "the heavens declare no other glory than that of Hipparchus, of Kepler, of Newton, and of all those who have cooperated in the establishment of laws."[12] In particular, what for Comte

11. Auguste Comte, *Système de politique positive,* vol. 1, in *Oeuvres de Auguste Comte* (Paris, 1964), 7:511.

12. August Comte, *Cours de philosophie positive* (Paris: Ancienne libraire Schleicher, 1924), 2, 24.

shows the unacceptability of theology is the realization that the Earth, rather than being the center of the universe, is only a secondary body circling the Sun "of which the inhabitants have entirely as much reason to claim a monopoly of the solar system which is itself almost imperceptible in the universe."[13] This suggests that Comte's own philosophy of astronomy may not as yet have transcended even the antitheological stage.

RALPH WALDO EMERSON (1803–1882)

One of the leading literary and religious figures of nineteenth-century America, Ralph Waldo Emerson, was specifically known as the leader of American Transcendentalism. He was so deeply interested in astronomy that Emerson scholars have described him as "obsessed with astronomy throughout his life."[14] Part of his passion for astronomy came from the power he believed it could and should exert on our thought, particularly in regard to spiritual matters.

A graduate of Harvard Divinity School, Emerson served for a period as minister to Boston's Second Unitarian Church. Among the most famous events in American religious and literary history during the first half of the nineteenth century was Emerson's announcement in September, 1832 that he was resigning his pastorate on the grounds that he could no longer reconcile his beliefs with administering the Lord's Supper. What led him to this decision? A possible answer (as well as much information on the religious significance of astronomy and its doctrine of a plurality of worlds) can be derived from a sermon titled "Astronomy," which Emerson delivered to his congregation on May 27, 1832, just a matter of days before he first revealed his difficulties with the Lord's Supper to his congregation.[15] The religious implications of astronomy had been much on Emerson's mind in the week before he delivered that sermon. In fact, he summarized the main theme of the sermon in a brief rhetorical question he raised in his journal for May 23. The question is, given modern astronomy, "Who can be a Calvinist or who an Atheist[?]"[16] The message he delivered in his sermon suggests that with this question he meant to urge that the revelations of astronomy force any thinking person not only to reject atheism,

---

13. Comte, *Cours*, 2, 130.

14. Sherman Paul, *Emerson's Angle of Vision* (Cambridge, Mass.: Harvard University Press, 1952), 95.

15. Ralph Waldo Emerson, *The Journals and Miscellaneous Notebooks of Ralph Waldo Emerson*, ed. Alfred R. Ferguson (Cambridge, Mass.: Harvard University Press, 1952), 4:27.

16. Emerson, *Journals*, 4:25.

but also to forsake the fundamental teachings of Christianity, in particular, belief in the divinity and redemptive actions of Jesus Christ.

Ralph Waldo Emerson, "Astronomy," in *Young Emerson Speaks: Unpublished Discourses on Many Subjects*, ed. Arthur Cushman McGiffert, Jr. (Boston: Houghton Mifflin Company, 1938), 170–79.

ASTRONOMY

God that made the world and all things therein, seeing that he is Lord of heaven and earth, dwelleth not in temples made with hands; neither is worshipped with men's hands as though he needed anything, seeing he giveth to all life and breath and all things . . . Forasmuch, then, as we are the offspring of God we ought not to think that the Godhead is like to gold or silver or stone graven by art and man's device. And the times of this ignorance God overlooked, but now commandeth all men every where to repent, because he hath appointed a day in which he will judge the world in righteousness by that man whom he hath ordained; whereof he hath given assurance unto all men, in that he hath raised him from the dead. Acts XVII, 24ff.

I QUOTE all these memorable words of Saint Paul[17] because they may serve as one out of many passages to show that the Scriptures claim to come from the same Being that made the heavens and the earth; that the God of nature and the God of the Bible are affirmed to be the same; that the Father of Jesus Christ is the Divine Providence in whose wisdom and love all beings are embosomed.

Since this is so,—since the records of the divine dealings with men claim no other origin than the author of nature, we may expect that they are to be read by the light of nature; that more knowledge of his works will enable us better to

17. [The quotation from *Acts* on which Emerson is commenting is from a speech by Paul that is quoted by Luke in *Acts*. MJC]

understand his word; and that religion will become purer and truer by the progress of science.

This consideration ought to secure our interest in the book of nature. The lover of truth will look at all the facts which every year science is bringing to light with curiosity as the commentary and exposition, say rather, the sequel of the revelation which our Creator is giving us of himself.

With this view I am led to offer you some reflexions that are suggested by the present state of the science of Astronomy, some thoughts upon the influence which the wonderful discoveries men have made of the extent and plan of the universe have had upon religious opinion.

• • •

An important result of the study of astronomy has been to correct and exalt our views of God, and humble our view of ourselves.

In all ancient speculation men were accustomed, of course, to take man for the type of the highest beings, and suppose whatever is intelligent and good among God's creatures must resemble human nature. Even God himself, the infant religion of all nations has clothed in human form, and idolatry imputed to him the passions as well as the person of man. Astronomy corrects all these boastful dreams, and demonstrates that whatever beings inhabit Saturn, Jupiter, Herschel,[18] and Mercury, even in this little family of social worlds that journey like us around the sun, they must have an organization wholly different from man. The human race could not breathe in the rare atmosphere of the moon; nor the human blood circulate in the climate of Uranus; nor the strength of men suffice to raise his own foot from the ground in the dense gravity of Jupiter.

Each of the eleven globes, therefore, that revolve round the sun must be inhabited by a race of different structure.

18. [For some decades after William Herschel discovered in 1781 the planet now called Uranus writers about astronomy referred to it under three different names: Herschel, Uranus, or Georgium Sidus. MJC]

And to suppose that the constitution of the race of yesterday that now plants the fields of this particular planet, should be the pattern for all the orders that people the huge globes in the heaven is too improbable to be entertained.

Rather believe that the benignant Power which has assigned each creature to its own element, the fish to the sea, the bird to the air, the beast to the field, has not less nicely adjusted elsewhere his creatures to their habitation and has enriched other seats of his love with other and perhaps far more excellent endowments than he has granted to mankind.

In the next place, the science of astronomy has had an irresistible effect in modifying and enlarging the doctrines of theology. It is known to all to whom I speak that until a few hundred years ago it was the settled opinion of all men that the earth was stationary in space, and that the sun and stars actually moved round it every day as they appear to do to the eye. The host of heaven were esteemed so many lanthorns to illuminate and set off the residence of man. It is only since the time of Galileo and Newton it was learned that the little ball on which we live spins upon its own axis to produce this appearance, and that it is at such a dizzy distance from the stars which were supposed thus revolving for its ornament, that it is not visible from them. And not only the earth but the whole system also to which it belongs, with the great sun in the centre, are perhaps too minute for observation from those remote luminaries.

• • •

When the solar system had been correctly explained to us we found ourselves journeying in a comparatively small opaque planet around a single star and quite too inconsiderable to be noticed amid the millions of burning suns which the telescope revealed. It was the effect of this new knowledge, to make an equal revolution in religious opinion, as in science, for it was impossible to regard the earth any longer, as the only object in the care of Providence.

It had been the belief of many generations that God from all eternity had foreseen the fall of man and had devised in his councils a method by which man might be saved. The second being in the universe, it was represented, undertook to save them, and in the vain imagination of man the scheme of his redemption, as it was called, occupied the attention of God and of angels, as if there were nothing in being but men. 'The earth,' in the strange language of an old divine, 'was the scaffold of the divine vengeance.' Now this system of theology was every way suited to the ancient system of the heavens. It could not but happen that the telescope should be fatal to both. I regard it as the irresistible effect of the Copernican astronomy to have made the theological *scheme of Redemption* absolutely incredible. The great geniuses who studied the mechanism of the heavens became unbelievers in the popular doctrine. Newton became a Unitarian.

In spite of the awful exhibition of wisdom and might disclosed to their eyes—the present God,—in spite of the natural expectation which dictated the sentiment, 'The undevout astronomer is mad'—the incongruity between what they beheld and the gross creeds which were called religion and Christianity by their fellow countrymen so revolted them, the profound astronomers of France rejected the hope and consolation of man and in the face of that divine mechanism which they explored denied a cause and adopted the belief of an eternal Necessity, as if that very external necessity were anything else than God, an intelligent cause.

• • •

And finally, what is the effect upon the doctrine of the New Testament which these contemplations produce? It is not contradiction but correction. It is not denial but purification. It proves the sublime doctrine of One God, whose offspring we all are and whose care we all are. On the other hand, it throws into the shade all temporary, all indifferent, all local provisions. Here is neither tithe nor priest nor Jerusalem nor Mount Gerizim. Here is no mystic sacrifice, no atoning blood.

But does it take one charm from the lowly grace of Christ? Does it take away any authority from his lips? It abridges what belongs to persons, to places and to times but it does not touch moral truth. We are assured in any speculation we may indulge concerning the tenants of other regions, in the wide commonwealth of God, that if we could carry the New Testament to the inhabitants of other worlds we might need to leave Jewish Christianity and Roman Christianity, Paul and Apollos and Cephas and Luther, and Socinus, but the moral law, justice and mercy would be at home in every climate and world where life is; that we can go nowhere but wisdom will not be valuable and justice, venerable and humility, suitable and diligence, useful and truth, sacred, and charity divine.

The largest consideration the human mind can give to the subject, makes moral distinctions still more important, and positive distinctions less. It will not teach any expiation by Jesus; it will not teach any mysterious relations to him. It will teach that he only is a mediator, as he brings us truth, and we accept it, and live by it; that he only saves us, by inducing us to save ourselves; that God now commands all men, all spirits, every where to repent; and that such principles as Jesus Christ inculcated must forevermore be the standard by which actions shall be judged.

• • •

It is, brethren, a glorious confirmation that is brought to our faith, the observation that it agrees well with all the new and astonishing facts in the book of nature. It is good to perceive that the beatitudes of the sermon on the mount will be such to all intelligent creatures. The Scriptures were written by human hands. God intends by giving us access to this original writing of his hand to correct the human errors that have crept into them. Let us yield ourselves with a grateful heart to the instruction that comes from this source and not repine to find that God is a greater, wider and more tender Parent than we were wont to worship. We shall not less distinctly see Jesus to be the gracious instrument of his bounty to instruct men in the character of God and the true nature of spiritual good, the teacher, and, by his teaching, the redeemer of men. But we shall fulfil the intent of Jesus by rendering the praise to God. The hour will already have arrived in our hearts, when means and instruments shall have done their office and when God shall be over all and through all and in all.

JOSEPH SMITH (1805–1844)

That Joseph Smith, the "First Prophet, Seer, and Revelator" of the Church of Jesus Christ of Latter-day Saints, also known as the Mormon Church, had a remarkable role in American religious history is widely known. It is less well known that among the doctrines that he bestowed on his new church were various beliefs about extraterrestrial intelligent life. Although these doctrines do not appear in the most famous Mormon scripture, the *Book of Mormon* (first published in 1830), they are prominent in some later scriptures: Smith's *The Doctrines and Convenants,* which was first published in 1835 but with supplements added later, and Smith's *The Pearl of Great Price,* which consists primarily of the "Book of Moses" and the "Book of Abraham." This was first published in 1851 but contains many writings based on earlier visions, the accounts of which had in some cases appeared in Mormon periodicals.

The scholar who most carefully studied the history of Mormon doctrines regarding extraterrestrials was E. Robert Paul, who was both a practicing Mormon and a widely respected historian of astronomy, chiefly known for his research on the history of statistical astronomy. In his 1992 book on Mormon cosmology, Paul divided Mormon beliefs in this area into two categories: the first category consists of five ideas deriving from the Copernican revolution and widely accepted in Europe and America in the early nineteenth century, and the second consists of nine other ideas formulated by Smith (although not necessarily for the first time) but not typically held by others who wrote in this area. The five ideas in the first group as specified by Paul are:

> (1) other planets of our solar system are inhabited by living, sentient, and rational beings; (2) the closed world of medieval cosmology was replaced with an infinite universe; (3) fixed stars are suns similar to our own and surrounded by planetary systems; (4) these planets are inhabited by conscious beings; and (5) an infinite number of solar systems exist.[19]

19. Erich Robert Paul, *Science, Religion, and Mormon Cosmology* (Urbana: University of Illinois Press, 1992), 85.

Paul lists and references the nine additional ideas advocated by Smith and that, at least as a whole, distinguish the beliefs of Mormons from those held by their contemporaries in the following way:

> (6) worlds have passed away and others have and are being formed (Moses 1:35, 38); (7) worlds are governed in a hierarchical relationship (Abraham 3:8–9); (8) every system has its own laws and bounds (D&C 88:36–38); (9) Christ made and/or makes all worlds (D&C 76:24; 93:9–10); (10) different kinds of people inhabit different worlds (D&C 76:112); (11) the Earth has been the most wicked of all worlds (Moses 7:36); (12) resurrected beings also reside on worlds (D&C 88:36–38); . . . (13) worlds exist in space and time (Moses 1:35, 38; D&C 88:36–38, 42–47; 93:9–10); [and thus] (14) Christ's redemption is universal.[20]

To get a feel for what Smith presented to his followers, one needs to encounter his formulations of these doctrines in the format in which they first appeared, in particular, in the solemn and lofty language that he selected to convey the messages that he claimed came from visions sent by the Divine. For example, in his *Doctrine and Convenants*, Smith revealed that "if God rewarded every one according to [his or her] deeds . . . , the term heaven, as intended for the Saints' eternal home, must include more kingdoms than one."[21] A slightly later section clarifies the role of Christ within this universe by stating that Christ not only sits at the right hand of God, but also "That by [Christ], and through him, and of him, the worlds were created, and the inhabitants thereof are begotten sons and daughters unto God" (76:24; also 93:10). And later he expanded on the great number of inhabited worlds in his universe: "And there are many kingdoms; for there is no space in which there is no kingdom; and there is no kingdom in which there is no space, either a greater or a lesser kingdom" (88:37). Drawing on a revelation that came to Smith on 2 April 1843, he elaborated on the role of angels in his cosmos:

> 4. In answer to the question—Is not the reckoning of God's time, angel's time, prophet's time, and man's time, according to the planet on which they reside?

20. Paul, *Mormon Cosmology*, 85–86. The abbreviations that appear in this quotation refer to the writings mentioned in the opening paragraph, e.g., "D&C" is Paul's abbreviation of *The Doctrine and Covenants*.

21. Joseph Smith, *The Doctrine and Covenants of the Church of Jesus Christ of Latter-day Saints: Containing Revelations Given to Joseph Smith, The Prophet* (Salt Lake City, Utah: Church of Jesus Christ of Latter-day Saints, 1957), section 76, introductory paragraph.

5. I answer, Yes. But there are no angels who minister to this earth but those who do belong or have belonged to it.

6. The angels do not reside on a planet like this earth;

7. But they reside in the presence of God, on a globe like a sea of glass and fire, where all things for their glory are manifest, past, present, and future, and are continually before the Lord (130:4–7).

Additional doctrines appeared in the Mormon scripture known as *The Pearl of Great Price*. For example, a major revelation, which is part of the "Book of Moses" section, extended the doctrine of the plurality of worlds into the past and future. Smith attributed this revelation to June 1830, when he learned:

29. And he [Moses] beheld many lands; and each land was called earth; and there were inhabitants on the face thereof. . . .

33. And worlds without number have I [God] created; and I also created them, for mine own purpose; and by the Son I created them, which is mine Only Begotten. . . .

35. But only an account of this earth, and the inhabitants thereof, give I unto you. For behold, there are many worlds that have passed away by the word of my power. And there are many that now stand, and innumerable are they unto man. . . .

38. And as one earth shall pass away, and the heavens thereof, even so shall another come; and there is no end to my works, neither to my words.[22]

Another section of the "Book of Moses" clarified the issues of the number of worlds and of humanity's place in this cosmos:

And were it possible that man could number the particles of the earth, yea millions of earths like this, it would not be a beginning to the number of thy creations (7:30). . . . Wherefore, I can stretch forth mine hands and hold all the creations which I have made; and among all the workmanship of mine hands there has not been so great wickedness as among thy brethren. (7:36)

The "Book of Abraham" section of *The Pearl of Great Price* purports to be a translation of some papyri that Smith had examined in 1833 and also presents vari-

22. *The Pearl of Great Price: A Selection from the Revelations, Translations, and Narrations of Joseph Smith* (Salt Lake City, Utah: Church of Jesus Christ of Latter-day Saints, 1957), Ch. I. "Visions of Moses."

ous cosmological ideas. For example, its third chapter proposes a hierarchical arrangement among the stars, with a particular star, Kolob, being dominant and especially near God's throne.

> 2. And I saw the stars, that they were very great, and that one of them was nearest unto the throne of God; and there were many great ones which were near unto it;
>
> 3. And the Lord said unto me: These are the governing ones; and the name of the great one is Kolob, because it is near unto me, for I am the Lord thy God; I have set this one to govern all those which belong to the same order as that upon which thou standest.

Another feature of Mormon beliefs regarding extraterrestrials is that Christ's redemptive actions have universal influence. Joseph Smith made this explicit in the last years of his life by stating:

> And I heard a great voice bearing record from Heav'n,
> He's the Saviour, and the only Begotten of God—
> By him, of him, and through him, the worlds were all made,
> Even all that careen in the heavens so broad.
> Whose inhabitants, too, from the first to the last,
> And sav'd by the very same Saviour as ours;
> And, of course, are begotten God's daughters and sons,
> By the very same truths, and the very same pow'rs.[23]

Smith's efforts to include ideas of extraterrestrial life in the founding documents of his new denomination no doubt enhanced its appeal to his contemporaries. Moreover, this feature of Mormon teaching continued to be prominent when after Smith's death leadership passed to Brigham Young and others and as the church grew in membership and spread throughout the world.[24] One evidence of the major role that pluralism has in the Mormon religion is a hymn that one of Smith's earliest disciples, William Wine Phelps, composed and that is still sung in Mormon congregations. This hymn incorporates a number of the features of Mormon cosmology seen in the earlier selections:

---

23. As quoted in Paul, *Mormon Cosmology*, 88.

24. On the history of Mormon views regarding ideas of extraterrestrial life, see Paul, *Mormon Cosmology*, esp. chaps. 4 and 5.

If you could hie to Kolob,
In th' twinkling of an eye,
And then continue onward,
With that same speed to fly—

D'ye think that you could ever,
Through all eternity,
Find out the generation
Where Gods began to be?

Or see the grand beginning,
Where space did not extend?
Or view the last creation,
Where Gods and matter end?

Methinks the Spirit whispers,
"No man has found 'pure space,'"
Nor seen the outside curtains
Where nothing has a place.

The works of God continue,
And worlds and lives abound;
Improvement and progression
Have one eternal round.

There is no end to matter,
There is no end to space,
There is no end to "spirit"
There is no end to race.[25]

ELLEN G. WHITE (1827–1915)

It is sometimes suggested that the discovery of extraterrestrial life would cause great consternation in religious denominations. The reality is that some denominations

25. Brigham H. Roberts, *A Comprehensive History of the Church of Jesus Christ of Latter-day Saints* (Provo, Utah: Brigham Young University Press, 1965), 2:388.

would view such a discovery not as a disruption of their beliefs, but rather as a confirmation. Such a group is the Seventh-day Adventist Church, which was founded in mid-nineteenth century America, with Ellen G. White as its central figure. Born Ellen G. Harmon, in 1846 she married James White (1821–1881), a minister who was also instrumental in the establishment of the new denomination, which now has approximately twelve million members worldwide.

The Seventh-day Adventist Church was an off-shoot of the Millerite movement in the first half of the nineteenth century in the United States. The Millerites, followers of William Miller, were convinced by their interpretation of various biblical passages that the end of the world and Christ's second coming would occur in 1843 or 1844. When this did not happen, the movement fractured, with different sects adopting diverse interpretations of the "Great Disappointment." In late 1844, Ellen Harmon began to experience visions, including one mandating that she spread the gospel. In this process, she met Miller and also a former sea captain, Joseph Bates (1792–1872). At first, Bates was skeptical of her visions. In 1846, however, Bates, who possessed some knowledge of astronomy, became convinced when Ellen experienced a vision that involved extraterrestrials. According to a contemporary report,

> Sister White was in very feeble health, and while prayers were offered in her behalf, the Spirit of God rested upon us. We soon noticed that she was insensible to earthly things. This was her first view of the planetary world. After counting aloud the moons of Jupiter, and soon after those of Saturn, she gave a beautiful description of the rings of the latter. She then said, "The inhabitants are a tall, majestic people, so unlike the inhabitants of earth. Sin has never entered here." It was evident from Brother Bates' smiling face that his past doubts in regard to the source of her visions were fast leaving him.[26]

Ellen White reported many hundreds of visions, including others involving extraterrestrials. For example, she revealed in 1849:

> The Lord has given me a view of other worlds. Wings were given me, and an angel attended me from the city to a place that was bright and glorious. The grass of the place was living green, and the birds there warbled a sweet song. The inhabitants of the place were of all sizes; they were noble, majestic, and

26. As quoted in J. N. Loughborough, *The Great Second Advent Movement: Its Rise and Progress* (New York: Arno, 1972, repr. of New York: Review and Herald Publishing Association, 1905), 258.

lovely. . . . I asked one of them why they were so much more lovely than those on the earth. The reply was, "We have lived in strict obedience to the commandments of God, and have not fallen by disobedience, like those on the earth." . . . Then I was taken to a world which had seven moons. . . . I could not bear the thought of coming back to this dark world again. Then the angel said, "You must go back, and if you are faithful, you with the 144,000 shall have the privilege of visiting all the worlds."[27]

By the 1860s, the Whites and Bates had adopted the name "Seventh-day Adventist" for their new denomination, which by then had grown to a membership of about three thousand. Ellen White was not only the main prophetess of the group, but also its most prolific spokesperson. Dozens of books flowed from her pen. In these, she set out the doctrines of the new denomination. One of her most dramatic teachings portrayed the universe as the scene of a "great controversy" between Christ and Satan. The fallen angels, according to her account, had tried to spread evil throughout the universe, but succeeded on only one planet, our Earth, where Adam's sin created a need for Christ's redemptive actions. In a section of her *The Story of Patriarchs and Prophets,* she vividly described in cosmic terms what this entailed.

Ellen G. White, *The Story of Patriarchs and Prophets* (Mountain View, Calif.: Pacific Press Publishing Association, 1913), 68–70.

But the plan of redemption had a yet broader and deeper purpose than the salvation of man. It was not for this alone that Christ came to the earth; it was not merely that the inhabitants of this little world might regard the law of God as it should be regarded; but it was to vindicate the character of God before the universe. To this result of his great sacrifice — its influence upon the intelligences of other worlds, as well as upon man — the Saviour looked forward when just before his crucifixion he said: "Now is the judgment of this world; now shall the prince of this world be cast out. And I, if I be lifted

27. Ellen G. White, *Early Writings* (Washington, D.C.: Review and Herald Publishing Association, 1945), 40–41.

up from the earth, will draw all unto me."[28] The act of Christ in dying for the salvation of man would not only make heaven accessible to men, but before all the universe it would justify God and his Son in their dealing with the rebellion of Satan. It would establish the perpetuity of the law of God, and would reveal the nature and the results of sin.

From the first, the great controversy had been upon the law of God. Satan had sought to prove that God was unjust, that his law was faulty, and that the good of the universe required it to be changed. In attacking the law, he aimed to overthrow the authority of its Author. In the controversy it was to be shown whether the divine statutes were defective and subject to change, or perfect and immutable.

When Satan was thrust out of heaven, he determined to make the earth his kingdom. When he tempted and overcame Adam and Eve, he thought that he had gained possession of this world; "because," said he, "they have chosen me as their ruler." He claimed that it was impossible that forgiveness should be granted to the sinner, and therefore the fallen race were his rightful subjects, and the world was his. But God gave his own dear Son—one equal with himself—to bear the penalty of transgression, and thus he provided a way by which they might be restored to his favor, and brought back to their Eden home. Christ undertook to redeem man, and to rescue the world from the grasp of Satan. The great controversy begun in heaven was to be decided in the very world, on the very same field, that Satan claimed as his.

It was the marvel of all the universe that Christ should humble himself to save fallen man. That he who had passed from star to star, from world to world, superintending all, by his providence supplying the needs of every order of being in his vast creation,—that he should consent to leave his glory and take upon himself human nature, was a mystery which the sinless intelligences of other worlds desired to understand. When Christ came to our world in the form of humanity, all were intensely interested in following him as he traversed, step

28. John 12:31, 32.

by step, the blood-stained path from the manger to Calvary. Heaven marked the insult and mockery that he received, and knew that it was at Satan's instigation. They marked the work of counter-agencies going forward; Satan constantly pressing darkness, sorrow, and suffering upon the race, and Christ counteracting it. They watched the battle between light and darkness as it waxed stronger. And as Christ in his expiring agony upon the cross cried out, "It is finished," a shout of triumph rung through every world, and through heaven itself. The great contest that had been so long in progress in this world was now decided, and Christ was conqueror. His death had answered the question whether the Father and the Son had sufficient love for man to exercise self-denial and a spirit of sacrifice. Satan had revealed his true character as a liar and a murderer. It was seen that the very same spirit with which he had ruled the children of men who were under his power, he would have manifested if permitted to control the intelligences of heaven. With one voice the loyal universe united in extolling the divine administration.

If the law could be changed, man might have been saved without the sacrifice of Christ; but the fact that it was necessary for Christ to give his life for the fallen race, proves that the law of God will not release the sinner from its claims upon him. It is demonstrated that the wages of sin is death. When Christ died, the destruction of Satan was made certain. But if the law was abolished at the cross, as many claim, then the agony and death of God's dear Son were endured only to give to Satan just what he asked; then the prince of evil triumphed, his charges against the divine government were sustained. The very fact that Christ bore the penalty of man's transgression, is a mighty argument to all created intelligences, that the law is changeless; that God is righteous, merciful, and self-denying; and that infinite justice and mercy unite in the administration of his government.

ELEVEN

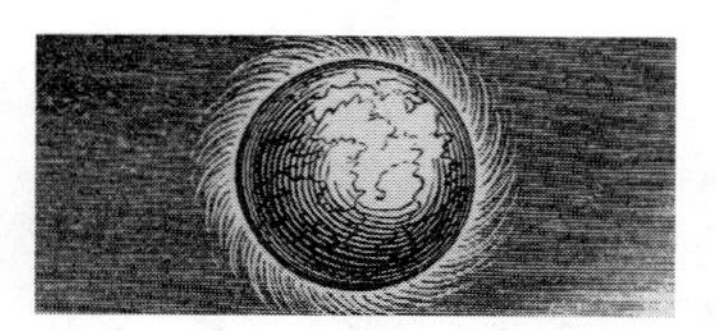

WHEWELL'S *OF THE PLURALITY OF WORLDS* AND RESPONSES

WHEWELL'S *ESSAY*

In late 1853, Rev. William Whewell (1794–1866), who by then had become distinguished as Master of Trinity College, the most famous of the colleges of Cambridge University, published anonymously a book entitled *Of the Plurality of Worlds: An Essay,* in which he shocked his contemporaries by calling into question the traditional doctrine of the plurality of worlds. The fact that Whewell had endorsed that doctrine in his *Astronomy and General Physics* (1833) raises the question why in the intervening two decades he had so radically altered his position. Briefly put, the answer seems to be that in the period around 1850 he had come to the conclusion that belief in extraterrestrial life can only with difficulty be reconciled with central doctrines of Christianity.[1] Although this is not the main line of argument in his book, this conviction seems to have motivated him to a fresh examination of the scientific,

1. For a presentation of this claim, see M. J. Crowe, *The Extraterrestrial Life Debate, 1750–1900: The Idea of a Plurality of Worlds from Kant to Lowell* (Cambridge: Cambridge University Press, 1986; new ed., Mineola, N.Y.: Dover Publications, 1999), esp. 277–82.

Fig. 10. Portrait of William Whewell made in 1836 by G. F. Joseph. (Courtesy of the Master and Fellows of Trinity College, Cambridge)

theological, and philosophical arguments that had been presented in favor of the doctrine—and these he decided were seriously defective. In fact, far and away the largest part of his book consists of his scientific and philosophical arguments in opposition to the pluralist position.

The shock felt by readers of the *Essay* was all the greater as word rapidly spread that its anonymous author was Whewell, a man known for his vast scientific and philosophical learning, who was also an ordained priest in the Church of England and recognized as a defender of orthodoxy. The book went through at least five English and two American editions. Moreover, it produced over seventy published responses, including nearly twenty books. One of these was Whewell's own *Dialogue on the Plurality of Worlds*, which he published in 1854 as a response to some of his critics. Samples of these responses are given in the second part of this chapter.

The following discussion of the contents of Whewell's book consists of summaries, chapter by chapter, of some of the main themes and high points of the book as well as selections from the work itself. These selections have been drawn from the fourth edition of that book, the text of which is nearly identical to that of the first edition, except that Whewell's *Dialogue on the Plurality of Worlds* was inserted at its beginning.

Summary of Whewell's *Of the Plurality of Worlds: An Essay*

Preface: Whewell makes two especially important statements in his preface. First, he writes, "It will be a curious, but not a very wonderful event, if it should now be deemed as blameable to doubt the existence of inhabitants of the Planets and Stars as, three centuries ago, it was held heretical to teach that doctrine" (13). Second, after stressing that scripture says nothing of the doctrine of a plurality of worlds and that Christianity for most of its history felt no need for it, Whewell states that "at the present day . . . many persons have so mingled this assumption with their religious belief, that they regard it as an essential part of Natural Religion" (13–14).

Chapter 1—"Astronomical Discovery": Whewell first cites the portion of Psalm 8 that Chalmers had taken as his text: "When I consider the heavens, the work of thy fingers, the moon and the stars, which thou has ordained; What is man, that thou art mindful of him? and the son of man, that thou visitest him?" He then stresses that the situation is now quite different from that of Old Testament times, that modern astronomy points to the need to ask this question in a new way. He then delineates some of the results of astronomy that have led people to suppose that the planets may be inhabited and that inhabited planets circle other suns.

Chapter 2—"The Astronomical Objection to Religion": Whewell's statement of this objection is: "if this world be merely one of innumerable worlds . . . occupied by intelligent creatures, [then] to hold that this world has been the scene of God's care and kindness, and still more, of his special interpositions, communications, and personal dealings with its individual inhabitants . . . is, the objector is conceived to maintain, extravagant and incredible" (115). The majority of the chapter is devoted to explicating that objection, Whewell expressing the hope that he will be able to deal with it in a manner sufficient to satisfy the friends of religion.

Chapter 3—"The Answer from the Microscope": Whewell first summarizes one of Chalmers's chief arguments. The Scottish divine had attempted to relieve the tensions resulting from the astronomical objection to religion by noting that the microscope has revealed millions of tiny organisms. Because we accept that these are under God's care, it should not seem problematic that God's solicitude can extend to innumerable worlds scattered throughout space. Although praising Chalmers, Whewell suggests that this argument would satisfactorily deal with the difficulty only if the microscope revealed intelligent organisms, comparable in some way to human beings. But this it does not do.

Chapter 4—"Further Statement of the Difficulty": Whewell here turns directly to the religious problem that arises if we assume that other planets are inhabited by "creatures analogous to man;—intellectual creatures, living . . . under a moral law, responsible for transgression, the subjects of a Providential Government." He then makes an extended argument for the claim that whenever persons speak of extraterrestrial beings, they conceive of such beings as essentially identical or at least very similar to humans. After suggesting that this claim is "an act of invention and imagination which may be as coherent as a fairy tale, but which, without further proof, must be as purely imaginary and arbitrary," he notes that humanity's intellectual development seems to have been directed by an overseeing Divine intelligence. This leads to one of the key sections of the book.

[William Whewell], *Of the Plurality of Worlds: An Essay; Also, A Dialogue on the Same Subject*, 4th ed. (London: John W. Parker and Son, 1855), 143–52.

12 But there is one step more, which we still have to make, in order to bring out this difficulty in its full force. As we have said, the moral law has been, to a certain extent, established, developed and enforced among men. But, as I have also said, looking carefully at the law, and at the degree of man's obedience to it, and at the operation of the sanctions

by which it is supported, we cannot help seeing, that man's knowledge of the law is imperfect, his conviction of its authority feeble, his transgressions habitual, their punishment and consequences obscure. When, therefore, we regard God as the Lawgiver and Judge of man, it will not appear strange to us, that he should have taken some mode of promulgating his Law, and announcing his Judgments, in addition to that ordinary operation of the faculties of man, of which we have spoken. Revealed Religion teaches us that he has done so: that from the first placing of the race of man upon the earth, it was his purpose to do so: that by his dealings with the race of man in the earlier times, and at various intervals, he made preparation for the mission of a special Messenger, whom, in the fulness of time, he sent upon the earth in the form of a man: and who both taught men the Law of God in a purer and clearer form than any in which it had yet been given; and revealed His purpose, of rewards for obedience, and punishments for disobedience, to be executed in a state of being to which this human life is only an introduction; and established the means by which the spirit of man, when alienated from God by transgression, may be again reconciled to Him. The arrival of this especial Message of Holiness, Judgment, and Redemption, forms the great event in the history of the earth, considered in a religious view, as the abode of God's servants. It was attended with the sufferings and cruel death of the Divine Messenger thus sent; was preceded by prophetic announcements of his coming; and the history of the world, for the two thousand years that have since elapsed, has been in a great measure occupied with the consequences of that advent. Such a proceeding shows, of course, that God has an especial care for the race of man. The earth, thus selected as the theatre of such a scheme of Teaching and of Redemption, cannot, in the eyes of any one who accepts this Christian faith, be regarded as being on a level with any other domiciles. It is the Stage of the Great Drama of God's Mercy and Man's Salvation; the Sanctuary of the Universe; the Holy Land of Creation; the Royal Abode, for a time at least, of the Eternal King. This being the character which has thus been conferred upon it, how can we assent to the assertions of Astronomers, when they tell us that it is only one among millions of similar habitations, not distinguishable from them, except that it is smaller than most of them that we can measure: confused and rude in its materials like them? Or if we believe the Astronomers, will not such a belief lead us to doubt the truth of the great scheme of Christianity, which thus makes the earth the scene of a special dispensation?

13 This is the form in which Chalmers has taken up the argument. This is the difficulty which he proposes to solve; or rather, (such being, as I have said, the mode in which he presents the subject,) the objection which he proposes to refute. It is the bearing of the Astronomical discoveries of modern times, not upon the doctrines of Natural Religion, but upon the scheme of Christianity, which he discusses. And the question which he supposes his opponent to propound, as an object to the Christian scheme, is;—How is it consistent with the dignity, the impartiality, the comprehensiveness, the analogy of God's proceedings, that he should make so special and preeminent a provision for the salvation of the inhabitants of this Earth, when there are such myriads of other worlds, all of which may require the like provision, and all of which have an equal claim to their Creator's care?

14 The answer which Chalmers gives to this objection, is one drawn, in the first instance, from our ignorance. He urges that, when the objector asserts that other worlds may have the like need with our own, of a special provision for the rescue of their inhabitants from the consequences of the transgression of God's laws, he is really making an assertion without the slightest foundation. Not only does Science not give us any information on such subjects, but the whole spirit of the scientific procedure, which has led to the knowledge which we possess, concerning other planets and other systems is utterly opposed to our making such assumptions respecting other worlds, as the objection involves. Modern Science, in proportion as she is confident when she has good grounds of proof, however strange may be the doctrines proved, is not only diffident, but is utterly silent, and abstains even from guessing,

when she has no grounds of proof. Chalmers takes Newton's reasoning, as offering a special example of this mixed temper, of courage in following the evidence, and temperance in not advancing when there is no evidence. He puts, in opposition to this, the example of the true philosophical temper,—a supposed rash theorist, who should make unwarranted suppositions and assumptions, concerning matters to which our scientific evidence does not reach;—the animals and plants, for instance, which are to be found in the planet Jupiter. No one, he says, would more utterly reject and condemn such speculations than Newton, who first rightly explained the motion of Jupiter and of his attendant satellites, about which Science can pronounce her truths. And thus, nothing can be more opposite to the real spirit of modern science, and astronomy in particular, than arguments, such as we have stated, professing to be drawn from science and from astronomy. Since we know nothing about the inhabitants of Jupiter, true science requires that we say and suppose nothing about them; still more requires that we should not, on the ground of assumptions made with regard to them, and other supposed groups of living creatures, reject a belief, founded on direct and positive proofs, such as is the belief in the truths of Natural and of Revealed Religion.

15 To this argument of Chalmers, we may not only give our full assent, but we may venture to suggest, in accordance with what we have already said, that the argument, when so put, is not stated in all its legitimate force. The assertion that the inhabitants of Jupiter have the same need as we have, of a special dispensation for their preservation from moral ruin, is not only as merely arbitrary an assumption, as any assertion could be, founded on a supposed knowledge of an analogy between the botany of Jupiter, and the botany of the earth; but it is a great deal more so. There may be circumstances which may afford some reason to believe that something of the nature of vegetables grows on the surface of Jupiter; for instance, if we find that he is a solid globe surrounded by an atmosphere, vapour, clouds, showers. But, as we have already said, there is an immeasurable distance between the existence of

unprogressive tribes of organized creatures,—plants or even animals,—and the existence of a progressive creature, which can pass through the conditions of receiving, discerning, disobeying, and obeying a moral law; which can be estranged from God, and then reconciled to him. To assume, without further proof, that there are, in Jupiter, creatures of such a nature that these descriptions apply to them, is a far bolder and more unphilosophical assumption, than any that the objector could make concerning the botany of Jupiter; and therefore, the objection thus supposed to be drawn from our supposed knowledge, is very properly answered by an appeal to our really utter ignorance, as to the points on which the argument rests.

16 This appeal to our ignorance is the main feature in Chalmers's reasonings, so far as the argument on the one side or the other has reference to science. Chalmers, indeed, pursues the argument into other fields of speculation. He urges, that not only we have no right to assume that other worlds require a redemption of the same kind as that provided for man, but that the very reverse may be the case. Man may be the only transgressor; and this, the only world that needed so great a provision for its salvation. We read in Scripture, expressions which imply that other beings, besides man, take an interest in the salvation of man. May not this be true of the inhabitants of other worlds if such inhabitants there be? These speculations he pursues to a considerable length, with great richness of imagination, and great eloquence. But the suppositions on which they proceed are too loosely connected with the results of science, to make it safe for us to dwell upon them here.

17 I conceive, as I have said, that the argument with which Chalmers thus deals, admits of answers, also drawn from modern science, which to many persons will seem more complete than that which is thus drawn from our ignorance. But before I proceed to bring forward these answers, which will require several steps of explanation, I have one or two remarks still to make.

18 Undoubtedly they who believe firmly both that the earth has been the scene of a Divine Plan for the benefit of

man, and also that other bodies in the universe are inhabited by creatures who may have an interest in such a Plan, are naturally led to conjectures and imaginations as to the nature and extent of that interest. The religious poet,[2] in his *Night Thoughts*, interrogates the inhabitants of a distant star, whether their race too has in its history, events resembling the fall of man, and the redemption of man.

> Enjoy your happy realms their golden age?
> And had your Eden an abstemious Eve?
> Or, if your mother fell are you redeemed?
> And if redeemed, is your Redeemer scorned?

And such imaginations may be readily allowed to the preacher or the poet, to be employed in order to impress upon man the conviction of his privileges, his thanklessness, his inconsistency, and the like. But every form in which such reflexions can be put shows how intimately they depend upon the nature and history of man. And when such reflexions are made the source of difficulty or objection in the way of religious thought, and when these difficulties and objections are represented as derived from astronomical discoveries, it cannot be superfluous to inquire whether astronomy has really discovered any ground for such objections. To some minds it may be more grateful to remedy one assumption by another: the assumption of moral agents in other worlds, by the assumption of some operation of the Divine Plan in other worlds. But since many persons find great difficulty in conceiving such an operation of the Divine Plan in a satisfactory way; and many persons also think that to make such unauthorised and fanciful assumptions with regard to the Divine Plans for the government of God's creatures is a violation of the humility, submission of mind, and spirit of reverence which religion requires; it may be useful if we can show that such assumptions, with regard to the Divine Plans, are called

2. [Edward Young, whose *Night Thoughts* is quoted elsewhere in this book. MJC]

forth by assumptions equally gratuitous on the other side: that Astronomy no more reveals to us extra-terrestrial[3] moral agents, than Religion reveals to us extra-terrestrial Plans of Divine government. Chalmers has spoken of the *rashness* of making assumptions on such subjects without proof; leaving it, however, to be supposed, that though astronomy does not supply proof of intelligent inhabitants of other parts of the universe, she yet does offer strong analogies in favour of such an opinion. But such a procedure is more than rash: when astronomical doctrines are presented in the form in which they have been already laid before the reader, which is the ordinary and popular mode of apprehending them, the analogies in favour of 'other worlds,' are (to say the least) greatly exaggerated. And by taking into account what astronomy really teaches us, and what we learn also from other sciences, I shall attempt to reduce such 'analogies' to their true value.

In paragraph 19, Whewell stresses that human beings have a spiritual character, suggesting thereby that they have claim to a special place in creation.

20 We have given a view of the peculiar character of man's condition, which seems to claim for him a nature and place, unique and incapable of repetition, in the scheme of the universe; and to this view, astronomy, exhibiting to us the habitation of man as only one among many similar abodes, offers an objection. We are, therefore, now called upon, I conceive, to proceed to exhibit the answer which a somewhat different view of modern science suggests to this difficulty or objection.

---

3. [For some years, I suspected that this is the earliest appearance of the word "extraterrestrial." That Whewell was the creator of the word would not be surprising in that he was known for coining new scientific terms. To check that suspicion, I consulted the *Oxford English Dictionary*, finding that the earliest usage cited there was in fact later. Consequently, around 1995 I wrote the editor of the *OED* about my discovery. The answer to my letter contained a surprise—that the *OED* had found an earlier instance and that they had located it by doing an electronic search of the *OED* itself. There, under the term "Howardite," which is a chemical found in meteorites by chemist Edward Howard, they located a passage from a paper by C. U. Shepard in the *American Journal of Science* 6 (1848): 253, where Shepard used the term in a discussion of Howard's work on meteorites. MJC]

For this purpose, we must begin by regarding the Earth in another point of view, different from that hitherto considered by us.

Summary of Whewell's *Of the Plurality of Worlds: An Essay* (continued)

Chapter 5—"Geology": Whewell reviews some of the main results attained by geology, emphasizing the point that geological studies have shown the great age of the Earth and that humanity appeared only late in Earth history.

Chapter 6—"The Argument from Geology": This chapter contains another key argument formulated by Whewell, and one of his most original ideas. He sets out to show that a knowledge of geology refutes one of the key arguments for extraterrestrial beings. Proponents of life elsewhere in the universe had repeatedly argued that were the vast reaches of space vacant of life, this would indicate that God's efforts had been wasted in creating them. This claim was based on the assumption that God would never act in such a manner that divine efforts would be wasted.

Whewell's response to this argument is that geology has shown that throughout the great majority of Earth history, the Earth lacked intelligent life. Were an argument analogous to the "waste" argument of the pluralists applied to this situation, it would entail the conclusion that God's efforts in creating the Earth were wasted for most of the Earth's past. Whewell has, consequently, used a sort of "reduction to absurdity" argument against the pluralists' claim; he has shown that if they accept the facts of geology, they must by a reasoning analogous to that which they apply to space, conclude that God acts wastefully. Put differently, Whewell is pointing out that geology suggests that God does in fact act in such a way that within our limited understanding we may perceive parts of the universe as in a wasteful state. He puts the core of his argument rather colorfully in the following passage:

> Here then we are brought to the view, which, it would seem, offers a complete reply to the difficulty which astronomical discoveries appeared to place in the way of religion:—the difficulty of the opinion that man, occupying this speck of earth, which is but an atom in the Universe, surrounded by millions of other globes, larger, and to appearances, nobler than that which he inhabits, should be the object of the peculiar care and guardianship, of the favor and government, of the creator of all, in the way in which Religion teaches us that He is. For we find that man, (the human race, from its first origin till now,) has occupied but an atom of time, as he has occupied but an atom of space:—that as

> he is surrounded by myriads of globes which may, like this, be the habitations of living things, so he has been preceded, on this earth, by myriads of generations of living things, not possibly or probably only, but certainly; and yet that, comparing his history with theirs, he has been, certainly has been fitted to be, the object of the care and guardianship, of the favor and government, of the Master and Governor of All, in a manner entirely different from anything which it is possible to believe with regard to the countless generations of brute creatures which had gone before him. . . . If the earth, as the habitation of man, is a speck in the midst of an infinity of space, the earth, as the habitation of man, is also a speck at the end of the infinity of time. (206–7)

Chapter 7—"The Nebulae": With this chapter, Whewell turns directly to some of the controversies current in the astronomy of his day. One of these concerned whether dim nebulous patches of light seen in the heavens are, in fact, other universes, "island universes" as they were called, comparable in size to our own Milky Way, but appearing dim because of their great distance. The relevance of this question is that if the nebulae were interpreted as other universes, they would create further opportunities for pluralists to postulate extraterrestrial life. Whewell's position on this issue was that "we appear to have good reason to believe that nebulae are vast masses of incoherent or gaseous matter, of immense tenuity, diffused in forms more or less irregular, but all of them destitute of any regular system of solid moving bodies" (251). Among his arguments for this claim, the most important derived from John Herschel's report that within the Magellanic Clouds one can see nebulae and clusters immediately adjacent to stars of comparable magnitude. As Whewell puts it, "There are such things as nebulae side by side with stars, and with clusters of stars. Nebulous matter resolvable occurs close to nebulous matter irresolvable. [Consequently these] are different kinds of things in themselves, not merely different to us" (230). In other words, if in looking at the Magellanic Clouds one sees a nebulous patch adjacent to a star of comparable magnitude, it makes no sense to assume that the nebula is composed of thousands of stars awaiting resolution. Nebulae and stars, in this case at least, must be comparable in size, although distinct in character (in other words, nebulae are not just clusters of stars). Although Whewell's contemporaries were not for the most part favorably disposed to this argument, by the end of the nineteenth century it had become widely accepted.

Chapter 8—"The Fixed Stars": In discussing the question of whether stars have habitable planets encircling them, Whewell, like all his contemporaries, was hindered by the fact that the distances of only a handful of stars had been determined by the 1850s. Nonetheless, Whewell makes a number of arguments. For example, he points out that there is no observational evidence that any star except the Sun has planets. Moreover, he notes that variable stars will not support inhabited satellites and that double star systems have greatly varying gravitational fields, which precludes the existence of life on planets within such systems.

Chapter 9—"The Planets": Turning to the solar system, Whewell begins with Neptune, which, because it is so much farther from the Sun than we are, seems ill provided with the heat and light necessary for life. Nonetheless, he admits that it might be possible to devise inhabitants for it, if it were found to be the rule that the objects of the solar system have inhabitants. This leads him to consider the Moon as the body most favorably situated both for life and for telescopic observation. Drawing on recent researches, he shows that the Moon not only lacks life but also water and an atmosphere. From this he concludes that it is unnecessary to postulate life on the planets. He then analyzes Jupiter and the question of its habitability in a section of his book that attracted widespread attention and that demonstrates Whewell's attention to detail.

[William Whewell], *Of the Plurality of Worlds*, 4th ed., 297–303.

14 Now having obtained these views from an examination of the Earth, let us apply them to other planets, as bearing upon the question of their being inhabited; and in the first place, to Jupiter. We can, as we have said, easily compare the mass of Jupiter and of the Earth; for both of them have Satellites. It is ascertained, by this means, that the mass or weight of Jupiter is about 333 times the weight of the Earth: but as his diameter is also 11 times that of the earth, his bulk is 1331 times that of the earth: (the *cube* of 11 is 1331;) and, therefore, the density of Jupiter is to that of the earth, only as 333 to 1331, or about 1 to 4. Thus the density of Jupiter, taken as a whole, is about a quarter of the earth's density; less than that of any of the stones which form the crust of the earth; and not much greater than the density of water. Indeed, it is tolerably certain, that the density of Jupiter is not greater than it would be if his entire globe were composed of water; making allowance for the compression which the interior parts would suffer by the pressure of the parts superincumbent

upon them. We might, therefore, offer it as a conjecture not quite arbitrary, that Jupiter is a mere sphere of water.

A paragraph has been omitted at this point in which Whewell cites observations indicating that Jupiter, rather than being spherical in shape, is "oblate, like an orange." Moreover, the amount of oblateness agrees well with what mechanics predicts for a fluid body rotating at Jupiter's rate of rotation.

16 But there are other circumstances in the appearances of Jupiter, which still further confirm this conjecture of his watery constitution. His belts,—certain bands of darker and lighter colour, which run parallel to his equator, and which, in some degree change their form, and breadth, and place, from time to time,—have been conjectured, by almost all astronomers, to arise from lines of cloud, alternating with tracts comparatively clear, and having their direction determined by currents analogous to our trade-winds, but of a much more steady and decided character, in consequence of the greater rotatory velocity.[4] Now vapours, supplying the materials of such masses of cloud, would naturally be raised from such a watery sphere as we have supposed, by the action of the Sun; would form such lines; and would change their form from slight causes of irregularity, as the belts are seen to do. The existence of these lines of cloud does of itself shew that there is much water on Jupiter's surface, and is quite consistent with our conjecture, that his whole mass is water.[5]

---

4. Herschel, [*Outlines of Astronomy*,] 513.

5. A difficulty may be raised, founded on what we may suppose to be the fact, as to the extreme cold of those regions of the Solar System. It may be supposed that water under such a temperature could exist in no other form than ice. And that the cold must there be intense, according to our notions, there is strong reason to believe. In the outer regions of our atmosphere, the cold is probably very many degrees below freezing; and in the blank and airless void beyond, it may be colder still. It has been calculated by physical philosophers, on grounds which seem to be solid, that the cold of the space beyond our atmosphere is 100° below zero. But when we emerge out of this frigorific void, and come into contact with a material planet, the materials of the planet, and the creatures, if there be any, which live upon it, must receive and retain the heat transmitted from the source of heat, the Sun, according to the same laws which

A paragraph in which Whewell argues that it is fully proper to view the law of gravitation as extending to Jupiter has been deleted at this point.

18 As bearing upon the question of life in Jupiter, there is another point which requires to be considered; the force of gravity at his surface. Though, equal bulk for equal bulk, he is lighter than the earth, yet his bulk is so great that, as we have seen, he is altogether much heavier than the earth. This, his greater mass, makes bodies, at equal distances from the centers, ponderate proportionally more to him than they would do to the earth. And though his surface is 11 times further from his center than the earth's is, and therefore the gravity at the surface is thereby diminished, yet, even after this deduction, gravity at the surface of Jupiter is nearly two and a half times that on the earth.[6] And thus a man transferred to the surface of Jupiter would feel a stone, carried in his hands, and would feel his own limbs also, (for his muscular power would not be altered by the transfer,) become 2 1/2 times as heavy, as difficult to raise, as they were before. Under such circumstances animals of large dimensions would be oppressed with their own weight. In the smaller creatures on the earth, as in insects, the muscular power bears a great proportion to the weight, and they might continue to run and to leap, even if gravity were tripled or quadrupled. But an elephant could not trot with two or three elephants

---

prevail in other parts of the Solar System, and which are made known to us on the Earth. Therefore the heat upon the surface of Jupiter must be such as corresponds to a heating power 50 times less than that which warms the surface of the Earth. If we suppose that this great cold must convert the watery mass of Jupiter into a mass of solid ice, we may still remark, that if there be a free surface, there will be vapour produced by the Sun's heat; and if there be air, there will be clouds. We may add, that so far as we have reason to believe, below the freezing point, no accession of cold produces any material change in ice. Even in the expeditions of our Arctic navigators, a cold of 40° below zero was experienced, and ice was still ice, and there were vapours and clouds as in our climate. It is quite an arbitrary assumption to suppose that any cold which may exist in Jupiter, would prevent the state of things which we suppose.

6. Herschel, 508.

placed upon his back. A lion or tiger could not spring, with twice or thrice his own weight hung about his neck. Such an increase of gravity would be inconsistent then, with the present constitution and life of the larger terrestrial animals; and if we are to suppose planets inhabited, in which gravity is much more energetic than it is upon the earth, we must suppose classes of animals which are adapted to such a different mechanical condition.

19 Taking into account then, these circumstances in Jupiter's state; his (probably) bottomless waters; his light, if any, solid materials; the strong hand with which gravity presses down such materials as there are; the small amount of light and heat which reaches him at five times the earth's distance from the sun; what kind of inhabitants shall we be led to assign to him? Can they have skeletons, where no substance so dense as bone is found, at least in large masses? It would seem not probable.[7] And it would seem they must be dwellers in the waters, for against the existence there of solid land, we have much evidence. They must, with so little of light and heat, have a low degree of vitality. They must then, it would seem, be cartilaginous and glutinous masses; peopling the waters with minute forms: perhaps also with larger monsters; for the weight of a bulky creature, floating in the fluid, would be much more easily sustained than on solid ground. If we are resolved to have such a population, and that they shall live by food, we must suppose that the waters contain at least so much solid matter as is requisite for the sustenance of the lowest classes; for the higher classes of animals will probably find their food in consuming the lower. I do not know whether the advocates of peopled worlds will think such a population as this worth contending for: but I think the only doubt can be, between such a population, and none. If Jupiter be a mere

7. It may be thought fanciful to suppose that because there is little or no solid matter (of any kind known to us) in Jupiter, his animals are not likely to have solid skeletons. The analogy is not very strong; but also, the weight assigned to it in the argument is small. *Valeat quantum valere debet.*

mass of water, with perhaps a few cinders at the center, and an envelope of clouds around it, it seems very possible that he may not be the seat of life at all. But if life be there, it does not seem in any way likely, that the living things can be anything higher in the scale of being, then such boneless, watery, pulpy creatures as I have imagined.[8]

20 Perhaps it may occur to some one to ask, if this planet, which presents so glorious an aspect to our eyes, be thus the abode only of such imperfect and embryotic lumps of vitality as I have described; to what purpose was all that gorgeous array of satellites appended to him, which would present, to intelligent spectators on his surface, a spectacle far more splendid than any that our skies offer to us: four moons, some as great, and others hardly less, than our moon, performing their regular revolutions in the vault of heaven. To which it will suffice at present to reply, that the use of those moons, under such a supposition, would be precisely the same, as the use of our moon, during the myriads of years which elapsed while the earth was tenanted by corals and madrepores, shell-fish and belemnites, the cartilaginous fishes of the Old Red Sandstone, or thesaurian monsters of the Lias; and in short, through all the countless ages which elapsed, before the last few thousand years:—before man was placed upon the earth, 'to eye the blue vault, and bless the *useful* light:'—to reckon by it his months and years:—to discover by means of it, the structure of the universe, and perhaps, the special care of his Creator for him alone of all his creatures. The moons of Jupiter, may in this way, be of use, as our own moon is. Indeed we know that they have been turned to most important purposes in astronomy and navigation. And knowing this, we may be content not to know how, either the satellites of Jupiter, or the satellite of the Earth, tend to the advantage of the brute inhabitants of the waters.

8. [Whewell's Jupiterians had by 1857 surfaced in one of the most popular novels of the period. See the selection from Anthony Trollope's *Barchester Towers* that is provided near the end of this chapter. MJC]

A paragraph in which Whewell discusses the question of how much of Jupiter exists in the form of ice has been deleted at this point. Whewell leaves the question open, but stresses that Jupiter receives only one twenty-fifth as much heat and light per unit surface area as the Earth.

Summary of Whewell's *Of the Plurality of Worlds: An Essay* (continued)

Chapter 9—"The Planets" (continued): Having disposed of Jupiter, Whewell had little difficulty dismissing as habitable Saturn and the planets beyond it because all are much farther than Jupiter from the Sun. Mars, however, having an orbit relatively near the Earth's, was a harder case. Whewell suggests that Mars may be comparable to the Earth long ago and hence may be populated by dinosaurs and such. Whewell does not seriously consider the asteroids as abodes of life, but Venus and Mercury receive mention, although astronomers had little to report concerning them. Their closeness to the Sun makes them unlikely locales for life. If Venus is tenanted at all, it is probably by "microscopic creatures, with siliceous coverings" protecting them from heat. In short, Whewell was able to present well-formulated arguments against life elsewhere in the solar system. And by the early twentieth century, it was realized that his overall position is in fact correct: that within our solar system, only the Earth is inhabited by intelligent beings.

Chapter 10—"Theory of the Solar System": Whewell presents a new and highly speculative theory of the solar system, which had some affinities with the nebular hypothesis. One feature of Whewell's theory was that as part of it he developed the notion that the Earth exists in what he called "the Temperate Zone of the Solar System," the only portion of the solar system possessing conditions compatible with life.[9] He rather rhetorically summarizes his position by stating:

> The Earth is really the domestic hearth of this Solar System; adjusted between the hot and fiery haze on one side, the cold and watery vapour on the other. This region only is fit to be a domestic hearth, a seat of habitation; and in this region is placed the largest solid globe of our system; and on this globe, by a series of

9. Recent discussions in astrobiology make extensive use of the notion of "habitable zone," but rarely note that Whewell originated this idea. For information on the history of this idea, see http://www.centauri-dreams.org?p=428.

> creative operations, entirely different from any of those which separated the solid from the vaporous, the cold from the hot, the moist from the dry, have been established, in succession, plants, and animals, and man. . . . the Earth alone . . . has become a World. (324–25)

Chapter 11—"The Argument from Design": This chapter is focused on showing how his antipluralist position can be reconciled with natural theology. Whewell begins by lauding the teleological design argument, that based on a notion of final causality. Nonetheless, he notes that problems have arisen for it. While exulting in the adaptation of the wing of a bird and the arm of a man to their functions, it fails to explain why those appendages, so different in function, should have strikingly similar skeletons. It is no more successful in explaining the functions of such bodily parts as nipples on the male. Whewell proposes to remove these problems by the consideration, which he owed largely to the comparative anatomist Richard Owen, that God acts according to overall patterns, modified to fit individual circumstances. Thus the homologies of the arm and wing indicate that God designed them on the same plan. This view does not destroy the argument from design. As he puts it,

> If the general law supplies the elements, still a special adaptation is needed to make the elements serve such a purpose; and what is this adaptation, but design? The radius and ulna, the carpal and metacarpal bones, are all in the general type of the vertebrate skeleton. But does this fact make it less wonderful, that man's arm and hand and fingers should be constructed so that he can make and use the spade, the plough, . . . the lute, the telescope. . . ? (336)

Moreover, this use of overall patterns, of general laws, may evidence "some other feature of the operation of the Creative Mind." We should avoid placing too much confidence in our teleological explanations of nature's adaptations, but when these explanations fail, we should see how readily these features can be explained as the result of God working by means of general laws and patterns. The relevance of this is that the planets and stars may be explained, not by God having created them for living beings, but as resulting from a general plan of creation of which the most noteworthy result is our inhabited planet. To the argument that if the other planets are not inhabited, then God created them in vain, Whewell is now able to respond that God, because he works in general patterns, frequently appears to have worked in vain. All around us, after all, we see "embryos which are never vivified, germs which are not developed. . . ." And he adds: "Of the vegetable seeds which are produced, what an infinitely small portion ever grow into plants! Of animal ova, how exceedingly few become animals, in proportion to those that do not; and that are wasted, if this be waste!" On this analogy, it is possible to view the Earth as the only "fertile seed of creation," as the "one fertile flower [of] the Solar System. . . . One such fertile result as the Earth, with all its hosts of plants and animals, and especially with Man . . . is a worthy and sufficient produce . . . of all the Universal Scheme." If a person be troubled by the thought that this would leave great portions of the universe without intelligent life, let that person reflect that this was the situation on Earth for most of its history (his argument from geology).

How are we to reconcile the splendors of the heavens with the thought that human beings may be the sole intelligent being in the universe? On one level, his colorfully worded answer is:

> The planets and the stars are the lumps which have flown from the potter's wheel of the Great Worker; the shred-coils which . . . sprang from His mighty lathe:—the sparks which darted from His awful anvil when the solar system lay incandescent thereon;—the curls of vapour which rose from the great cauldron of creation when the elements were separated. If even these superfluous portions are marked with universal traces of regularity and order, this shows that universal rules are his implements, and that Order is the first and universal Law of the heavenly work. (369–70)

On another level, he urges that the heavens are nothing compared to humanity:

> The majesty of God does not reside in planets and stars . . . which are . . . only stone and vapour, materials and means. . . . the material world must be put in an inferior place, compared with the world of mind. If there be a World of Mind, that . . . must have been better worth creating . . . than thousands and millions of stars and planets, even if they were occupied by a myriad times as many species of brute animals as have lived upon the earth since its vivification. (370)

After citing poets to the same point, Whewell again stresses that humanity has such potencies that its existence serves to justify the universe. Not only humanity, but "one soul created never to die . . . outweighs the whole unintelligent creation." He also suggests:

> That the human race possesses a worth in the eyes of Reason beyond that which any material structure, or any brute population can possess, might be maintained on still higher and stronger grounds; namely, on religious grounds: but we do not intend here to dwell on that part of the subject. If man be, not merely (and he alone of all animals) capable of Virtue and Duty, of Universal

> Love and Self-Devotion, but be also immortal; if his being be of infinite duration, his soul created never to die; then, indeed, we may well say that one soul outweighs the whole unintelligent creation. And if the Earth have been the scene of an action of Love and Self Devotion for the incalculable benefit of the whole human race, in comparison with which the death of Socrates fades into a mere act of cheerful resignation to the common lot of humanity; and if this action, and its consequences to the whole race of man, in his temporal and eternal destiny, and in his history on earth before and after it, were the main object for which man was created, the cardinal point round which the capacities and the fortunes of the race were to turn; then indeed we see that the Earth has a pre-eminence in the scheme of creation, which may well reconcile us to regard all the material splendour which surrounds it, all the array of mere visible luminaries and masses which accompany it, as no unfitting appendages to such a drama. The elevation of millions of intellectual, moral, religious, spiritual creatures, to a destiny so prepared, consummated, and developed, is no unworthy occupation of all the capacities of space, time, and matter. And, so far as any one has yet shown, to regard this great scheme as other than the central point of the divine plan; to consider it as one part among other parts, similar, co-ordinate, or superior; involves those who so speculate, in difficulties, even with regard to the plan itself, which they strive in vain to reconcile; while the assumption of the subjects of such a plan, in other regions of the universe, is at variance with all which we, looking at the analogies of space and time, of earth and stars, of life in brutes and in man, have found reason to deem in any degree probable. (372–73)

These considerations, Whewell urges, show that natural theology has little to lose by forsaking the idea of a plurality of worlds.

Chapter 12—"The Unity of the World": This chapter begins with the statement: "The two doctrines which we have here to weigh against each other are the Plurality of Worlds, and the Unity of the World." Whewell notes that by world he means one inhabited by *intelligent* beings. In discussing these contrasting views, Whewell suggests that we first look inward at the nature of humanity to see whether we find some characteristics of its nature that lessen the difficulty that arises from considering the Earth as being, in a unique and special manner, the province of God. In this context, Whewell claims that the human "mind is, in some measure, as the Divine Mind." Plato and other ancients held this doctrine, but it is also supported by recent scientific thought, in particular, that embodied by Richard Owen in his *On the Nature of Limbs*. Whewell quotes from Owen who had claimed that God, in creating the world, used "an Idea and Exemplar." Owen recognizes "an ideal exemplar for the vertebrated animals," which takes on many modifications, these corresponding to the diverse species of animals. Whewell then develops this idealist position somewhat further, but backs off from Newton's notion that space is the sensorium of God, preferring a Kantian view of space. Then Whewell asserts that "God has constituted *man*, so that *he* can apprehend the works of creation, only as existing in time and space" (380). If human beings can partake of the Divine Mind, then the Earth may be the only locale where intelligent beings live. Whewell adds: "The remotest planet is not devoid of life, for God lives there." Whewell's point emerges especially clearly when he states: "One school of moral discipline, one theatre of moral actions, one arena of moral contests for the highest prizes, is a sufficient center for innumerable hosts of stars and planets, globes of fire and earth, water and air, whether or not tenanted by corals and madrepores, fishes and creeping things." Moreover, if humans are immortal, we need no other array of creatures to give dignity to the scheme of creation. Furthermore, if humanity improves, this would do more for the nobility of the universe than any number of stars and planets.

Whewell goes on to deal with the claim that if the existence of humanity glorifies God, then intelligent life on all the planets would give greater glory to God. His response, which refers back to the scientific evidence presented in prior chapters, is that "we cannot think ourselves authorized to assert cosmological doctrines, selected arbitrarily by ourselves, on the ground of their exalting our sentiments of admiration and reverence for the Deity, *when the weight of all the evidence which we can obtain respecting the constitution of the universe is against them*" (387–88). And he adds that one "great scheme of moral and religious government . . . may well suffice for the religious sentiments of man in the present age." Whewell admits that a plurality of inhabited worlds is a possibility, but stresses that the uniqueness of life on the Earth is also possible. Moreover, the idea of a plurality of worlds is no more than a conjecture. Some of the stars may be centers of solar systems, but this is "founded upon the single fact, shown to be highly ambiguous, of the stars being self-luminous; and to this possibility, we oppose all the considerations flowing from moral, historical and religious views, which represent the human race as unique." Some may respond to the view that the human race has only recently appeared on Earth, and all the past has but led up to this development, by saying that such will not be true in the future. Human beings may be around as long as the age that preceded their appearance. To this, Whewell replies that it is not legitimate to base such an argument on a future possibility.

Chapter 13—"The Future": In this chapter, Whewell discusses the degree to which the future can be predicted. Because the claims that he makes have little if any direct bearing on the question of extraterrestrial life, they need not be summarized.

SOME BRITISH RESPONSES TO WHEWELL'S *ESSAY*

Anonymous

The following appeared in the *London Daily News* in October 1854:

> We scarcely expected that in the middle of the nineteenth century, a serious attempt would have been made to restore the exploded idea of man's supremacy over all other creatures in the universe; and still less that such an attempt would have been made by one whose mind was stored with scientific truths. Nevertheless a champion has actually appeared, who boldly dares to combat against all the rational inhabitants of other spheres; and though as yet he wears his vizor down, his dominant bearing, and the peculiar dexterity and power with which he wields his arms, indicate that this knight-errant of nursery notions can be no other than the Master of Trinity College, Cambridge.[10]

Sir David Brewster (1781–1868)

David Brewster was a leading Scottish scientist, who had been a close associate of Thomas Chalmers and who on a number of previous occasions had crossed swords with Whewell. These two facts help explain why Brewster reacted very strongly when he was asked to write a review of Whewell's *Essay*. In fact, his reaction was so strong that he ended up writing not only the review, but also a long book, *More Worlds than One: The Creed of the Philosopher and the Hope of the Christian*, castigating Whewell and the anti-pluralist ideas he had endorsed.

The following excerpts are from Brewster's review of Whewell's *Essay* in *North British Review* 21 (1854): 2–44.

> To conceive any one material globe, whether a gigantic clod slumbering in space, or a noble planet equipped like our own, and duly performing its appointed tasks, to have no living occupants, or not in a state of preparation to be occupied, seems to us one of those notions which could be harboured only in the ill-educated and ill-regulated mind,—a mind without faith and without hope; but to conceive a whole universe of moving and revolving worlds in such a category, indicates, in our apprehension, a mind dead to feeling and shorn of

10. As quoted in [Samuel Warren], "Speculators among the Stars, Part II," *Blackwood's Edinburgh Magazine* 76 (Oct. 1854): 372.

> reason. . . . That such a work [as Whewell's *Essay*] could have been written in the present day by a man of high mental attainment, and professing the Christian faith, is to us one of the most marvellous events in these marvellous times; and did we believe in the proximity of the millennial age, we should rank it among the lying wonders which are to characterize the latter times. (10)

> After a careful perusal of [Whewell's chapter entitled "Further Statement of the Difficulty"], we must acknowledge our inability either fully to understand its meaning, or to see its bearing on the real question of a plurality of worlds. It is a mere display of ingenuity, obliterating metaphysically the brightness of our perceptions, and coming over our minds like an Eastern fog on a spring morning or like the tail of a comet over a cluster of stars. (16)

Next we have a series of excerpts from Brewster's *More Worlds than One: The Creed of the Philosopher and the Hope of the Christian* (London: John Camden Hotten, 1870 edition of the 1854 original):

> There is no subject within the whole range of knowledge so universally interesting as that of a Plurality of Worlds. (1)

> Hence the size or bulk of Jupiter is 1300 times greater than that of the Earth, and this alone is a proof that it must have been made for some *grand* and *useful* purpose [that is] of being the seat of animal and intellectual life. (65–67)

> [Brewster asserts that because the Sun is] a domain so extensive, so blessed with eternal light, it is difficult to claim that it is not occupied by the highest order of intelligences. . . . Universal life upon universal matter is an idea to which the mind instinctively clings. (102–3)

> When our Saviour died, the influence of His death extended backwards, in the past, to millions who never heard his name, and forwards, in the future, to millions who will never hear it. . . . [D]istance in time and distance in place did not diminish its healing virtue. . . . [I]t was a force which did not vary with any function of distance. . . . Emanating from the middle planet of the system, because, perhaps, it most required it, why may it not have extended . . . to all the planetary races? (149–50)

> [Referring to Whewell's argument from geology] . . . our author has recourse to what we consider the most shallow piece of sophistry which we have ever encountered in modern dialectics. (202)

> In concluding his chapter on the Fixed Stars, our Essayist [Whewell] utters sentiments and throws out conjectures so insulting to Astronomy, and casting such ridicule even on the subject of his own work, that we can ascribe them only to some morbid condition of the mental powers, which feeds upon paradox and delights in doing violence to sentiments deeply cherished and to opinions universally believed. (227)

> But while the astronomer ponders the wonderful structures of the spheres, . . . the Christian contemplates them with a warmer and more affectionate interest. From their past and present history his eager eye turns to the future of the sidereal systems, and he looks to them as the hallowed spots in which his immortal existence is to run. Scripture has not spoken with an articulate voice of the future locality of the blest, but Reason has combined with the scattered utterances of Inspiration, and with a voice, almost oracular, has declared that He who made the worlds, will in the worlds which he has made, place the beings of His choice. . . . [I]t is impossible for intellectual man with the light of revelation as his guide, to doubt for a moment that on the celestial spheres his future is to be spent. (258–59)

Frederick William Cronhelm

Sparked apparently by the enthusiasm for Whewell's book shown by Dr. Charles Musgrave, who was Cronhelm's pastor, Cronhelm wrote a short book siding with Whewell. He wrote in his *Thoughts on the Controversy As to a Plurality of Worlds* (London: Rivingtons, 1858):

> If . . . we seek to extend the hope of redemption to the fallen races, among the intelligent inhabitants of innumerable worlds, we must suppose the Second Person of the blessed Trinity to go on successive missions of salvation from world to world, assuming one after another the nature of every fallen race. . . . Is there a Bethlehem in Venus, a Gethsemane in Jupiter, a Calvary in Saturn? (17)

James David Forbes (1809–1868)

Forbes was a leading Scottish scientist, noted primarily for his work on the theory of glaciers. He was the Professor of Natural Philosophy at Edinburgh University. He stated in his review of Whewell's *Essay* in *Littell's Living Age* 41 (1854), referring to Jupiter:

> Alas! for the imagined seat of higher intelligences; alas! for the glories of the most majestic planet of our heavens, the stern will of the ruthless destroyer has dissipated with no sparing hand the threads on which we hung the net-work of our imagery. No unsentimental housemaid ever made with relentless broom a cleaner sweep of a geometrical cobweb! (53–54)

Sir John Herschel (1792–1871)

On 3 January 1854,[11] Whewell sent a copy of his book to his long time friend John Herschel, who was then Master of the British Mint, asking Herschel for his views on it. Whewell, however, did not reveal to Herschel that he had himself authored it. In February,[12] Herschel, shocked and taken aback by many of Whewell's assertions, sent the following response:[13]

> My dear Whewell
>
> Only continual occupation has prevented me from replying to yours of the 3d ult[14] & telling you what I think of the "Plurality" which by the bye, people very oddly persist in attributing to yourself. In common with these people, I find myself obliged to admit that I should not have thought there was so much to be said on the non-plurality side of the question. True, Humboldt drew attention to the fact of the Classification of the planets into heavy & light and shewed that the little ones are heavy & the large ones light.—But people's thoughts (most people's) are sluggish—and really though somewhere I have myself stated that taken in a lump Saturn might be regarded as made of Cork[15]—it *never did* occur to me to draw the conclusion that *ergo* the *surface* of Saturn must be of

11. For Whewell's letter, see Isaac Todhunter, *William Whewell* (London: Macmillan, 1876), 2:399–400.

12. This dating is based on Herschel's use of the word "ult" (see next note) and on the fact that the letter is a response to Whewell's letter of 3 January 1854, in which Whewell somewhat disingenuously asked Herschel for an assessment of the views of a "friend," whose book (Whewell's *Plurality of Worlds*) presents ideas, which Whewell admitted to being "much at variance with opinions which you have countenanced."

13. Transcribed from the original at Trinity College Library (Cambridge) Whewell Papers Add. Ms.a. 207[90]; compared with the transcription at the Royal Society RS:HS.23.140.

14. [ult = ultimo, meaning last month, i.e., January. MJC]

15. [In his *Outline of Astronomy*, 3rd ed. (London: Longman, Brown, Green, and Longmans, 1850), #508, Herschel had noted that "the density of Saturn hardly exceeds one-eighth of the mean density of the Earth, so that it must consist of materials not much heavier than cork." MJC]

extreme tenuity—though I long ago came to the conclusion that the rings were fluid (for the same reason that others have done so—that if solid they would tear themselves to pieces)—and that the streaks on them were mere lines of cloudiness or other liquid streakinesses.

But to dispossess Jupiter of his solidity & make him a huge Arctic Ocean I confess never once occurred to my thoughts. Yet so it would seem it must be—(yet the Satellites have bright & dark halves).[16]

Yet in the interior depths of that sea What fishes! there may be What Crystal palaces[17] they may construct on the tabasheer[18]—nucleus of that huge aquatic globe (so comfortably warm in its interior—up to any given temperature) What water organs in the nature of Sirenes they may construct and to what a perfection they may have brought the science of hydrography & many other things—Who can conceive their configurations—They may exist as immense alga-like or medusa-like Creatures "floating many a rood"[19] & thinking observing feeling & operating at each of their infinitely multiplied extremities &c &c.

So *this* then is the best of all possible worlds—the *ne plus ultra* between which and the 7th heaven there is nothing intermediate. Oh dear! Oh dear! 'Tis a sad cutting down. Look only at the Russians & Turks.[20]—Look at the revelations of the Blue Books[21] & the Police Courts I can't give in my adhesion to the doctrine that *between* this and the angelic there are not some dozen or two grades of intellectual and moral creatures.

You say (I mean the Author says) that of millions of germs only a few are reproductive that for thousands of flowers there are hardly units of fruits (if he does not say so in words 'tis his argument). Ergo among all the stars there is (not a *few* but) *one* Sun. Among all the planets not *a* few but only *one* Earth.—

Dissentions & Protesting.—The whole theory is destroyed if there can be two cases produced in which the process has gone on to its completion in the production of that ne plus ultra—An Earth! inhabited by Men!! for if two why not 2000!

---

16. [In the original, Herschel used square brackets, which have here been replaced by parentheses. MJC]

17. [The Crystal Palace was featured at the London Great Exhibition of 1851. MJC]

18. [Tabasheer is a siliceous substance that forms in the joints of bamboo. Herschel's suggestion seems to be that a solid nucleus has formed in the planet. MJC]

19. [From John Milton, *Paradise Lost*, Book I, line 196. MJC]

20. [England was then on the verge of entering the Crimean War. MJC]

21. [Reports of Parliament and the Privy Council, which were issued with a blue paper cover, were colloquially referred to as "Blue Books." MJC]

However n'importe.[22] — The book is full of striking things. The geological argument is very handily put. The Magellanic Clouds[23] are very availably brought into action — Time & Space are duly & properly scorned and reduced to their true value.— The "deep hidden Law — the "Sacred riddle" the fact that it has pleased the Creator to work "by Law" when it almost seems to us that it is for Law's sake — is I should say the sort of conclusion that rises up in my recollection of that same book — more than anything else.

I hope Mrs. Whewell is well.— We are all well to do — My family are all in the country. My own existence is limited now to the one & only idea of *making money*[24] & getting through daily duties which allow *no moment* for any intellectual pursuit whatever. | Adieu | Yours very truly | JFW Herschel

Herschel wrote in his obituary of Whewell in *Proceedings of the Royal Society* 16 (1867–68):

> The essay on the 'Plurality of Worlds' . . . can hardly be regarded as expressing his deliberate opinion, and should rather be considered in the light of a *jeu d'esprit*, or possibly, as has been suggested, as a lighter composition on the principle of "audi alteram partem," [hear the other side] undertaken to divert his thoughts in a time of deep distress. Though it may have had the effect I have heard attributed to it, of "preventing a doctrine from crystallizing into a dogma," the argument it advances will hardly be allowed decisive preponderance against the general impression which the great facts of astronomy tend so naturally to produce. (lxi)

Thomas Henry Huxley (1825–1895)

One of the most prominent biological scientists of the Victorian period, he is most often remembered as the leading champion of Darwinian theory. In his review of Whewell's *Dialogue* in the *Westminster Review* (July 1854), Huxley wrote:

---

22. [French expression meaning "no matter." MJC]

23. [In his *Results of Observations Made . . . at the Cape of Good Hope* (London: Smith, Elder and Co., 1847), 147, Herschel had noted that in the Magellanic Clouds, nebulous regions exist side-by-side with stars and are of comparable brightness. Whewell drew upon this observation to argue that nebulae are not island universes composed of millions of stars, but rather consist of clouds of shining fluid, possibly condensing to form stars. Whewell's argument became a major reason for rejecting the "island universe" theory of nebulae. MJC]

24. [A pun on the fact that Herschel was at that time Master of the British Mint. MJC]

> It is not a little singular that just at the present time when the stirring events of the day might be supposed to furnish even philosophers with abundant subjects of direct and personal interest, a hot controversy should have broken out upon that most hyper-hypothetical of speculations, the "Plurality of Worlds." (242)

> But we are glad to leave these speculations with the concluding reflection, that the fact of so eminent a man writing so ill upon it, strengthens our conviction that the subject is essentially unfitted for discussion. Surely there are sufficiently wide fields of investigation whose cultivation will yield results which *can* be tested; surely, our scientific Alexanders are not yet justified in crying for other worlds to conquer. (244)

Hugh Miller (1802–1856)

Originally a stonemason, Miller, a Scotsman, established a reputation as a writer on geology and on the relations between science and religion. In his *Geology versus Astronomy* (Glasgow: James R. MacNair, 1855), he wrote:

> Our globe may be the great nursery, not of the solar system only, but of the whole material universe. (31)

> [T]hough only one planet and one race may have furnished the point of union between the Divine and the created nature, the effects of that junction may extend to *all* created nature. . . . If it was necessary that the point of junction be somewhere, why not here? (33)

John Henry Cardinal Newman (1801–1890)

Oxford educated, Newman was a central figure in the Oxford Movement. In 1845, he left the Church of England to become a Roman Catholic. Newman was one of the leading intellectuals of the Victorian period and the most prominent Catholic theologian in nineteenth-century Britain.

In his 13 April 1858 letter to E. B. Pusey, Newman stated:

> But in the whole scientific world men seem going ahead most recklessly with their usurpations on the domain of religion. Here is Dr. Brewster, I think, saying that 'more worlds than one is the hope the Christian —' and, as it seems to me, building Christianity more or less upon astronomy. I seem to wish that divine

and human science might each be suffered in peace to take its own line, the one not interfering with the other. Their circles scarcely intersect each other.[25]

Furthermore in his *Grammar of Assent* (1870), he wrote:

Facts cannot be proved by presumptions, yet it is remarkable that in cases where nothing stronger than presumption was even professed, scientific men have sometimes acted as if they thought this kind of argument, taken by itself, decisive of a fact which was in debate. Thus in the controversy about the Plurality of worlds, it has been considered, on purely antecedent grounds, as far as I see, to be so necessary that the Creator should have filled with living beings the luminaries which we see in the sky, and the other cosmical bodies which we imagine there, that it almost amounts to a blasphemy to doubt it.[26]

Rev. Baden Powell (1796–1860)

Powell was simultaneously the Savilian Professor of Geometry at Oxford and one of the chief spokesmen for liberal Anglicanism. He endorsed Darwin's *Origin of Species* in an essay appearing in a multiauthored and highly controversial volume entitled *Essays and Reviews*. His son of the same name founded the Boy Scouts. In his *Essays on the Spirit of the Inductive Philosophy, The Unity of Worlds, and the Philosophy of Creation* (London: Longman, Brown, Green, and Longmans, 1855), he wrote:

If it be an inscrutable mystery *wholly beyond human comprehension* that God should send His Son to redeem this world, it cannot be a *more* inscrutable mystery . . . that He should send His Son to redeem ten thousand other worlds. (291)

Looking at the subject solely as a question of plausible philosophical conjecture, and guided as we should be by the pure light of inductive analogy, all astronomical presumption, taking the truths of geology into account, seems to be in favour of progressive order, advancing from the inorganic to the organic, and from the insensible to the intellectual and moral in all parts of the material world alike, though not necessarily in all at the same time or with the

25. As quoted in *The Letters and Diaries of John Henry Newman*, ed. Charles Stephen Dessain (London: Thomas Nelson and Sons, Ltd., 1968), 18:322.

26. As reprinted (Notre Dame, Ind.: University of Notre Dame Press, 1979), 298–99.

> same rapidity; in some worlds one stage being reached, while in others only a comparatively small advance may have been made. (231)

Rev. Adam Sedgwick (1785–1873)

Cambridge educated, Sedgwick became Woodwardian Professor of Geology at Cambridge. He served for a time as Vice-Master of Trinity College, Cambridge, with Whewell as Master. In his letter of 17 February 1854 to Sir John Herschel, he wrote:

> What wonderful health [Whewell] has! And indeed he ought to be strong, to destroy a plurality of worlds, as he is trying to do. Have you seen the big pestle and mortar by which he has pounded 500,000 worlds into comet-tail-dust, and the big snuffers by which he has put out the *lights* of all *livers* above and below the earth? I was much amused by it, but not convinced.[27]

H. J. S. Smith (1826–1883)

Oxford educated, Smith succeeded Powell as Savilian Professor of Geometry at Oxford. He stated in his "The Plurality of Worlds":

> We cannot imagine a more painful spectacle of human presumption than that which would be afforded by a man who would sit down to arrange, 'in a satisfactory way,' a scheme for the extension of Divine mercy to some distant planet, and who, when he found 'great difficulty in conceiving' such an extension . . . instead of desisting from his vain attempt, should . . . infer that no such scheme can exist because he fails to discover a *modus operandi* for it.[28]

Sir James Stephen (1789–1859)

After serving in the British government and rising to become undersecretary for the colonies, Stephen in 1849 returned to Cambridge as professor of Modern History.

From his letter of 13 October 1853 commenting on the manuscript of Whewell's *Essay*:

---

27. As quoted in John W. Clark and Thomas M. Hughes, *Life and Letters of Reverend Adam Sedgwick* (Cambridge: Cambridge University Press, 1890), 2:269.

28. In *Oxford Essays for 1855* (London: John W. Parker and Sons, n. d.), 117.

> [Y]ou should somewhere indicate (briefly but dogmatically of course) what are the rules of logic applicable to this debate, for nothing is more evident to me than that in this particular subject those rules are almost universally unheeded;—I suppose because they are almost universally unknown.[29]

From his letter of 15 October 1853 commenting on the manuscript of Whewell's *Essay*:

> Can it really be that this world is the best product of omnipotence, guided by omniscience and animated by Love?—that the Deity has called into existence one race of rational beings only, and that one race corrupt from the very dawn of its appearance?—that of this solitary family "many are called but few are chosen"?—for the vast majority of them, as far as we can judge, it had been infinitely better that they had never been born.[30]

From his letter of 18 October 1853 commenting on the manuscript of Whewell's *Essay*:

> On behalf of my clients, the inhabitants of the extraterrestrial universe, I am pleading like some of my old Bar associates—"Gentlemen of the Jury, before you give credit to that evidence, think on the misery that your verdict must inflict on the Prisoner's family." If the learned logician had said, not "be incredulous lest you should give pain," but "be cautious where your error might lead to such painful consequences," he would not have been so far wrong. It is in this latter sense that I oppose your terrible artillery (for I acknowledge myself to be alarmed by it).[31]

From his letter of 10 November 1853 commenting on the manuscript of Whewell's *Essay*:

> [The doctrine of a plurality of worlds] aims formidable blows at the foundation of our faith in Christianity. The opposite doctrine aims blows scarcely less formidable at the foundation of our faith in natural religion. . . . If one or the

29. Stephen's letters to Whewell are preserved in Whewell's correspondence at Trinity College, Cambridge University. The identification mark for this letter is (Add. Ms. a. 216[129]).

30. Trinity College Whewell letters (Add. Ms. a. 216[130]).

31. Trinity College Whewell letters (Add. Ms. a. 216[131]).

other of the two must be abandoned, it is impossible not to see that [men will tend] . . . to disbelieve the Evangelists, rather than to disbelieve the Natural Theologians.[32]

Alfred Lord Tennyson (1809–1892)

As we have seen in chapter 10 (and as we shall see again in chapter 13), the greatest of the Victorian poets was very involved with ideas of extraterrestrial life. For this reason and because Whewell had been his tutor at Cambridge University, Tennyson read Whewell's book with both interest and distress as his son and biographer, Hallam Tennyson, recorded in his *Alfred Lord Tennyson: A Memoir* (New York: Macmillan, 1897):

> Whewell's *Plurality of Worlds* he [Lord Tennyson] . . . carefully studied. "It is to me anything but a satisfactory book. It is inconceivable that the whole Universe was merely created for us who live in this third-rate planet of a third-rate sun." (1:379)

Anthony Trollope (1815–1882)

While working as a post office official, Trollope began to compose novels, eventually publishing over fifty and attaining a reputation not far below that of Charles Dickens. The following exchange appears in Trollope's *Barchester Towers* (1857).

> "Are you a Whewellite or a Brewsterite, or a t'othermanite, Mrs. Bold?" said Charlotte, who knew a little about everything, and had read about a third of each of the books to which she alluded.
>
> "Oh," said Eleanor; "I have not read any of the books, but I feel sure that there is one man in the moon at least, if not more."
>
> "You don't believe in the pulpy gelatinous matter?" said Bertie.
>
> "I heard about that," said Eleanor; "and I really think it's almost wicked to talk in such a manner. How can we argue about God's power in the other stars from the laws which he has given for our rule in this one?"
>
> "How indeed!" said Bertie, "Why shouldn't there be a race of salamanders in Venus? and even if there be nothing but fish in Jupiter, why shouldn't the fish there be as wide awake as the men and women here?"

32. Trinity College Whewell letters (Add. Ms. a. 216[141]).

"That would be saying very little for them," said Charlotte. "I am for Dr. Whewell myself; for I do not think that men and women are worth being repeated in such countless worlds. There may be souls in other stars, but I doubt there having bodies attached to them. But come, Mrs. Bold, let us put our bonnets on and walk round the close. If we are to discuss sidereal questions, we shall do much better under the towers of the cathedral, than stuck in this narrow window."[33]

Samuel Warren (1807–1877)

After studying medicine and law for a period, Warren became a prominent lawyer and legal writer. He also was known for a number of novels. In his review of the books of Whewell and of Brewster in *Blackwood's Edinburgh Magazine* 75 (1854), he stated:

> If Dr. Whewell may be regarded as . . . a sort of Star-Smasher, his opponent is in very truth a Star-Peopler. Though he admits "there are some difficulties to be removed, and some additional analogies to be adduced, before the mind can admit the startling proposition that the Sun, Moon, and all the satellites, are inhabited spheres"—yet he believes that they are: that all the planets of their respective systems are so; as well as all these single stars, double stars, and nebulae, with all planets and satellites circling about them!—though "our *faltering reason utterly fails us!*" he owns, "when called on to believe that even the *Nebulae* must be surrendered to life and reason! wherever there is matter *there must* be life!" One can by this time almost pardon the excitement, the alarm rather, and anger, with which Sir David ruefully beheld Dr Whewell go forth on this exterminating expedition through Infinitude! It was like a father gazing on the ruthless slaughter of his offspring. Planet after planet, satellite after satellite, star after star, sun after sun, single suns and double suns, system after system, nebula after nebula, all disappeared before this sidereal Quixote! . . . This could be borne no longer; so thus Sir David pours forth the grief and indignation of the Soul Astronomic. (380)

33. In Trollope, *Barchester Towers and The Warden* (New York: Random House, 1950), 378–79.

PART FOUR

FROM 1860 TO 1915

Fig. 11. Illustration by Henrique Alvim Corréa for the 1906 French edition of H. G. Wells, *War of the Worlds*

TWELVE

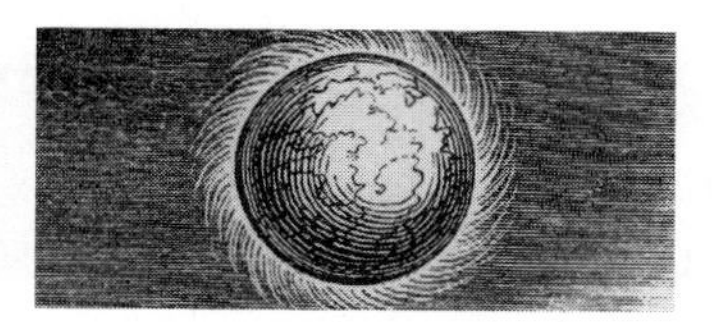

NEW APPROACHES TO AN ANCIENT QUESTION

This chapter focuses on four authors: Charles Darwin, Richard Proctor, Camille Flammarion, and Alfred Russel Wallace. One of its recurring themes concerns evolution and extraterrestrials. Significant as the publication in 1859 of Darwin's *Origin of Species* was, it is important to realize that evolutionary views broadly considered were already widespread before 1859. Earlier readings from Kant as well as consideration of the nebular hypothesis of the origin of the solar system show that in astronomy evolutionary ideas were present long before 1859. Similarly, geologists for more than a century had been coming to see the Earth as evolving in many aspects. Likewise in the life sciences: an array of pre-Darwinian biologists including Buffon, Lamarck, and Chambers had raised a variety of evolutionary issues. Darwin's impact derived not from his introducing the idea of biological evolution, but rather from his presenting biological evolution in a manner that forced major thinkers to take it seriously.

Before we turn directly to Darwin, mention should be made of one crucial development from the period around 1860 in physical science, including astronomy. This development was spectrum analysis. By means of the spectroscope, astronomers could determine in certain cases the chemical compositions of celestial bodies.

For example, glowing gases in the Sun or stars radiate bright-line spectra, the lines of which reveal the chemical composition of those glowing gases. Similarly, nonradiating, cooler gases such as those in a planetary atmosphere absorb rays of light at particular wavelengths producing dark-line spectra, the lines of which indicate the chemical composition of the gases. Important as spectroscopy was as an astronomical technique, it was also deeply significant in providing the clearest evidence ever obtained of the chemical homogeneity of the universe; it showed that the chemical elements making up the Earth are present throughout the universe, which implies that the laws governing terrestrial chemical reactions hold sovereignty throughout the universe. For these reasons, astronomers, possessing this powerful new technique and aware of its universal message, began to speak of the "New Astronomy" and to refer to themselves as astrochemists and astrophysicists. And not surprisingly, spectrum analysis intersected at many points with the extraterrestrial life debate.

Sir William Huggins (1824–1910) was the leading pioneer of astronomical spectroscopy. Among his achievements in this area was the discovery in the mid-1860s that various nebulous appearing objects exhibit bright-line spectra, indicating that they consist of glowing gases and thereby showing that they cannot be island universes. He also developed spectroscopic methods for measuring radial motion of celestial objects—that is, the motion of the object toward or away from the Earth. Some of Huggins's early papers were published in collaboration with William Allen Miller. Their "On the Spectra of Some of the Fixed Stars" (1865) contains the following endorsement of the idea of extraterrestrial life:

> It is remarkable that the elements most widely diffused through the host of stars are some of those most closely connected with the constitution of the living organisms of our globe, including hydrogen, sodium, magnesium, and iron. . . . On the whole we believe that the foregoing spectrum observations on the stars contribute something towards an experimental basis on which a conclusion hitherto but a pure speculation, may rest—viz. that at least the brighter stars are, like our sun, upholding and energising centres of systems of worlds adapted to be the abode of living beings.[1]

When editing this paper, which became a classic in the history of astronomical spectroscopy, for his *Scientific Papers*, Huggins added a footnote stating that al-

1. William Huggins, *The Scientific Papers of Sir William Huggins*, ed. Sir William Huggins and Lady Huggins (London: W. Wesley and Son, 1909), 60.

though Miller persisted in this belief, he (Huggins) had by 1866 freed himself "from the dogmatic fetters of my early theological education,"[2] which had motivated this statement. As the above passage suggests, the involvement of spectrum analysis with ideas of extraterrestrial life was quite substantial.[3] Nonetheless, because of the delicacy of many of the observations, astronomers in attempting to determine, for example, whether the Martian atmosphere contains oxygen and water vapor, at times found contradictory results or reported results that were later shown to be spurious.

CHARLES DARWIN (1809–1882)

What is the most important (and also most controversial) theory ever put forward in the life sciences? Who put it forward? And what bearing does this theory have on the extraterrestrial life debate? The answer to the first question is almost universally recognized to be the theory of evolution by means of natural selection. Nearly everyone knows the answer to the question "Who created this theory?" but in fact in most cases the answer given is incorrect. Darwin was not the discoverer of this theory but, as will be shown shortly, the codiscoverer—Alfred Russel Wallace having simultaneously and independently formulated the same theory. The third question has been answered many times by different authors, but the responses so often differ among themselves, partly because of the complexity of the issues, that some would say that this question even now awaits a definitive answer.

It was in the year 1858 that the public first learned of the theory of evolution by means of natural selection. The dramatic story of this first encounter is less widely known, but is well worth recounting in the present context. Charles Darwin in early 1858 was hard at work on a massive volume presenting the theory of evolution by means of natural selection, a theory he by then had been secretly developing for almost two decades. Darwin correctly believed it needed to be bolstered by all the evidence he could muster. An interesting aspect of the development of Darwin's theory is that in 1842 and then more fully in 1844 Darwin drafted (but withheld from publication) essays in which he presented his new theory of evolution

2. Huggins, *Papers*, 60.

3. See M. J. Crowe, *The Extraterrestrial Life Debate, 1750–1900: The Idea of a Plurality of Worlds from Kant to Lowell* (Cambridge: Cambridge University Press, 1986; new ed. Mineola, N.Y.: Dover Publications, 1999), 359–67.

by means of natural selection. In his 1844 essay the great biologist made a fascinating comment that implies that the idea of extraterrestrial life may have helped Darwin in his attempt to explain by purely naturalistic means the evolution of terrestrial life forms. In particular, he remarked: "It is derogatory that the Creator of countless Universes should have made by individual acts of His will the myriads of creeping parasites and worms, which since the earliest dawn of life have swarmed over the land."[4]

In devoting so much effort to the massive volume that he was preparing, Darwin was sensitive to both the controversial character and the importance of what he was undertaking. Thus he was intent on making his first presentation of his theory as perfect as possible. To that end, he had secretly shared some of his key ideas and arguments with a few co-workers. Among these were the eminent geologist Charles Lyell and the distinguished botanists Joseph Dalton Hooker and Asa Gray, all of whom realized the importance of Darwin's theory and the controversy it would bring forth. Lyell, for one, had noted a danger in this seemingly prudent strategy; he had warned Darwin of the danger that another naturalist might independently discover his theory and by publishing it first gain priority over Darwin.

That Lyell's warning was not groundless, Darwin learned to his great distress in June of 1858. Moreover, what he learned at that time caused Darwin to bring his theory before the public in a form that he realized was "*most* imperfect." The thunderbolt that forced this on Darwin struck on 18 June 1858 in the form of a letter from the Malay States and from the pen of a respected younger naturalist, Alfred Russel Wallace (1823–1913), who sent Darwin an essay asking him to forward it to Charles Lyell. What was so striking and troubling to Darwin is that after reading Wallace's essay, Darwin realized that Wallace in essentially total independence from Darwin had arrived at the same theory. Darwin, clearly devastated, immediately wrote to Lyell and after mentioning that Wallace had sent an essay commented:

> Your words have come true with a vengeance—that I should be forestalled. . . . I never saw a more striking coincidence; if Wallace had my MS. sketch written out in 1842, he could not have made a better short abstract! Even his terms now stand as heads of my chapters. Please return me the MS., which

4. Charles Darwin, *The Foundation of the Origin of Species: Two Essays Written in 1842 and 1844*, ed. Francis Darwin (Cambridge: Cambridge University Press, 1909), 254.

he does not say he wishes me to publish, but I shall of course at once write and offer to send to any journal. So all my originality, whatever it may amount to, will be smashed.[5]

Darwin's distress was greatly relieved when Hooker and Lyell developed a plan that they hoped would do justice to both these brilliant scientists. They arranged that at a 1 July 1858 meeting of the Linnean Society in London the theories of both men would be presented. In particular, the first document presented consisted of two extracts from Darwin's book manuscript, which extracts Darwin had shown to Hooker and Lyell long before Wallace's letter. The second Darwin document was an October 1857 letter Darwin had sent to Asa Gray. These were followed by Wallace's essay. This swift and judicious action led to a situation where experts have recognized that each man deserved a major share of the credit for formulating the theory of evolution by natural selection.

Given below are the materials Darwin published in the *Journal of the Proceedings of the Linnean Society* and that revealed to the world what is widely recognized as the paramount theory in the life sciences. One can assess the theory's importance from a remark made by the famous geneticist Theodosius Dobzhansky: "Nothing in biology makes sense except in the light of evolution."[6] Although no mention is made in these materials of issues regarding extraterrestrial life, the theory they present had such major and extensive implications regarding ideas of extraterrestrials that it deserves inclusion in this collection. Although Wallace's paper[7] is not included, we shall see later in this chapter that Wallace eventually entered the extraterrestrial life debate in a major way.

Charles Darwin, "On the Tendency of Species to Form Varieties; and on the Perpetuation of Varieties and Species by Natural Means of Selection," *Journal of the Proceedings of the Linnean Society of London: Zoology* 3 (1858): 46–50.

5. Charles Darwin, *Darwin: A Norton Critical Edition*, ed. Philip Appleman (New York: Norton, 1970), 57.

6. Theodosius Dobshansky, "Nothing in Biology Makes Sense except in the Light of Evolution," *American Biology Teacher* 35 (1973): 125–29.

7. Wallace's paper is available in many forms, for example, at http://www.life.umd.edu/emeritus/reveal/pbio/darwin/dw05.html. Accessed on 6 July 2007.

On the Tendency of Species to form Varieties; and on the Perpetuation of Varieties and Species by Natural Means of Selection. By CHARLES DARWIN, Esq., F.R.S., F.L.S., & F.G.S., and ALFRED WALLACE, Esq. Communicated by Sir CHARLES LYELL, F.R.S., F.L.S., and J.D. HOOKER, Esq., M.D., V.P.R.S., F.L.S., &c.

[Read July 1st, 1858.]

London, June 30th, 1858.

MY DEAR SIR,—The accompanying papers, which we have the honour of communicating to the Linnean Society, and which all related to the same subject, viz. the Laws which affect the Production of Varieties, Races, and Species, contain the results of the investigations of two indefatigable naturalists, Mr. Charles Darwin and Mr. Alfred Wallace.

The gentlemen having, independently and unknown to one another, conceived the same very ingenious theory to account for the appearance and perpetuation of varieties and of specific forms on our planet, and both fairly claim the merit of being original thinkers in this important line of inquiry; but neither of them having published his views, though Mr. Darwin has for many years past been repeatedly urged by us to do so, and both authors having now unreservedly placed their papers in our hands, we think it would best promote the interests of science that a selection from them should be laid before the Linnean Society.

Taken in the order of their dates, they consist of:—

1. Extracts from a MS. work on Species[8], by Mr. Darwin, which was sketched in 1839, and copied in 1844, when the copy was read by Dr. Hooker, and its contents afterwards communicated to Sir Charles Lyell. The first Part is devoted to "The Variation of Organic Beings under Domestication and in their Natural State;" and the second chapter of that Part, from which we propose to read to the Society the extracts referred

8. This MS. work was never intended for publication, and therefore was not written with care.—C. D. 1858.

to, is headed, "On the Variation of Organic Beings in a state of Nature; on the Natural Means of Selection; on the Comparison of Domestic Races and true Species."

2. An abstract of a private letter addressed to Professor Asa Gray, of Boston, U.S., in October 1857, by Mr. Darwin, in which he repeats his views, and which shows that these remained unaltered from 1839 to 1857.

3. An Essay by Mr. Wallace, entitled "On the Tendency of Varieties to depart indefinitely from the Original Type." This was written at Ternate in February 1858, for the perusal of his friend and correspondent Mr. Darwin, and sent to him with the expressed wish that it should be forwarded to Sir Charles Lyell, if Mr. Darwin thought it sufficiently novel and interesting. So highly did Mr. Darwin appreciate the value of the views therein set forth, that he proposed, in a letter to Sir Charles Lyell, to obtain Mr. Wallace's consent to allow the Essay to be published as soon as possible. Of this step we highly approved, provided Mr. Darwin did not withhold from the public, as he was strongly inclined to do (in favour of Mr. Wallace), the memoir which he had himself written on the same subject, and which, as before stated, one of us had perused in 1844, and the contents of which we had both of us been privy to for many years. On representing this to Mr. Darwin, he gave us permission to make what use we thought proper of his memoir, &c.; and in adopting our present course, of presenting it to the Linnean Society, we have explained to him that we are not solely considering the relative claims to priority of himself and his friend, but the interests of science generally; for we feel it to be desirable that views founded on a wide deduction from facts and matured by years of reflection, should constitute at once a goal from which others may start, and that, while the scientific world is waiting for the appearance of Mr. Darwin's complete work, some of the leading results of his labours, as well as those of his able correspondent, should together be laid before the public.

We have the honour to be yours very obediently,

CHARLES LYELL.
JOS. D. HOOKER.

J. J. Bennett, Esq.,
Secretary of the Linnean Society.

I. *Extract from an unpublished Work on Species, by* C. DARWIN, Esq., *consisting of a portion of a Chapter entitled, "On the Variation of Organic Beings in a state of Nature; on the Natural Means of Selection; on the Comparison of Domestic Races and true Species."*

De Candolle, in an eloquent passage, has declared that all nature is at war, one organism with another, or with external nature. Seeing the contented face of nature, this may at first well be doubted; but reflection will inevitably prove it to be true. The war, however, is not constant, but recurrent in a slight degree at short periods, and more severely at occasional more distant periods; and hence its effects are easily overlooked. It is the doctrine of Malthus applied in most cases with tenfold force. As in every climate there are seasons, for each of its inhabitants, of greater and less abundance, so all annually breed; and the moral restraint which in some small degree checks the increase of mankind is entirely lost. Even slow-breeding mankind has doubled in twenty-five years; and if he could increase his food with greater ease, he would double in less time. But for animals without artificial means, the amount of food for each species must, *on an average*, be constant, whereas the increase of all organisms tends to be geometrical, and in a vast majority of cases at an enormous ratio. Suppose in a certain spot there are eight pairs of birds, and that *only* four pairs of them annually (including double hatches) rear only four young, and that these go on rearing their young at the same rate, then at the end of seven years (a short life, excluding violent deaths, for any bird) there will be 2048 birds, instead of the original sixteen. As this increase is quite impossible, we must conclude either that birds do not rear nearly half their young, or that the average life of a bird is, from accident, not nearly seven years. Both checks probably concur. The same kind of calculation applied to all plants and animals affords results more or less striking, but in very few instances more striking than in man.

Many practical illustrations of this rapid tendency to increase are on record, among which, during peculiar seasons, are the extraordinary numbers of certain animals; for instance, during the years 1826 to 1828, in La Plata, when from drought some millions of cattle perished, the whole country actually swarmed with mice. Now I think it cannot be doubted that during the breeding-season all the mice (with the exception of a few males or females in excess) ordinarily pair, and therefore that this astounding increase during three years must be attributed to a greater number than usual surviving the first year, and then breeding, and so on till the third year, when their numbers were brought down to their usual limits on the return of wet weather. Where man has introduced plants and animals into a new and favourable country, there are many accounts in how surprisingly few years the whole country has become stocked with them. This increase would necessarily stop as soon as the country was fully stocked; and yet we have every reason to believe, from what is known of wild animals, that *all* would pair in the spring. In the majority of cases it is most difficult to imagine where the checks fall—though generally, no doubt, on the seeds, eggs, and young; but when we remember how impossible, even in mankind (so much better known than any other animal), it is to infer from repeated casual observations what the average duration of life is, or to discover the different percentage of deaths to births in different countries, we ought to feel no surprise at our being unable to discover where the check falls in any animal or plant. It should always be remembered, that in most cases the checks are recurrent yearly in a small, regular degree, and in an extreme degree during unusually cold, hot, dry, or wet years, according to the constitution of the being in question. Lighten any check in the least degree, and the geometrical powers of increase in every organism will almost instantly increase the average number of the favoured species. Nature may be compared to a surface on which rest ten thousand sharp wedges touching each other and driven inwards by incessant blows. Fully to realize these views much reflection is requisite. Malthus on man should be studied; and all such cases as those of the mice in La Plata, of the cattle and horses when first turned out in South America, of the birds by our calculation, &c., should be well considered. Reflect on the enormous multiplying power *inherent and annually in action* in all animals; reflect on the countless seeds scattered by a hundred ingenious contrivances, year after year, over the whole face of the land; and yet we have every reason to suppose that the average percentage of each of the inhabitants of a country usually remains constant. Finally, let it be borne in mind that this average number of individuals (the external conditions remaining the same) in each country is kept up by recurrent struggles against other species or against external nature (as on the borders of the Arctic regions, where the cold checks life), and that ordinarily each individual of every species holds its place, either by its own struggle and capacity of acquiring nourishment in some period of its life, from the egg upwards; or by the struggle of its parents (in short-lived organisms, when the main check occurs at longer intervals) with other individuals of the *same* or *different* species.

But let the external conditions of a country alter. If in a small degree, the relative proportions of the inhabitants will in most cases simply be slightly changed; but let the number of inhabitants be small, as on an island, and free access to it from other countries be circumscribed, and let the change of conditions continue progressing (forming new stations), in such a case the original inhabitants must cease to be as perfectly adapted to the changed conditions as they were originally. It has been shown in a former part of this work, that such changes of external conditions would, from their acting on the reproductive system, probably cause the organization of those beings which were most affected to become, as under domestication, plastic. Now, can it be doubted, from the struggle each individual has to obtain subsistence, that any minute variation in structure, habits, or instincts, adapting that individual better to the new conditions, would tell upon its vigour and health? In the struggle it would have a better *chance* of surviving; and those of its offspring which inherited the variation, be it ever so slight, would also have a better *chance*. Yearly more are bred than can survive; the smallest grain in the balance, in the long run, must tell on which death shall fall, and which shall survive. Let this work of selection on the one hand, and death on the other, go on for a thousand generations, who will pretend to affirm that it would produce no effect, when we remember what, in a few year, Bakewell effected in cattle, and Western in sheep, by this identical principle of selection?

To give an imaginary example from changes in progress on an island:—let the organization of a canine animal which preyed chiefly on rabbits, but sometimes on hares, become slightly plastic; let these same changes cause the number of rabbits very slowly to decrease, and the number of hares to increase; the effect of this would be that the fox or dog would be driven to try to catch more hares: his organization, however, being slightly plastic, those individuals with the lightest forms, longest limbs, and best eyesight, let the difference be ever so small, would be slightly favoured, and would tend to live longer, and to survive during that time of the year when food was scarcest; they would also rear more young, which would tend to inherit these slight peculiarities. The less fleet ones would be rigidly destroyed. I can see no more reason to doubt that these causes in a thousand generations would produce a marked effect, and adapt the form of the fox or dog to the catching of hares instead of rabbits, than that greyhounds can be improved by selection and careful breeding. So would it be with plants under similar circumstances. If the number of individuals of a species with plumed seeds could be increased by greater powers of dissemination within its own area (that is, if the check to increase fell chiefly on the seeds), those seeds which were provided with ever so little more down, would in the long run be most disseminated; hence a greater number of seeds thus formed would germinate, and would tend to produce plants inheriting the slightly better-adapted down.[9]

Besides this natural means of selection, by which those individuals are preserved, whether in their egg, or larval, or

9. I can see no more difficulty in this, than in the planter improving his varieties of the cotton plant.—C. D. 1858.

mature state, which are best adapted to the place they fill in nature, there is a second agency at work in most unisexual animals, tending to produce the same effect, namely, the struggle of the males for the females. These struggles are generally decided by the law of battle, but in the case of birds, apparently, by the charms of their song, by their beauty or their power of courtship, as in the dancing rock-thrush of Guiana. The most vigorous and healthy males, implying perfect adaptation, must generally gain the victory in their contests. This kind of selection, however, is less rigorous than the other; it does not require the death of the less successful, but gives to them fewer descendants. The struggle falls, moreover, at a time of year when food is generally abundant, and perhaps the effect chiefly produced would be the modification of the secondary sexual characters, which are not related to the power of obtaining food, or to defence from enemies, but to fighting with or rivalling other males. The result of this struggle amongst the males may be compared in some respects to that produced by those agriculturists who pay less attention to the careful selection of all their young animals, and more to the occasional use of a choice mate.

II. *Abstract of a Letter from* C. DARWIN, Esq., to Prof. ASA GRAY, *Boston, U.S., dated Down, September 5th,* 1857.

1. It is wonderful what the principle of selection by man, that is the picking out of individuals with any desired quality, and breeding from them, and again picking out, can do. Even breeders have been astounded at their own results. They can act on differences inappreciable to an uneducated eye. Selection has been *methodically* followed in *Europe* for only the last half century; but it was occasionally, and even in some degree methodically, followed in the most ancient times. There must have been also a kind of unconscious selection from a remote period, namely in the preservation of the individual animals (without any thought of their offspring) most useful to each race of man in his particular circumstances. The "roguing," as nurserymen call the destroying of varieties which depart from

their type, is a kind of selection. I am convinced that intentional and occasional selection has been the main agent in the production of our domestic races; but however this may be, its great power of modification has been indisputably shown in later times. Selection acts only by the accumulation of slight or greater variations, caused by external conditions, or by the mere fact that in generation the child is not absolutely similar to its parent. Man, by this power of accumulating variations, adapts living beings to his wants—may be said to make the wool of one sheep good for carpets, of another for cloth, &c.

2. Now suppose there were a being who did not judge by mere external appearances, but who could study the whole internal organization, who was never capricious, and should go on selecting for one object during millions of generations; who will say what he might not effect? In nature we have some slight variation occasionally in all parts; and I think it can be shown that changed conditions of existence is the main cause of the child not exactly resembling its parents; and in nature geology shows us what changes have taken place, and are taking place. We have almost unlimited time; no one but a practical geologist can fully appreciate this. Think of the Glacial period, during the whole of which the same species at least of shells have existed; there must have been during this period millions on millions of generations.

3. I think it can be shown that there is such an unerring power at work in *Natural Selection* (the title of my book), which selects exclusively for the good of each organic being. The elder De Candolle, W. Herbert, and Lyell have written excellently on the struggle for life; but even they have not written strongly enough. Reflect that every being (even the elephant) breeds at such a rate, that in a few years, or at most a few centuries, the surface of the earth would not hold the progeny of one pair. I have found it hard constantly to bear in mind that the increase of every single species is checked during some part of its life, or during some shortly recurrent generation. Only a few of those annually born can live to propagate their kind. What a trifling difference must often determine which shall survive, and which perish!

4. Now take the case of a country undergoing some change. This will tend to cause some of its inhabitants to vary slightly—not but that I believe most beings vary at all times enough for selection to act on them. Some of its inhabitants will be exterminated; and the remainder will be exposed to the mutual action of a different set of inhabitants, which I believe to be far more important to the life of each being than mere climate. Considering the infinitely various methods which living beings follow to obtain food by struggling with other organisms, to escape danger at various times of life, to have their eggs or seeds disseminated, &c. &c., I cannot doubt that during millions of generations individuals of a species will be occasionally born with some slight variation, profitable to some part of their economy. Such individuals will have a better chance of surviving, and of propagating their new and slightly different structure; and the modification may be slowly increased by the accumulative action of natural selection to any profitable extent. The variety thus formed will either coexist with, or, more commonly, will exterminate its parent form. An organic being, like the woodpecker or misseltoe, may thus come to be adapted to a score of contingencies—natural selection accumulating those slight variations in all parts of its structure, which are in any way useful to it during any part of its life.

5. Multiform difficulties will occur to every one, with respect to this theory. Many can, I think, be satisfactorily answered. *Natura non facit saltum* answers some of the most obvious. The slowness of the change, and only a very few individuals undergoing change at any one time, answers others. The extreme imperfection of our geological records answers others.

6. Another principle, which may be called the principle of divergence, plays, I believe, an important part in the origin of species. The same spot will support more life if occupied by very diverse forms. We see this in the many generic forms in a square yard of turf, and in the plants or insects on any little uniform islet, belonging almost invariably to as many genera and families as species. We can understand the meaning of this fact amongst the higher animals, whose habits we

understand. We know that it has been experimentally shown that a plot of land will yield a greater weight if sown with several species and genera of grasses, than if sown with only two or three species. Now, every organic being, by propagating so rapidly, may be said to be striving its utmost to increase in numbers. So it will be with the offspring of any species after it has become diversified into varieties, or subspecies, or true species. And it follows, I think, from the foregoing facts, that the varying offspring of each species will try (only few will succeed) to seize on as many and as diverse places in the economy of nature as possible. Each new variety or species, when formed, will generally take the place of, and thus exterminate its less well-fitted parent. This I believe to be the origin of the classification and affinities of organic beings at all times; for organic beings always *seem* to branch and sub-branch like the limbs of a tree from a common trunk, the flourishing and diverging twigs destroying the less vigorous — the dead and lost branches rudely representing extinct genera and families.

This sketch is *most* imperfect; but in so short a space I cannot make it better. Your imagination must fill up very wide blanks.

C. DARWIN

Despite being bereft of any mention of extraterrestrials, these publications by Darwin presented scientists with a powerful theory with many ramifications for the extraterrestrial life debate. Moreover, Darwinian theory gave major support to a naturalistic understanding of the universe, which supported many advocates of extraterrestrials. On the other hand, Darwinism challenged the teleological approach to life, thereby creating problems for those who used teleological arguments for extraterrestrials.

Working out the ramifications of Darwinism for life beyond the Earth has involved numerous scientists over many years; in fact, major issues and controversies remain. One of the most influential positions on these issues was presented (as we shall soon see) by A. R. Wallace himself. Before leaving Darwin's writings, however, it will be useful to ponder four crucial questions about the Darwin-Wallace theory of evolution by natural selection.

1. Is it correct that Darwin and Wallace had only one mechanism for evolution, in particular, evolution by natural selection?
2. Is it correct that the direction taken by evolution is always toward improvement in species?
3. Do humans continue to evolve?
4. It is widely assumed that a major test of good scientific theories is that they are predictive. Is this a quality possessed by the Darwin-Wallace theory?

Darwin showed restraint in giving his famous book the modest title *The Origin of Species*, rather than such a title as *The Origin of Life*. Nonetheless, there is a record of Darwin's views on the origin of life, although it does not come from his published writings. It appears rather in a 1 February 1871 letter Darwin wrote to Joseph Dalton Hooker.

> It is often said that all the conditions for the first production of a living organism are now present, which could ever have been present. But if (and oh! what a big if!) we could conceive in some warm little pond, with all sorts of ammonia and phosphoric salts, lights, heat, electricity, &c. present, that a proteine compound was chemically formed ready to undergo still more complex changes, at the present day such matter would be instantly devoured or absorbed, which would not have been the case before living creatures were formed.[10]

And Darwin's "warm little pond" remains to this day one of the competitors in the continuing discussion of the origin of life, a topic that has great bearing on the extraterrestrial life debate.

RICHARD ANTHONY PROCTOR (1837–1888)

When Richard Proctor died in 1888, the London *Times* praised him for having "probably done more than any other man during the present century to promote an interest among the ordinary reading public in scientific subjects."[11] From across

10. Francis Darwin (ed.), *The Life and Letters of Charles Darwin; Including an Autobiographical Chapter* (London: John Murray, 1888), 3:18.

11. *Times* (London), Sept. 14, 1888, 5.

the Atlantic, the American astronomer C. A. Young agreed: "As an expounder and popularizer of science he stands, I think, unrivaled in English literature."[12] This claim is surely justified, at least if its scope is limited to popularization of astronomy in the English-speaking world. In the twenty-five years between Proctor's first publication and his sudden death in 1888, he authored fifty-seven books, mainly on astronomy, most of these consisting of essays previously published in periodicals. Yet so prolific was Proctor that these republished essays represented, as he once stated, less than one-fourth of his total output.[13] Proctor, moreover, gave lecture tours in Britain, the United States, Canada, Australia, and New Zealand. After marrying an American widow and taking up residence in the United States around 1881, he also edited, first from St. Joseph, Missouri, then from Orange Lake, Florida, the London-based scientific periodical *Knowledge*, which he had founded in 1881. Proctor was not only prolific; he was also the most widely read author in the pluralist debate in Britain and America during the period from 1870 to 1890.

A graduate of St. John's College, Cambridge, Proctor, forced by the financial need resulting from losing his inheritance in a bank crash, attempted in the 1860s to carve out a career as a popularizer of science. His first book, *Saturn and Its System* (1865), was a financial failure, but won praise from the astronomical community. Realizing that if he were to succeed in reaching the public, he would need more attractive fare, he centered one of his next books, *Other Worlds than Ours* (1870), on ideas of extraterrestrial life. His motivation for this is clear from a remark he made in 1878 regarding the reading public: "The interest with which astronomy is studied by many who care little or nothing for other sciences is due chiefly to the thoughts which the celestial bodies suggest respecting life in other worlds than ours."[14] This approach succeeded so well that his *Other Worlds than Ours* remained in print for nearly forty years and probably outsold any of his other publications. Moreover, having discovered the public's appreciation of pluralist publications, Proctor included discussions of extraterrestrial life ideas in a dozen later books. By 1873, the situation had dramatically changed; in that year, an American author stated: "Ten years ago, the name of Richard Anthony Proctor was absolutely unknown; five years later, it was familiar

12. As quoted in Charlotte R. Willard, "Richard A. Proctor," *Popular Astronomy* 1 (1894): 319.

13. R. A. Proctor, *The Borderland of Science* (London: Wyman & Sons, 1882), v.

14. R. A. Proctor, "Other Worlds and Other Universes," in *Myths and Marvels of Astronomy*, new ed. (London: Chatto & Windus, 1880), 135; see also 137.

in scientific circles in London; and to-day it is familiar as household words to every educated man in England, and to many thousands in this country."[15]

In formulating his position regarding extraterrestrial life in *Other Worlds than Ours*, Proctor distanced himself from the anti-pluralist position that William Whewell had advocated. For example, in his discussion of Jupiter, Proctor remarked: "Surely no astronomer worthy the name can regard this grand orb as the cinder-centered globe of watery matter so contemptuously dealt with by one who, be it remembered thankfully, was not an astronomer."[16] Nonetheless, the position Proctor advocated regarding Jupiter, and Saturn as well, was far from the traditional pluralist claim for their habitability. In particular, Proctor denied that Jupiter is "at present a fit abode for living creatures," suggesting instead that it is "in a sense a sun . . . a source of heat" serving its four satellites on which "life — even such forms of life as we are familiar with — may still exist."[17] Not content with making Jupiter a quasi-sun, he urged that Jupiter "must be intended to be one day the abode of noble races."[18]

This point is doubly important. First, it suggests that Proctor, more than most earlier pluralists, inclined toward viewing the universe in evolutionary terms; he saw the planets as entities that have evolved over the course of the history of the universe. Second, and more significantly, the gap that Proctor had opened up in his 1870 book between himself and mainline pluralism, which typically included arguments for the habitability of all the planets, increasingly widened during the next five years to the point that the twenty-nine reprintings of his *Other Worlds than Ours* that appeared between 1870 and 1909 presented views that Proctor had substantially abandoned by 1875. In fact, by century's end, and due to some extent to Proctor's influence, claims for life in our solar system were quite frequently restricted to the Earth and to Mars, which planet became the last best hope of the pluralists for detecting extraterrestrial life in our solar system. And Mars itself became at century's end the subject of an international controversy.

Proctor expressed his changing views regarding extraterrestrial life in a series of essays that appeared between 1870 and 1875. In one such essay, titled "A Whewellite Essay on the Planet Mars" Proctor argued that "Neither animal nor vegetable forms of life known to us could exist on Mars" and that if living beings exist

15. John Fraser as quoted in "Proctor the Astronomer," *English Mechanic* 18 (Dec. 12, 1873): 322.

16. Richard Proctor, *Other Worlds than Ours* (London: Longmans, Green, 1870), 4.

17. Proctor, *Other Worlds*, 141.

18. Proctor, *Other Worlds*, 145.

on Mars, they "must differ so remarkably from what is known on earth, that to reasoning beings on Mars the idea of life on our earth must appear wild and fanciful."[19] Readers shocked by the reversal evident in this claim must have been even further taken aback by essays that appeared in Proctor's *Our Place among Infinities* and *Science Byways,* both published in 1875. A selection from Proctor's most important essay from the former book constitutes the next reading.

After Proctor's sudden death in 1888, efforts were made to provide a memorial for him. There was an ironic twist to this. Proctor, whose views had so changed that in 1888 he had asserted that "we must regard it as at the very least highly probable that on Mars . . . few of the higher forms of life were (or have been) developed,"[20] was made the focus in 1896 of efforts both to establish an international Proctor Memorial Association and to erect in California a telescope in his honor. The proposed telescope of one hundred foot aperture (!) was to be so powerful that earthlings, as the *New York Times* headline proclaimed, "Will See Men on Mars."[21]

Richard Proctor, *Our Place among Infinities,* 2nd ed. (London: Henry S. King & Co, 1876), 45–51, 52–70.

A NEW THEORY OF LIFE IN OTHER WORLDS

Two opposite views have been entertained respecting life in other worlds. One is the theory which Brewster somewhat strongly described as the creed of the philosopher and the hope of the Christian, that nearly all the orbs which people space are the abode of life. Brewster, Chalmers, Dick, and a host of other writers, have adopted and enforced this view, Brewster going so far as to maintain the probability that life may exist upon the moon, dead though her surface seems, or beneath the glowing photosphere of the sun. But even where so extreme an opinion has not been entertained, the believers in the theory of a plurality of worlds have maintained that all the celestial orbs have been created to be, and are at

19. Proctor, *Borderland of Science,* 156–57.
20. R. A. Proctor, "Varied Life in Other Worlds," *Open Court* 1 (1888): 599.
21. Anonymous, "Will See Men on Mars," *New York Times,* June 21, 1896, 22.

this present time, the abodes of life, or else minister to the wants of creatures living in other orbs. It is worthy of notice that this view has been entertained even by astronomers, who, like the Herschels, have devoted their lives to the scientific study of the heavens. So completely has the theory been identified, as it were, with modern astronomy, that we find the astronomer passing from a statement respecting some observed fact about a planet, to the consideration of the bearing of the fact on the requirements of living creatures on the planet's surface, without expressing any doubts whatever as to the existence of such creatures. For example, Sir John Herschel, writing about the rings of Saturn, after discussing Lardner's supposed demonstration that the eclipses caused by the rings would last but for a short time;[22] says, 'This will not prevent, however, some considerable regions of Saturn from suffering very long total intervention of the solar beams, affording to our ideas but an inhospitable asylum to animated beings, ill compensated by the feeble light of the satellites; but we shall do wrong to judge of the fitness or unfitness of their condition from what we see around us, when perhaps the very combinations which convey to our minds only images of horror may be, in reality, theatres of the most striking and glorious displays of beneficent contrivance.' And many other such cases might be cited.

Before passing to the opposite view of life in other worlds, a view commonly associated with the name of the late Dr. Whewell, I shall venture to quote a few passages from his Bridgewater Treatise on Astronomy and General Physics, in which he writes very much like a supporter of the theory he subsequently opposed in his 'Plurality of Worlds.' Thus speaking of the satellites in the solar system, he says,—'There is one fact which immediately arrests our attention; the number

---

22. This is disapproved, and the justice of Herschel's views demonstrated in chapter vii. of my treatise on Saturn, in which work I give a table of the climatic relations in Saturn (for I also once adopted the theory criticized above) the time and place of sunrise and sunset in Saturnian latitudes in Saturnian Spring, Summer, Autumn, and Winter, and so on. Labour wasted, I fear, except as practice in Geometrical Astronomy.

of these attendant bodies appears to increase as we proceed to planets farther and farther from the sun. Such, at least, is the general rule. Mercury and Venus, the planets near the sun, have no attendants; the earth has but one. Mars, indeed, who is still further removed, has none, nor have the minor planets, so that the rule is only approximately verified. But Jupiter, who is at five times the earth's distance, has four satellites; and Saturn, who is again at a distance nearly twice as great, has seven' (now eight) 'besides that most extraordinary phenomenon, his ring, which for purposes of illumination is equivalent to many thousand satellites. Of Uranus it is difficult to speak, for his great distance renders it almost impossible to observe the smaller circumstances of his condition. It does not appear at all probable that he has a ring like Saturn; but he has at least four satellites which are visible to us, at the enormous distance of 900 millions of miles, and I believe that the astronomer will hardly deny that he may possibly have thousands of smaller ones circulating about him. But leaving conjecture, and taking only the ascertained cases of Venus, the Earth, Jupiter, and Saturn, we conceive that a person of common understanding will be strongly impressed with the persuasion that the satellites are placed in the system *with a view to compensate for the diminished light of the sun at greater distances.*' Then he presently adds, after considering the exceptional case of Mars,—'No one familiar with such contemplations will, by one anomaly, be driven from the persuasion that the end which *the arrangements of the satellites seem suited to answer is really one of the ends of their creation.*' Here is the theory of life in other worlds definitely adopted, and moreover presented in company with the extremest form of the teleological argument, and that, too, by Whewell, whose name afterwards became associated with the extremest development of the doctrine of the paucity of worlds!

The Whewellite theory is tolerably well known, though certainly it is not held in very great favour. For my own part, I used, at one time, to think that Whewell only advanced it in jest; but now (perhaps because my own researches and study have led me to regard the Brewsterian theory as untenable) I recognise in Whewell's later views the result of longer and more careful study than he had given to the subject, when (nearly a quarter of a century earlier) he wrote his Bridgewater Treatise.

Whatever opinion we form as to the theory advanced in the 'Plurality of Worlds,' we must admit that Whewell did good service to science in breaking the chains of old-fashioned ideas, and inaugurating freedom of discussion. The stock writers on astronomy had been repeating so often the imperfect analogies on which astronomers had earlier insisted, that the suggestions based on such analogies had come to be regarded as so many scientific facts. The Earth is a planet, and Mars is a planet, therefore what we know about the Earth may be inferred respecting Mars, no account being taken of the known difference in the condition of the two planets: accordingly, not only are the white spots at the Martian poles to be regarded as snow-covered regions, and the blue markings on his surface as seas, but we are to infer a similarity of climatic conditions and other habitudes, without entering into any close consideration of the probable extent of the planet's atmosphere, the heat received from the Sun by Mars, and a variety of other relations respecting which we are at least as well informed as we are respecting the analogies in question. Jupiter, again, is a planet, and though he is so much larger than the Earth that we might be disposed at the outset to regard him as a body of another order, we must be so guided by analogies (which, after all, may be imaginary) as to consider that his size only renders him so much the nobler an abode for such life as we are familiar with: and instead of being struck by the fact that Jupiter, unlike Mars, shows no polar snowcaps, we are to direct our attention to his belts, and to regard them as cloud-belts analogous to the tropical cloud-zone of the Earth. Nor are we to enquire too closely whether the aspect of his equatorial belt, to say nothing of his other belts, corresponds in any degree with that which the cloud-zone of our Earth would present to observers on another planet:—Let it suffice to note a few analogies, as thus—"The Earth is a planet, Jupiter is a planet; the Earth rotates and therefore has a day, Jupiter rotates and has a day; the Earth has a year, Jupiter has a year; the Earth has clouds, Jupiter has clouds;

the Earth has a moon, Jupiter has four moons: this done, every other consideration may be conveniently overlooked, and we may proceed to descant on the wonderful extent and dignity of this distant world, with as little question of its being inhabited as though we had seen with our own eyes the creatures which exist upon the planet's surface. So with Saturn, and the rest."

Whewell broke through all these old-fashioned methods. He dealt with the several planets on the true scientific principle long since enunciated by Descartes, taking nothing for granted that had not been proved. He showed how unlike the conditions prevailing in the other planets must be to those existing on the Earth, and without pretending to demonstrate absolutely that none of the higher forms of life can exist on certain planets, he showed that at any rate the probabilities are in favour of that hypothesis. Passing on to the stars, he did good service by showing how much had been taken for granted by astronomers in their assumptions respecting these orbs; nor is the value of his work, in this field, by any means diminished, by the circumstance that during recent years evidence which was wanting when Whewell wrote has been obtained, and the stars have been shown demonstratively to be suns. And lastly, he dealt in an independent, and therefore instructive manner, with the star-cloudlets or nebulae, giving many strong reasons for doubting the views which were at that time repeated in every text-book of astronomy.

The conclusions to which Whewell was led were (1) that no sufficient reason exists for believing in other worlds than ours; and (2) if the other planets are inhabited, it can only be, in all probability, by creatures belonging to the lowest orders of animated existence.

In the remainder of this long paragraph, Proctor contrasted Whewell's "philosophic, calm, and dispassionate force of reasoning" in the debate over extraterrestrials with Brewster's "impassioned outbursts," which were "almost ludicrous."

Here, however, are two theories — opposed to each other, and not admitting of being reconciled. If we are to make a selection between them, to which shall we turn in preference? The balance of evidence is on the whole in favour of Whewell's, (so at least the matter presents itself to me after careful and long-continued study); but certainly Brewster's is the theory which commends itself most favourably to the mind which would believe that God "hath done all things well," and that nothing that He has made was made in vain. Even those who, like myself, are indisposed to admit that the ways and works of God are to be judged by our conceptions of the fitness of things, (thought we may be altogether certain that all things are made in wisdom and fitness), would prefer to accept the Brewsterian theory, if decision were to be made between the two. For, what amount of evidence could reconcile us to the belief (even though it forced this belief upon us) that our Earth alone of all the countless orbs which people space, is the abode of reasoning creatures, capable of recognising the glories of the universe, and of lauding the Creator of those wonders and of their own selves? Nevertheless we must be guided in these matters by evidence, not by sentiment—by facts, not by our feelings. It is well, therefore, to note that the decision does not lie between the two theories which have just been dealt with. Another theory, holding a position intermediate between those two, and combining in my judgment the evidence which favours one theory with the fitness characterising the other, remains yet to be presented. The last essay was intended to prepare the way for this theory.

I propose to take, as the basis of the new theory of life in other worlds, the analogy which has commonly been regarded as affording the strongest evidence in favour of the Brewsterian theory,—only I shall take a more extended view of the subject than has been customary.

Before introducing that Brewsterian argument, I may remark that the mere fact that our Earth is an inhabited world is not in itself sufficient even to render probable the theory that there is life in other worlds than ours. An equally strong argument might be derived against that theory from the study of our Moon,—the only other planet of which we have obtained reliable information,—for few can suppose that the

Moon is fit to be the abode of life. Since then of the two planets we can examine, one — the Earth — is inhabited, while the other — the Moon — is probably not inhabited, the only evidence we have is almost equally divided between the Whewellite and Brewsterian theories, whatever balance remains in favour of the latter being too slight to afford any sufficient basis for a conclusion.

But while this reasoning is just, as applied to the mere fact that the Earth is inhabited, it is by no means capable of overthrowing the evidence which is derived from the manner in which life exists on the Earth. When we consider the various conditions under which life is found to prevail, that no difference of climatic relations or of elevation, of land or of air or of water, of soil in hand, of freshness or saltness in water, of density in air, appears (so far as our researches have extended) to render life impossible, we are compelled to infer that the power of supporting life is a quality which has an exceedingly wide range in nature. I refrain, it will be noticed from using here the usual expression, and saying, as of yore, that 'the great end and aim of all the workings of nature is to afford scope and room for the support of life,' because this mode of speaking may be misunderstood. We can see what nature actually does, and we may infer, if we so please, that such or such is the end and aim of the God of nature; nevertheless we must remember that the evidence we have belongs to the former relation not to the latter. I am careful to dwell on this point because the longer I study such matters the more clearly do I recognise the necessity of most studiously limiting our statements to that which the evidence before us really establishes.

Passing beyond the evidence which the Earth at present affords, we find that during many ages the Earth has presented a similar scene. 'Geology,' I wrote four years ago, 'teaches us of days when this Earth was peopled with strange creatures such as now are not found upon its surface. We turn our thoughts to the epochs when these monsters throve and multiplied, and picture to ourselves the appearance which our Earth then presented. Strange forms of vegetation clothe the scene which the mind's eye dwells upon. The air is heavily laden with moisture to nourish the abundant flora; hideous reptiles crawl over their slimy domain, battling with each other, or with the denizens of the forest; huge bat-like creatures sweep through the dusky twilight which constituted the primaeval day; weird monsters pursue their prey amid the depths of ocean: and we forget, as we dwell upon the strange forms which existed in those long past ages, that the scene now presented by the Earth is no less wonderful, and that the records of our time may, perhaps, seem one day as perplexing as we now find those of the geological eras.' In the past, then, as in the present, this Earth was inhabited by countless millions of living creatures, and during the enormous period which has elapsed since life first appeared on the surface of the Earth, myriads if not millions, of orders of living creatures have appeared, have lived the life appointed to their order, and have vanished, or exist only under modified forms. As each individual has had its period of life, so also has each race, and we may say with the poet (noting always that the personification of nature is but a poetical idea, and does not present any real substantive truth),—

> Are God and Nature then at strife,
> That Nature lends such evil dreams?
> So careful of the type she seems
> So careless of the single life
> * * * * * * * *
> 'So careful of the type?' but no,
> From scarped cliff and quarried stone
> She cries, 'A thousand types are gone;
> I care for nothing, all shall go.'[23]

Abundant life, in ever-varying forms, and under all-various conditions, continuing age after age during hundreds of thousands of years, such is what our Earth presents to us when we turn our thoughts to its past history. And looking forward,

23. [These lines are from Alfred Lord Tennyson's famous poem "In Memoriam." MJC]

a similar scene is presented to our contemplation. For many a long century, probably for hundreds of thousands of years, life will continue on the Earth, unless some catastrophe (the occurrence of which we have as yet no reason to anticipate) should destroy life suddenly from off her surface.

So viewing this Earth, we seem to find forced upon us the belief that the support of life is the object for which the Earth was created, and thus we are led to regard the other orbs which, like her, circle around a central Sun, as intended to be the abode of life. The only object which, so far as we can see, the Earth has fulfilled during an indefinitely long period has been to present a field, so to speak, for the support of life, nor can we recognise any other purpose which she will fulfil in the future. If we admit this, and if we also believe that God made nothing without some purpose, of course we have no choice but to admit that the purpose with which the Earth was made was the support of life. And reasoning from Analogy, we infer that the other planets, as well those of our own system as those which we believe to exist, 'wheeling in perpetual round,' as attendants upon other Suns, were similarly created to be the abode of life.[24]

But, before we infer from the strength of this reasoning that the other planets are inhabited worlds, let us look somewhat more closely into the circumstances, or rather, instead of examining only a portion of the evidence, let us take a wider survey and examine all the evidence we possess. It may appear, at a first view, that already we are dealing with periods which, to our conceptions, are practically infinite. How long, compared with the brief span of human life, are the eras with which history deals! how enormous, even by comparison with these eras, appears the range of time (tens of thousands, if not hundreds of thousands of years), since man first appeared upon this earth! and, according to the teachings of geology, we have to deal with a yet higher order of time in passing to

24. [A long footnote has been omitted at this point in which Proctor discussed a religiously based argument for extraterrestrials put forward by Dr. Richard Bentley, an associate of Isaac Newton. MJC]

the beginning of life upon our globe. From one million of years to ten millions! It is between such limits, say the most experienced geologists, that the choice lies. Surely we may be content with periods such as these, periods as utterly beyond our powers of conception as the duration of the pyramids would be to creatures like the ephemeron, did such creatures possess the power of reason!

And yet, why should we stop at the beginning of life upon this Earth? We have passed to higher and higher orders of time-intervals, but the series has no limit that we know of, while it possesses terms, recognisable by us, of higher order than those we have been dealing with. We know that in the far-off times before life appeared,

> The solid Earth whereon we tread
> In tracts of fluent heat began,
> And grew to seeming-random forms,
> The seeming prey of cyclic storms.[25]

Let us look back at that part of the Earth's history, and see whether the long periods which we have contemplated may not be matched and more than matched by the aeons which preceded them. When we thus

> Contemplate all this work of Time
> The giant labouring in his youth,

we see how far we have been from recognising the true breadth of the mighty waves on one of which the life upon this Earth has been borne, we see that as yet we have not

> Come on that which is, and caught
> The deep pulsations of the world,—
> Æonian music measuring out
> The steps of time.

25. [These lines as well as the lines of poetry quoted later in this paragraph are from Alfred Lord Tennyson's "In Memoriam." MJC]

Taking as the extremest span of the past existence of life upon the Earth ten millions of years, we learn from the researches of physicists that the age preceding that of life (the age during which the world was a mass of molten rock), lasted more than thirty-five times as long, since Bischoff has shown that the Earth would require 350 millions of years to cool down from a temperature of 2,000° Centigrade to 200°. But far back beyond the commencement of that vast era, our Earth existed as a nebulous mass, nor can we form even a conjecture as yet respecting the length of time during which that earlier stage of the Earth's existence continued.

So much for the past. Of the future we know less. But still we recognise, not indistinctly, a time when all life will have ceased upon the Earth. Whether by the process of refrigeration which is going on, or by the gradual exhaustion of the forces which at present reside in the Earth, or by the change in the length of the day which we know to be slowly taking place, a time must come when the condition of our earth will no longer be suited for the support of life. Or it may be that Stani[s]las Meunier is right in his theory that as a planet grows older, the oceans, and even the atmosphere, are gradually withdrawn into the interior of the planet's globe, where space is formed for them by the cooling and contracting of the solid frame of the planet. But apart from all such considerations, we know that a process of exhaustion is taking place, even in the Sun himself, whence all that exists upon the Earth derives its life and daily nourishment. So that indirectly by the dying out of the source of life, if not directly by the dying out of life, this Earth must one day become as bleak and desolate a scene as we believe the Moon to be at this present time.

It is easy to recognise the bearing of these considerations upon the question of life in other worlds. We had been led by the contemplation of the long continuance of life upon this Earth, to regard the support of life as in a sense the object of planetary existence, and therefore to view the other planets as the abode of life. But we now see that the time during which life has existed on the earth, has been a mere wavelet in the sea of our Earth's lifetime, this sea itself being but a minute portion of the infinite ocean of time, while, as Tyndall has well remarked in that infinite ocean, the history of man (the sole creature known to us that can appreciate the wonders of creation) is but the merest ripple. We learn, then, from the Earth's history, a lesson the very reverse of that which before we had seemed so clearly to read there. It is not the chief, but only a minute portion of the Earth's existence which has been characterised by the existence of life upon our globe; and if we adopted the teaching now brought before us, as readily as before we learned that other lesson, we should say, 'It is not the chief, but only an utterly subordinate part of nature's purpose, to provide for the existence and support of life.'

We have been led by the study of the probable past history of the earth, and by the consideration of her probably future fortunes, to the conclusion, that although life has existed on her surface for an enormously long period, and will continue for a corresponding period in the future, yet the whole duration of life must be regarded but as a wave on the vast ocean of time, while the duration of the life of creatures capable of reasoning upon the wonders which surround them, is but a ripple upon the surface of such a wave. It matters little then whether we take life itself, without distinction of kind or order, or whether we take only the life of man, we still find a disproportion which must be regarded as practically infinite, between the duration of such life, and the duration of the preceding and following periods when there has been and will be no such life upon the earth.

But yet, in passing, I cannot but point to the fact that in considering the usual arguments for life in others worlds, I might limit myself to the existence of rational beings. It would be difficult to show that mere life, without the power which man possesses of appreciating the wonders of the universe, is a more fitting final purpose in creation than the existence of lifeless but moving masses like the suns and their attendant planets. The insect or the fish, the bird or the mammal, the minutest microscopic animalcule or the mightiest cetacean, may afford suggestive indications of what we describe as beneficent contrivance; yet it is hard to see in what essential respect a universe of worlds beyond our own, inhabited only by such animals, would accord better with those ideas which the believers in the plurality of worlds entertain respecting the purpose of the Almighty, then a universe with none but vegetable life, or a universe with no life at all, yet replete with wonderful and wonderfully moving masses of matter. It is rational life alone to which the arguments of our Brewsters and Chalmers really relate. Nor would it be difficult to raise here another perplexing consideration, by inquiring what degree of cultivation of the intellect in human races accords with the 'argument from admiration' which the followers of Brewster delight to employ. The savage engaged in the mere effort to support life or to combat his foes, knows nothing of the glories whereof science tells us. The wonders of nature, so far as they affect him at all, tend to give ignoble and debasing ideas of the being or beings to whose power he attributes the occurrence of natural phenomena. Nor as we advance in the scale of civilization, do we quickly arrive at the stage where the admiration of nature begins to be an ordinary exercise even of a few minds. Still less do we arrive quickly, even in reviewing the progress of the most civilized races, at the stage when the generality of men give much of their thoughts to the natural wonders which surround them. Is it saying too much to assert that this stage has never yet been attained by any nation, even the most advanced and the most cultured? If we limit ourselves, however, to the existence merely of some few nations, amongst whom the study of nature has been more or less in vogue, how brief in the history of this earth has been the period when such nations have existed! how brief the continuance of those among such nations which belong to the past, and whose whole history is thus known to us! how few even in such nations the men who have been so deeply impressed with the wonders of nature, as to be led to the utterance of their thoughts! If the life of man is but as a ripple where life itself is as a wave on the ocean of time, surely the life of man as the student and admirer of nature, is but as the tiniest of wave-crests upon the ripple of human life.

How, then, does all this bear upon the question of life in other worlds? The answer will be manifest if we apply to these considerations the same argument which Brewster and Chalmers have applied to the evidence which indicates the enormous duration of life upon the earth. Since this enormous duration, taking life even in its most general aspect, has been shown to be as a mere nothing by comparison with the practically infinite duration of the earth without life, the argument as respects life in any other world (at least, in any world of which antecedently we know nothing) must be directly reversed. It is far more probable that that world is now passing through a part of the stage preceding the appearance of life, or of the stage following the appearance of life, than that this particular epoch belongs to the period when that particular world is inhabited. If, indeed, we had some special reason for believing that this epoch to which terrestrial life belongs has some special importance as respects the whole universe, we might feel unwilling to consider the question of life in any other world independently of preconceptions derived from our experience in this world. But I apprehend that we have no reason whatever for so believing. It appears to me that such a belief—that is, the belief that life in this earth corresponds with a period special for the universe itself—is as monstrous as the old belief that our earth is the centre of the universe. It is, in fact, a belief which bears precisely the same relation to time that the last-mentioned belief bears to space. According to one belief, the minute space occupied by our earth was regarded as the central and most important part of all space, and the only part which the Creator had specially in His plans, so to speak, in creating the universe; according to the other, the minute time occupied by the existence of life on the earth is the central and most important part of all time, and the only part during which the Creator intended that living creatures should exist anywhere. Both ideas are equally untenable, though one only has been formally discarded.

This present time, then, is a random selection, so to speak, regarded with reference to the existence of life in any other world, and being a random selection, it is much more likely to belong to the period when there is no life there. Let me illustrate my meaning by an example. Suppose I know that a friend of mine, living at a distance, will be at home for six minutes exactly, some time between noon and ten on any given day, but that I have no means of forming any opinion as to when the six minutes will be. Then, if at any given moment, say at three, I ask myself the question, 'Is my friend at home?' although I cannot know, I can form an opinion as to the probability of his being so. There are six hundred minutes between noon and ten and he is to be at home only six minutes, or the one-hundredth part of the time,—accordingly, the chance that he is at home is one in a hundred, or speaking in a general way it is much more likely that he is not at home than that he is. And so precisely with any given planet, apart from any evidence we may have as to its condition,—what we know about life on our earth teaches us that the probability is exceedingly minute that that planet is inhabited. The argument is the favourite argument from analogy. Thus: life on our earth lasts but a very short time compared with the duration of the earth's existence; therefore life in any given planet lasts but a very short time compared with the planet's existence; accordingly, the probability that that planet is inhabited at this present moment of time is exceedingly small, being, in fact, as the number of years of life to the number of years without life, or as one chance in many hundreds at the least.

This applies to the planets of our solar system only in so far as we are ignorant of their condition. We may know enough about some of them to infer either a much higher probability that life exists, or almost certainly that life cannot exist. Thus we may view the condition of Venus or Mars as perchance not differing so greatly from that of our earth as to preclude the probability that many forms of life may exist on those planets. Or on the other hand, we may believe from what we know about Jupiter and Saturn that both these planets are still passing through the fiery stages which belong to the youth of planet life; while in our moon we may see a world long since decrepit, and now utterly unfit to support any forms of animated existence. But even in the case of our solar system, though the evidence in some cases against the possibility of life is exceedingly strong, we do not meet with a single instance in which evidence of the contrary kind is forcible, still less decisive. So that in the solar system the evidence is almost as clear in favour of the conclusion above indicated as where we reason about worlds of whose actual condition we know nothing. As respects such worlds,—that is, as respects the members of those systems of worlds which circle, as we believe (from analogy), around other suns than ours,—the probability that any particular world is inhabited at this present time is exceedingly small.

But let us next consider what is the probability that there is life on *some member or other* of a scheme of worlds circling around any given sun. Here, again, the argument is from analogy being derived from what we have learned or consider probable in the case of our own system. And I think we may adopt as probable some such view as I shall now present. Each planet, according to its dimensions, has a certain length of planetary life, the youth and age of which include the following eras:—a sunlike state; a state like that of Jupiter or Saturn, when much heat but little light is evolved; a condition like that of our earth; and lastly, the stage through which our moon is passing, which may be regarded as planetary decrepitude. In each case of world-existences the various stages may be longer or shorter, as the whole existence is longer or shorter, so that speaking generally the period of habitability bears the same proportion in each world to the whole period of its existence; or perhaps there is no such uniform proportion, while, nevertheless, there exists in all cases that enormous excess of the period when no life is possible over the period of habitability. In either case, it is manifest that regarding the system as a whole, now one, now another planet (or more generally, now one, now another member of the system) would be the abode of life, the smaller and shorter-lived having their turn first, then larger and larger members, until life has existed on the mightiest of the planets, and even at length upon the central sun himself. We need not concern ourselves specially with the peculiarities affecting the succession of life in the case of subordinate systems, or of the members of the asteroidal family, or in other cases where we have little real knowledge to guide us: the general conclusion remains the same, that life would appear successively in planet after planet, step by step from the smaller to the larger, until the approach of the last scene of all, when life would have passed from all the planets, and our sun would alone remain to be in due time inhabited, and then in turn to pass (by time-intervals to us practically infinite) to decrepitude and death.

During all this progression, the intervals without life would in all probability be far longer than those when one or other planet was inhabited. In fact, the enormous excess of the lifeless periods for our earth over the period of habitability, renders the conclusion all but certain that the lifeless gaps in the history of the solar system must last very much longer than the periods of life (in this or that planet) with which they would alternate.

If we apply this conclusion to the case of any given star or sun with its scheme of dependent worlds, we see that even for a *solar system* so selected at random the probability of the existence of life is small. It is, of course, greater than for a single world taken at random;—just as if I had ten friends who were to be at home each for six minutes between noon and ten, the chance would be greater that *some* one of the number would be at home at a given moment of that interval than would be the chance that a *given* one of the number would be then at home; while yet even taking all the ten it would still be more likely than not that at that moment not one would be at home.

Thus when we look at any star, we may without improbability infer that *at the moment* that star is not supporting life in any one of those worlds which probably circle round it.

Have we then been led to the Whewellite theory that our earth is the sole abode of life? Far from it. For not only have we adopted a method of reasoning which teaches us to regard every planet in existence, every moon, every sun, every orb in fact in space, as having *its period* as the abode of life, but the very argument from probability which leads us to regard any given sun as not the centre of a scheme in which at this moment there is life, forces upon us the conclusion that among the millions on millions, nay, the millions of millions of suns

> which people space, millions have orbs circling round them which are at this present time the abode of living creatures. If the chance is one in a thousand in the case of each particular star, then in the whole number (practically infinite) of stars, one in a thousand has life in the system which it rules over: and what is this but saying that millions of stars are life-supporting orbs? There is then an infinity of life around us, although we recognise infinity of time as well as infinity of space as an attribute of the existence of life in the universe. And remembering that as life in each individual is finite, in each planet finite, in each solar system finite, and in each system of stars finite, so (to speak of no higher orders) the infinity of life itself demonstrates the infinity of barrenness, the infinity of habitable worlds implies the infinity of worlds not as yet habitable, or which have long since passed their period of inhabitability. Yet is there no waste, whether of time, of space, of matter, or of force, for waste implies a tending towards a limit, and therefore of these infinities, which are without limits, there can be no waste.

CAMILLE FLAMMARION (1842–1925)

The two most successful popularizers of astronomy in the latter half of the nineteenth century were Richard Proctor and Camille Flammarion. A measure of their productivity and popularity can be gained from a study of astronomical publications up to 1881, which found that as of that year, among all astronomical authors who had ever lived, Proctor ranked seventh in number of publications, Flammarion fifth. The figures for their rates of publication are even more impressive; in that ranking, Proctor came out third, his French contemporary first.[26] Moreover, whereas Proctor's career ended with his death in 1888, Flammarion in 1881 was at the end of only the second of the six decades of his career during all of which he wrote incessantly. The author of over seventy books, he also through numerous translations reached an audience that extended far beyond his native France.

---

26. C. Houzeau and A. Lancaster, *Bibliographie générale de l'astronomie . . . jusqu'en 1880* (London: Holland Press, 1964 reprint of the 1882 original), 2:lxxiv.

Not only were Proctor and Flammarion remarkably popular authors, they both first attracted an audience by writing a book on the question of a plurality of worlds. The story of this book in Flammarion's case is especially dramatic. In 1862, Flammarion published a fifty-four page booklet titled *La pluralité des mondes habités,* on the title page of which he had listed himself as "Ancien calculateur à l'observatoire imperial de Paris, professeur d'astronomie, membre de plusiers sociétés savantes, etc." In fact, he was a twenty year old in his fourth year at the Paris Observatory as an apprentice astronomer, who although without university training was not without enthusiasm for writing or for extraterrestrials. The book was an immediate sensation; as Flammarion himself put it, the book "at once made my reputation."[27] By 1864, he had expanded it to 570 pages, and by 1865, it had received at least twenty-four reviews. It continued in print in France until the 1920s, going through dozens of editions, and was translated into six or more foreign languages.[28] Like Proctor, Flammarion learned from the success of his first book that the public possesses great interest in this subject and continued to deal with this topic in dozens of later volumes. Having placed himself near the center of the extraterrestrial life debate in 1862, he relinquished that position only with his death in 1925. Despite the numerous similarities in the careers of Proctor and Flammarion, they adopted significantly different approaches to treating ideas of extraterrestrial life. This should be evident if the selection from Flammarion provided below is compared with the Proctor piece that precedes it.

Flammarion's *Plurality of Inhabited Worlds* has never been published in English, although many of his books have. Thus the selection provided here offers English readers for the first time a concise statement of his main arguments as he formulated them in his most historically significant book on the subject of extraterrestrials.

Camille Flammarion, *The Plurality of Inhabited Worlds,* unpublished translation by Robert L. Jones, Jr., from Flammarion's *La pluralité des mondes habités* (Paris: Mallet–Bachelier, 1862), 1, 5, 9–11, 22, 33–34, 39–45, 48–50, 51–53, 54–62, and 81–85.

27. As quoted in R. A. Sherard, "Flammarion the Astronomer," *McClure's* 2 (May 1894): 569.

28. For details, see M. J. Crowe, *The Extraterrestrial Life Debate, 1750–1900: The Idea of a Plurality of Worlds from Kant to Lowell* (Cambridge: Cambridge University Press, 1986; new ed., Mineola, N.Y.: Dover Publications, 1999), 378–79.

INTRODUCTION

The philosophical movement occurring in the intellectual world over the past few years has gone unnoticed by no one; after bowing their heads beneath an unwarranted shame caused by the skeptical philosophy of the last century, men have lifted themselves up once again, filled with the latent aspirations which the shame had suppressed. The cult of the Idea counts new and fervent worshippers. Political upheavals, financial uncertainties, and the indifference of most men to questions beyond material existence have not, despite what people say, deadened the human mind to the point of preventing it from reflecting occasionally even today on its *raison d'être* and its destiny; during these past several years the soldiers of philosophy have been successively roused to action by the appeal of a few eloquently uttered words and have rallied together in diverse groups beneath the standard of the modern Idea.

After this opening paragraph, Flammarion asserts that this new era is dawning because "philosophy is no longer confined to the circle of sects or systems, but marches side by side with science and concurrently with it, adopting the same methods of experimentation, the results of which bring the truth into broad daylight" (2). He notes that recently numerous authors have warned against an "exaggerated partiality of the terrestrial" and stressed that attention should be given to the "universality of humanities" (2–3). Although admitting that "certitude of the plurality of worlds does not yet exist," he promises that by employing the "Baconian method," he will "give a new foundation to the doctrine of a plurality of worlds" (4).

Now let us add, in order quickly to justify in your eyes, reader, the *raison d'être* of our publication, that besides being of current interest in light of recent endeavors of human thought, this chapter of natural philosophy is the living side, if it may thus be expressed, of the science of astronomy. Despite its magnificent discoveries, that science would be of only minor usefulness in the advancement of the human mind if it could not be approached from a philosophical point of view;

as philosophy, it must join forces with the other branches of science *in teaching us who and what we are.* The spectacle of the exterior universe is, in fact, the only measure by which we can know our true place in nature; without this sort of comparative study we would be living on the surface of an unknown world, not even knowing who or where we are in relation to the immense whole of created things.

In the remainder of the introduction, Flammarion indicates that his presentation will consist of three parts. The first section is a "Historical Study," which demonstrates the continuity of ideas about the plurality of worlds throughout history. Second is an "Astronomical Study," which summarizes what is known about the solar system and argues that, in comparison to the other planets, our Earth is unremarkable. The third section is a "Physiological Study," which cites evidence of the diversity of life on Earth to argue for the diversity of life throughout the solar system. Flammarion summarizes the polemical point of his text with his statement, "the ant in our fields would have infinitely more basis and reason to believe its anthill the only inhabited spot on the globe than we have to regard the heavens as an immense desert in which our world is the only oasis and man is the only and everlasting contemplator."

HISTORICAL STUDY

After reflecting on a quotation from the Roman philosopher Lucretius, who was an early advocate of the idea of a plurality of inhabited worlds, Flammarion explains the purpose of this first section of his text.

To locate the origin of this admirable doctrine and to know to which mortal we are indebted for this marvelous conception of human intelligence, it suffices for us to carry ourselves back in thought to those splendid nights where the soul, alone with nature, meditates, pensive and silent, under the immense dome of the starry sky. There, a thousand stars lost in the distant regions of the expanses bathe the Earth in a soft light which shows us our true place in the universe; there, the mysterious idea of the infinite which surrounds us isolates us from all terrestrial distractions and bears us unaware to those vast provinces inaccessible to the weakness of our senses.

Absorbed in a hazy dream, we contemplate those twinkling pearls which tremble in the melancholy azure, we follow those passing stars which from time to time streak the ethereal plains, and, removing ourselves with them into the immensity, we roam from world to world in the infinity of the heavens. But the admiration excited in us by the most moving scene of the spectacle of nature is soon transformed into an indescribable feeling of sadness, because we are strangers to those worlds ruled by an apparent solitude which cannot give birth to the immediate sensation by which life attaches us to the Earth. We feel the need to populate these globes apparently overlooked by Life; in these eternally deserted and silent provinces we search for gazes which respond to our own. So once did a hardy navigator explore the oceans for a long time in accordance with his dream, searching for the land which had been revealed to him, piercing the most vast distances with his eagle-like gazes and audaciously exceeding the limits of the known world, until finally he wandered onto the immense plains where the New World had been resting for centuries. He realized his dream. May ours break out of the mystery which yet envelops it so that, on the ship of thought, we may sail up to the heavens and search there for other earths.

This deep-seated belief, which shows us in the universe a vast empire where life develops under the most varied forms and where thousands of nations live simultaneously in the expanse of the heavens, seems to date from the very establishment of men on the Earth. It originated with the first mortal who, devoting himself with the good faith of a simple and studious soul to the gentle contemplation of the heavens, merited to comprehend that eloquent spectacle. All peoples, most notably the Indians, Chinese, and Arabs, have conserved down to our own day theogonic traditions which recognize among ancient dogmas that of the plurality of human inhabitations in the worlds which shine above our heads; going back to the first pages of the historical annals of humanity, we find this same idea, either in a religious context, as concerning the transmigration of souls and their future state, or in an astronomical context, as concerning simply the inhabitability of heavenly bodies.

In several paragraphs omitted at this point, Flammarion attributes belief in the plurality of inhabited worlds to many ancient Egyptians, Ionians, and Greeks. Though Aristotle was a notable exception to the apparently widespread belief in extraterrestrial life, Flammarion notes that "had [he] known the true system of the world" (14), the philosopher would have agreed with the majority. Flammarion's history extends through Bruno, Galileo, Newton, Kant, and other modern European natural philosophers.

Without mentioning our century, which would speak with still more eloquence than those preceding it in favor of our cause, we hope that this glorious series of names, forever famous in the history of science and philosophy from the most remote ancient history right up to the present day, will not be a vain and useless palladium in our hands; we will permit ourselves to think that if all of these illustrious men did not believe that they discredited their genius in proclaiming the plurality of worlds, then we will be able, we who need not fear such an accusation, to proclaim ourselves this beautiful doctrine and attempt to develop it and show all its grandeur.

Flammarion proceeds through a series of lengthy corroborating quotations from Kant, Laplace, John Herschel, and others. He concludes from his "Historical Study" that in regard to the idea of a plurality of worlds, "the eminent men of all ages who have been versed in the operations of nature have also been profoundly impressed by its prodigious fecundity and have understood the insanity of those who would limit that fecundity to our abode alone" (26). He adds that "The study of nature engenders and affirms in the mind of men the idea of the plurality worlds" (27) and promises that the "Astronomical Study" to follow "will elevate to philosophical certitude the probability of the Plurality of Worlds" (28).

ASTRONOMICAL STUDY

The blazing orb of day, inexhaustible source of light and heat with which it floods the immensity of space, unceasing rejuvenator of the beauty and youth of the planets forming its royal court, inexhaustible hearth of the life and fecundity that unfold in its empire, resides in glory at the center of our planetary system and presides over the celestial revolutions of its world. This immense globe is 1,400,000 times larger than the Earth.

After this poetic description of the Sun, Flammarion turns to its satellites, working outward from the tiny intermercurial planets then attributed to the Sun, through Mercury, Venus, and the Earth, arriving at the planet Mars.

About 50,000,000 miles farther circles the planet Mars, which also presents characteristics strikingly similar to the preceding planets. It is 144,282,928 miles distant from the central orb, completes its year in 687 days, and accomplishes its daily rotation in 24 hours and 39 minutes. The atmospheric envelopes which surround this planet and the preceding one, the snows which appear periodically at their poles, the clouds which form from time to time on their surfaces, the analogous geographical configuration of their continents and of their plains designated as maritime, and the variation of seasons and climates common to these two worlds, give us a firm basis for believing that these planets are both inhabited by beings whose physical structures must contain more than one analogous feature. By the same token, if one of them were doomed to solitude and emptiness, then the other, finding itself in the same condition, should have the same fate.

In several paragraphs omitted at this point, Flammarion continues to move further from the Sun, describing the asteroid belt between Mars and Jupiter as the remnants of a lost planet. When he reaches Saturn, Flammarion writes, "The alternating light and dark bands which appear on [Jupiter and Saturn] and which are a certain indication of atmospheric variations, the changing color of the polar and equatorial regions, the magnificence of the spectacle of creation viewed from Saturn, where the games played by nature among the mysterious rings must be of an unqualified splendor for its inhabitants, and the similar spectacle seen from Jupiter, where the most favorable conditions for life are united, all witness well enough how far the domain of life is from being limited to the small world which has given us life." (36) After describing Uranus and Neptune, Flammarion summarizes his view of the solar system, noting that Mercury, Venus, Earth, and Mars share "the same history, same appearance, same conditions and same role in the universe." (38)

In beginning a comparative study of the planets, the first point which demands our attention is the position occupied by the Earth in our system. Considering simply the number of planets and their respective distances from the radiant orb, the asteroids counting as a single planet, we note first of all that the Earth is the third of nine and is consequently distinguished neither by proximity nor long distance, nor by virtue of a median position. We assert next that the Earth is three times farther from the Sun than Mercury and thirty-six times nearer than Neptune, and it is not located at the midpoint of the radius of the planetary system, for that point falls between the orbits of Saturn and Uranus. We conclude, then, that from this initial point of view the Earth has no particular distinction among the planets.

In considering the quantity of heat and light which these spheres receive from the Sun, knowing that the intensity for each one varies, everything else being equal, in proportion to the inverse square of the distance, and taking the Earth as a reference point, we find that Mercury receives seven times more light and heat than our globe, Venus twice as much, Mars half as much, the asteroids seven times less, Jupiter 27 times less, Saturn 90 times less, Uranus 365 times less, and Neptune 1300 times less. This brief glance is sufficient to teach us that the Earth has received no particular distinction from this new point of view. We said "everything else being equal" because it is certain that in order to solve this problem we require data which will probably never be available to us. We would need to know, for example, the density and chemical composition of the surrounding atmospheres, for it is known that they allow varying amounts of solar radiation to pass through them to heat their planets and then act with varying degrees of efficiency to trap this heat and prevent it from radiating back into space; this property, suitably proportioned to the distances, would suffice to give the same average temperature to planets dispersed at varying distances from the Sun. We would need to know also the nature of the materials which compose the planetary bodies, the accidents of terrain and the particular circumstances with a notable effect on the heat absorbed or reflected, the general color and particular tints of various surfaces, the ordinary degree of dryness or humidity of the soil and the rate of evaporation of liquid masses, the height of mountains, and particularly the interior heat of each planet. Finally, we would need to know a thousand influential causes of which we have not the slightest inkling, since we judge all creation by terrestrial phenomena, not being able to observe anything else. Let it suffice for us to understand that all the objections based on the proximity or distance of the Sun, which seem to prohibit the existence of living beings on certain worlds because they would burn or on other worlds because they would freeze, have no validity at all when they are placed against the efficacious power of nature. Consequently, whether nature produces beings structured to withstand the normal conditions on these planets or whether she attenuates the extreme conditions unfavorable to the existence of the beings she wishes to place there, it remains sufficiently well established that from this new point of view the position of the Earth does not distinguish it from the other planetary worlds.

In considering the possibility that the satellites have been placed in the heavens to light our nights and to control the movements of atmospheric phenomena by the flux and reflux of the ocean and the atmosphere,[29] we observe that certain planets possess up to eight satellites and that the Earth is far from being favored in that regard. We have here an important observation to address to believers in final causes, who rightly admire the luminaries whose gentle and beneficial light replaces during the night the dazzling light of day but who wrongly pretend that the moon and the other satellites would serve no purpose if they did not render some small service to

29. One will note that we are not speaking, either here or elsewhere, of *water* or of *air*, because we have no proof that the planetary liquids and fluids are of a chemical composition analogous to that of terrestrial liquids and fluids. We believe to the contrary that those of the planets differ essentially from those of the Earth, because they found themselves at the time of their formation in conditions wholly other from those that presided over the formation of terrestrial matter. It is on this point that moderns who have written on the plurality of worlds have been so terribly mistaken and have been led to the most erroneous conclusions.

their planets, and that the rendering of this service is the satellites' only *raison d'être*. We will point out quite simply to them that their argument could be advantageously turned against them. In fact, the inhabitants of these little worlds certainly have a greater right than we do ourselves to consider themselves privileged and to maintain that the Earth and the planets, which reflect much more light, were formed expressly for the purpose of illuminating their very long nights; this right is supported all the more by the fact that the planets have reflecting surfaces so much more vast than those of the satellites. Hence, the Earth reflects thirteen times more light to the moon than it receives from that sphere, and despite the plurality of the satellites of Jupiter, Saturn, and Uranus, the difference is even greater for those worlds. From whichever side one examines this question, not only is the Earth less favored than the large planets, but it is favored even less so with respect to the satellites themselves. In order to dissipate completely the opposition of those who invoke final causality in such a manner, we observe along with Arago that, to satisfy their views, it would have been necessary that the planets have a greater number of satellites at their service the greater their distance from the Sun, which they do not; with Laplace we observe that, for a permanent illumination of nights on Earth, the moon would have had to have been four times farther from the Earth and always in opposition, and would have had to complete its revolution in one year in an orbit outside that of the Earth but in the same plane, which is not possible; with Comte, finally, we observe that the best arrangement would have been to have two satellites placed in such a manner that the rising of one coincided with the setting of the other, which would occur if the two circulated in the same orbit and remained constantly 180° apart from one another in longitude, which is no more possible than the idea of Laplace.

In considering the relative sizes and surface areas which distinguish the planets from one another, we will again make the same remark that the Earth has not been at all favored over the other celestial bodies and that it has neither the smallest, nor the median, nor the largest surface. While the

Earth's average diameter is less than 7900 miles,[30] Saturn's measures 71,050 miles and Jupiter's nearly 89,300 miles. This comparison brings to mind one of the most ingenious pages of Fontenelle's book, where the Countess takes it upon herself to ask him whether or not the inhabitants of Jupiter have been able to ascertain the existence of our little globe. The philosopher answered her,

> In all good faith, I fear that we are unknown to them: the Earth would appear one hundred times smaller to them than their planet appears to us; it is too little, they do not see it at all. Here is the best we can hope for ourselves: there may be astronomers on Jupiter who, after having toiled to fabricate excellent glasses and after having chosen the most beautiful nights for observing, will finally have discovered in the heavens a tiny planet that they had never seen. First of all the Journal of the Learned of that country will speak of it. The people of Jupiter either will not hear about it or they will only laugh. The philosophers whose opinions the discovery destroys decide that they will believe none of it; only exceptionally reasonable people will suspect the truth. Another observation is made, the little planet is seen again, men reassure themselves that it is not an illusion, and finally, thanks to all the pains which the wise men have taken upon themselves, the people of Jupiter know that the Earth exists. . . . But our Earth is not us—no one has the slightest suspicion that she might be inhabited, and if someone comes to imagine it, God knows how all Jupiter will ridicule him.[31]

30. The average terrestrial radius, that which falls near the center of France, is 6,366,407 meters [3,918 miles]; the average diameter of the globe is then 12,732,814 meters [7,836 miles], and its circumference is 4,000 myriameters or 10,000 metric *lieues* [24,800 miles]. One could note here that a voyage of circumnavigation, which is completed in three years on Earth would, under identical conditions, last over 28 years on Saturn, nearly 35 years on Jupiter, and over 110 years on the Sun. [Editor's Note: In this and subsequent references in this translation, translator's notes are indicated by square brackets. MJC]

31. *Entretiens sur la pluralité des mondes*, 132 ["Quatrième Soir"].

If, after having compared our globe to Saturn and Jupiter, we compared it to the Sun, we would establish that the diameter of the latter is 883,000 miles, and its surface area is 955 trillion, 130 billion square miles. Judging by our globe, whose surface of 789 million square miles nourishes nearly one billion inhabitants, the surface of the Sun, which is 12,000 times larger, could nourish 12,000 billion men.

> Now, if it could be proven, as it has been maintained, that the inhabitation of the Sun might be characterized by incessant delights and extreme longevity, what case can be made for the pretention of those who do not fear affirming, without any sure proof and even against the whole of the multiple witnesses of science, that all that we see was created for the sake of the Earth and the happiness of its inhabitants, whose life is so ephemeral, so disturbed, susceptible to so many sufferings, deceptions, and miseries? Indeed, if the globes of our world were formed for one another, is it not more natural to think that every type of advantage must rest with the greatest among them, the one which, placed at the center of the system, obliges the others to circle around it, governs them, masters them, dominates them with such power, and finally, lights and heats them with his benevolent rays?[32]

How much more right would the inhabitants of splendid Saturn and of magnificent Jupiter have to regard the other worlds as launched into space to teach them the laws of the universe and to make them admire its harmony, those beings whose years are counted by centuries and who see all of creation disposed in their favor? How much more basis do these inhabitants, privileged in the moral order as they are in the physical, have to regard themselves as monarchs of the world, they who are elevated so high above the puny human creatures who stammer about on the surface of our globe? So then,

32. Plisson, *les Mondes*, 242 [Quatrième Section, #88 (Paris, 1847)].

here as before, the Earth has received not the slightest distinction from Nature.

Next, Flammarion treats the differences in volume between the planets, concluding that the immense volumes of the outer planets would be a waste if only the tiny Earth is inhabited. He quotes Fontenelle, "Life is everywhere, and should the moon be nothing but a heap of rocks, I would rather have the inhabitants gnawing on the rocks than to think there were no inhabitants at all" (46). Flammarion moves to the subject of planetary density next, deciding once again that the Earth is nothing special, since its density "is neither the least, nor the median, nor the greatest" (48).

A study of the interesting question of the effects of gravity on the surface of the different globes of our system shows us that on the Sun objects are twenty-nine times heavier, and on Mars only half as heavy as on Earth. Consequently, a body which travels 4.90 meters in the first second of fall at the terrestrial surface will travel 143.91 meters on the Sun and only 2.16 meters at the surface of Mars. One can see from this that the force of gravity is not of an average intensity on Earth, and that if the physical organization of terrestrial beings is in harmony with that intensity due to an entirely fortuitous state of matter, one must conclude that nature was not greatly hindered in establishing on the other globes beings whose constitutions are equally in harmony with the worlds they inhabit. Let us conclude from this that the inhabitants of each planet are essentially different from one another, for the effects of gravity exert a considerable influence over the laws of organization. On our continents, for example, there could exist no animals much larger than the elephant, because since the additional weight would hinder a proportional increase in muscular activity, the movements of such enormous masses could not be effected with the same facility. Conversely, in the depths of the seas, the specific gravity of the animals' bodies permit them to swim with great agility in the environment for which they were born. Now, what observation demonstrates on Earth is extended by analogy to the other planetary worlds. A kilogram of terrestrial matter would be reduced to several grams if transported to the small planets, whereas it would increase to more than 30 kilograms on the solar globe; an Earthman of 198 pounds would be extremely light on the first, whereas on the Sun he would weigh more than 4400 pounds.

> He could most likely fall from five stories on Pallas without hurting himself any more than if he jumped off of a chair on Earth, while the slightest fall on the Sun, supposing he could even hold himself up in a standing position, would break his body into a thousand pieces as if he were ground by an iron mortar.[33]

As futile as they may seem, these last considerations are quite appropriate for enlightening us concerning the innumerable effects of a single natural force and for teaching us how far those we see on Earth are from being the only creatures in the universe.

A few lines omitted at this point discuss the comparative weights of the planets.

May the reader deduce for himself the conclusion which follows from the foregoing considerations, for we desire no further proof of the truth of our doctrine than the witness of his own judgment. Were he to follow the philosophical progress of modern astronomy, he would see that when the movement of the Earth and the volume of the Sun became known, astronomers and philosophers began to find it odd that such a magnificent star should be employed exclusively to illuminate and heat an imperceptible little world tucked away in the company of a large number of others under that star's supreme domination. The absurdity of such an opinion became even more striking when Venus was found to be a planet of the same dimensions as the Earth, with mountains and plains, seasons and years, and days and nights analogous to our own; that analogy was expanded to the following conclusion, that,

33. Plisson, *les Mondes*, 275.

since they are alike in their physical characteristics, they must be alike also in their role in the universe: if Venus were without population, then the Earth must be similarly lacking, and reciprocally, if the Earth were populated, Venus must be populated also. But later, when astronomers observed the gigantic worlds of Jupiter and Saturn surrounded by their splendid retinues, they were irrepressibly driven to deny living beings on the smaller preceding planets as long as the larger planets were denied living beings, and not just beings of our stature, but rather far superior to those on Venus or the Earth. And, in fact, is it not evident that the absurdity of the immobility of the Earth is perpetuated a thousand times more extravagantly in the final causality which places our globe in the highest rank of celestial bodies? Is it not evident that this world is tossed without distinction on the planetary rockheap, and that it is no better suited than the others to be the exclusive seat of life and intelligence? Draw yourself away in thought for an instant, reader, to a point in space from which you can embrace the whole of the solar system, and suppose that you are oblivious to the existence of the planet which gave you life! Keep in mind that, in order to give yourself freely and fully to this exercise, you must no longer think of the Earth as your homeland or favor it over the other abodes; now contemplate without prejudice and with the eye of an extraterrestrial the planetary worlds which circle about the hearth of life. If you suspected that there might be life, if you imagined that certain planets were inhabited, would you, in all honesty, grant a population to this lowly globe of Earth without first expecting that all the marvels of living creation had been established on the larger planets? Or if you wished to settle on a celestial body from which you could embrace the splendor of the heavens and on which you could enjoy the benefits of a rich and fruitful nature, would you choose for your abode this puny Earth which is eclipsed by so many resplendent spheres? For each response, reader, and this is the weakest conclusion we could possibly draw from the preceding considerations, let us assert that *no preeminence has been bestowed on Earth in the solar system to make it the only inhabited world*, and that *astronomically speaking, the other planets are as well disposed as the Earth to support life.*

PHYSIOLOGICAL STUDY

"But," some will object, "the preceding determinations are based only on astronomical evidence, unimpeachable it is true but still insufficient to establish a firm conviction favoring the inhabitability of worlds. Up to this point you have completely suppressed the physiological point of view, which should have comprised a goodly portion of the discussion of your thesis. Even if all the other planets are *in appearance* as well suited as Earth to harbor life, nothing proves to us that this appearance mirrors reality and that the planets were *in fact* given the conditions capable of supporting life or were given life itself. To the contrary, the considerable weight and hardness of the spheres on one hand and the lightness and incohesiveness of molecules on the other, the blazing heat and dazzling light of certain worlds and the glacial cold and eternal darkness of others, all appear invincibly to oppose the manifestation of the phenomena of life."

The physiological point of view is certainly very important to consider here, but the objections which it raises and which seem serious at first sight refute themselves as soon as we begin to analyze them thoroughly. In fact, not only is it unnecessary for us to strain our minds in order to perceive their invalidity and to understand the possibility of life forms entirely incompatible with those of the Earth, but moreover it suffices for us to cast a mere glance over our dwelling place in order to conceive of planets populated very differently, and even to convince ourselves that it is nearly impossible that they are inhabited by beings similar to those living on Earth.

What infinite variety, for example, between the joyful inhabitants who flit about in the air above the plains and those who ply the mobile regions of the oceans or who pass their life on the surface of the continents! What diversity in their manner of life and in their language! Who could count the degrees of this ladder of life which began with the zoophytes of primitive times and whose highest rung is occupied by man? And even within humanity itself, what difference of structures, of characters, of morals, of habits, and of physical and moral strength between the European whose will transforms empires and the Eskimo who is incapable of expressing his own thoughts! Without even mentioning the plant kingdom, the spectacle of the tremendous variety of zoological life suffices in itself to convince us of the impotence of biological obstacles which oppose themselves to the fecundity of nature. To choose an example relating to our subject, let us recall that during the primitive epoches of the globe, when the interior heat and the instability of the terrestrial surface prohibited the existence of present day animals and vegetables, another type of life suited to those early ages propagated itself in a prodigious manner. On one side, a powerful vegetation, the Cycadaceae, which measured no less than seven feet in diameter, arborescent ferns whose only living vestiges are conserved at the equator, spread far and wide on lands which were still swampy and which were forming, millions of years ago, the current oxygenated atmosphere and deposits of oil. On another side, microscopic animals were constructing, in conditions of extreme heat, mountains formed entirely of their debris, animals so small that 300 have been placed in a length of .078 of an inch, and so numerous that as many as 3,840,000 have been counted in a single ounce of water.[34] These beings, whose organic simplicity was in harmony with the newness of the globe, were succeeded by richer and more elegant vegetation which bore flowers, and by animals more elevated in the economy of life and endowed with a vitality so prodigious that these races were unaffected by the upheavals of the soil and even by the great revolutions in the Earth's crust which occurred so frequently at that time. From this epoch dates the creation of the Radiata and the Polyps which, broken and divided into diverse

34. De Humboldt, *Cosmos*, tome I, pl. 270. [The correct name is von Humboldt, or simply Humboldt. Although Flammarion misspells the name here, he spells it correctly in a later footnote. I believe the reference here is to a fact mentioned in the "Universality of Animal Life" section of volume 1 of *Cosmos*, 344 of the New York edition of 1851.]

parts, still live and reproduce today. Also created at this time were the Annulata, equally endowed with a great vital force, and later came the Crustaceans, whose bodies, protected by their shells, retained a last vestige of the foresight of Nature, which always acts according to the time and place. It is in primitive times, also, but during an epoch a bit closer to us, that the gigantic Saurians appeared, those colossal animals covered by scales and a tough, resistant hide who were at that time sole masters of living creation, who dominated for thousands of years the regions where one day man was destined to appear. Let us keep in mind that, between the infancy of the world and the appearance of the most recent created beings, multitudes of varieties of plants and animals have succeeded one another on Earth's surface according to the transformation in the physical state of the soil and atmosphere, being born, developing, and disappearing with the geological periods in order to make room for other beings which in turn reenacted the same process. Aware of these facts, we realize that the power of creation is infinite and that we can reasonably conceive no obstacle whatever to the manifestation of life, as long as such an obstacle does not formally contradict the laws which govern the world. Our only reason for sketching the picture of primitive times was to extend this principle to the domain of celestial bodies, and to show that, at least within broad limits, those worlds could be populated by all sorts of living things which could never survive on our world. But even in her present state, Nature abounds with evidence to prove to us, by the diversity of her terrestrial productions, the variety which she could have spread through the heavens. This diversity manifests itself everywhere. First, in terms of environments and their vital principles, we see innumerable species of aquatic animals sharing an existence incompatible with that of the other organisms on the globe (Cuvier), and we see amphibious reptiles such as alligators and snakes living in a medium which would be fatal to man and the higher animals (Humboldt). In terms of light, we see condors and eagles, which reside at higher altitudes amid dazzling snows, staring habitually at the Sun (Lenorman) while certain species of fish enjoy the benefits of light in the murky obscurity of the oceanic depths, where darkness, deeper than that of the blackest night on Earth's surface, reigns eternally (Biot).[35]

Finally, in terms of heat, climate, gravity, atmospheric pressure, etc., we see that certain Infusoria know neither heat nor cold, that the same species which live in China and Japan have been found in the Baltic Sea (J. Ross), that the diatoms which thrive in the hot springs of Canada are found also in the polar regions, and that those living on the surface of the sea have been found on lead weights sunk to a depth of 1800 feet, where they were subjected to a pressure of sixty atmospheres (Zimmerman); just as bodies do not possess an absolute weight, so neither does an absolute cold or hot exist in the universe, where all is only relative and all is in harmony. Now, if such is the lesson which nature teaches us here below, if her inexhaustible fecundity, against which no resistance ever did or ever will prevail, causes so much variety in the productions of the Earth, what more assurance do we need that no obstacle whatever could successfully prevent the manifestation of life on the planets and satellites, whose productions, moreover, could vary to infinity!? We say that these diverse productions can and must vary to infinity, and we are as far from admitting that the inhabitant of Mercury conforms to that of Neptune as we are assured that there is an infinity of different structures, not only from one world to another, but even within each world with its different ages, climates, and biological conditions. If we form in our minds an accurate idea of the efficacious power of Nature,[36] we will be forced to admit that the

35. Man himself can, by a prolonged exercise, render his eye so precisely sensitive to the slightest luminous sensation that he can read and write in places where anyone else would believe himself in the most absolute darkness. A prisoner of the Bastille had this sad experience. Imprisoned for forty years in an underground cell, apparently completely deprived of light, he came not only to write, but even to read. Nevertheless, his eye became so sensitive that when finally he was released, he solicited as a favor permission to reenter his prison, because it was impossible for him to become accustomed again to the light of day.—Valerius, *Les phenomenes de la nature*, 10.

36. In order that no one gives a pantheistic interpretation to this word "nature," which appears often in our study, we will state what *we consider nature, that is to say, the universality of created things and the laws which govern them, as the expression of the Divine Will.*

inhabitants of the planets farthest from the Sun receive no less light and heat, relative to their respective constitutions, than the inhabitants of Mercury or of Earth, and that one cannot logically rely on the proximity or distance of the planets to determine their inhabitability. We point out also that the elements inherent in the constitution of such or such a planet cannot be more contrary to its inhabitability than those which constitute the Earth are contrary to our own existence. Thus, when someone objects to us that water would take the form of steam in some worlds and of ice or snow in others, that minerals would be in a molten state in certain worlds and a state of extreme hardness in others, that agriculture and the arts would be impossible, or a thousand other equivalent arguments, such objections could apply only to the things of Earth transported to other heavenly bodies, which strips them of all but a shadow of scientific validity. It is certain that nature knows perfectly well how to appropriate the physical structure of living beings to that of the organic or inorganic beings in their environment, as well as to the vital principles peculiar to that environment itself.

We will go even further, daring to extend this principle to all celestial bodies in general, no matter how great the difference between their conditions and our own. In doing so we remain within the broad limits, for nature, with eternity behind and before her, can and must necessarily have worlds in the process of formation and destruction. We think therefore that certain biological conditions which seem incompatible with life could in reality be favorable to beings structured in a manner unknown to us, that the absence of an atmosphere, for example, and by the same token the absence of liquids at the surface of certain worlds does not *necessarily* negate the possibility of life. In fact, then, do the modern authors who admit the plurality of worlds only with this restriction judge nature incapable of forming living beings on other models than those she has established on Earth? Is this a valid reason, the fact that we cannot live without the crude fluid which envelops our globe, to believe that no possible being could ever inhabit the spheres deprived of this fluid? Or, given that water is essential to the nourishment of terrestrial life, must we then conclude that this same condition exists on every other world? Would the Creator have encircled the Earth with an oxygenated atmosphere composed precisely as it is had man necessarily had a different structure, or would He have placed man here below, structured as he is, if this atmosphere had not existed? What absurdity that the moderns should circumscribe the power of creation within such narrow limits, which human science itself would not dare to adopt forever! What folly to pretend that without a certain proportion of oxygen and nitrogen, omnipotent Nature could beget neither animal life nor plant life, or more precisely no type of being whatever, for even though terrestrial nature is divided into three kingdoms, there is no reason why nature could not appear in other worlds under forms incompatible with terrestrial forms!

In several paragraphs omitted at this point, Flammarion argues that, because contemporary thinkers have only a limited understanding of the conditions that make terrestrial life possible, it is vain to extend this limited understanding to declarations of the impossibility of extraterrestrial life. At this point, Flammarion draws two general conclusions: "(1) *No preeminence has been bestowed on Earth;* (2) *the other planets are inhabitable just as Earth is*" (64). After developing the traditional argument that if the planets were not inhabited, it would be difficult to explain why God created them, he considers the discoveries made possible by microscopes and the investigation of fossils, which demonstrate the abundance of life under a staggering variety of conditions and environments. In support of the claim that "even for human life, the earth is not the best of all possible worlds" (75), Flammarion discusses the fact that the plane of the Earth's equator is inclined at 23° to the plane of the Earth's orbital motion, which causes major climactic changes (the seasons), variations in the length of day, and other problems, some of which do not occur on other planets.

Let us embrace in one glance, if possible, the human population covering the Earth, and we shall affirm that this globe is far from being agreeable to man, that the sterility of his planet forces him to spend the greater part of his time in the acquisition of the means of subsistence. The plants which nourish him must be sown, cultivated, and prepared; the animals which he uses must be sheltered against the intemperance of the seasons, so he must build lodgings for them, prepare their food, and make himself their slave. Alone in the midst of nature, man does not receive from her the slightest aid, and if he finds the means of living on Earth, it is by virtue of continual work, not in the least by virtue of the benevolent dispositions of nature. Watch it, this same terrestrial nature, swallow every year thousands of men who go overseas in search of the food of progress! Watch it jolt and destroy in the blink of an eye the cities where men have established centers of civilization! Watch it dry up the produce of the earth with a torrid heat or inundate it with torrential rains and the overflowing of rivers! Contemplate those hardy multitudes bent over toward the ground, broken by a sterile labor, whose intelligence is closed to every aspiration of thought! Let us cast our probing gazes over the surface of the terrestrial world, everywhere encountering the same disheartening spectacle; should we encounter here and there palaces where luxury glitters, let our gazes but pierce the gold-lined walls and we will find there also eyes wet with tears. Then we will know that human intelligence with its lofty thoughts has in no way established its reign here below where everything obeys the dictates of matter, and that the vast majority of men toil mightily to provide a very small number with the commodities of life, while themselves remaining in dire misery.

If the preceding reflections are not enough, let us consider that apart from the hostility of nature on the exterior, an even more formidable hostility is imposed on us by the interior forces which govern our world. Our fields, our cities, and our inhabitants are borne along on a sea of incandescent matter which could at any moment open up and swallow us in its fiery depths. It is known that the solid crust of the globe is less than twenty-five miles thick and that it is constantly agitated by the incessant action of subterranean forces, so that an interior fluctuation could at a given moment raise the basins of the seas, flooding our countries and swallowing us up, while at the same time leaving the sea-beds dry and in effect forming new continents of them. A geological revolution could also, one fine day, break into a thousand fragments this fragile envelope on which we believe ourselves secure and disperse its debris throughout space. Now, following such considerations, could one still pretend that this globe is the best of all possible worlds, even for man? Or that numerous other celestial bodies could not be infinitely superior to our world and unite much better than our own the conditions favorable to the development and longevity of human life? Far from placing it above the other celestial bodies, one will be astonished that life has taken up residence here at all; one will admit that the world's present state of population must be due to Nature's prodigious fecundity, and that she begets life even in places where man would never have dared to conceive of it. One will understand that she peopled the Earth only because it is her essence to produce life everywhere there is matter to receive it, and far from thinking that her eternal spring has run dry in multiplying the beings on Earth's surface, one will find in the diversity and infinity of her productions an eloquent proof that she has not exhausted herself in decorating the other worlds with an innumerable multitude of creatures, since she continues to be able to produce them here.

Therefore, not only the astronomical position of the Earth in its orbit, but also the normal disposition of its nature and its particular geological constitution, prove to us that it is far from being the most favorably established world for the care of life. The differences in age, position, mass, density, size, environment, biological conditions, etc., place a large number of other worlds in the immense amphitheater of the heavens in a condition of inhabitability superior to that of Earth. It is in these worlds that humanity lives tranquil and glorious, protected by a benevolent sky and a perfectly constant temperature and enjoying in peace the friendly dispositions of nature. An eternal springtime, perhaps more diversified by ever new charms than the most disparate of our own seasons, adorns these blessed worlds where man is relieved of every material occupation and of all those needs inherent in our terrestrial constitution. There, instead of groveling for his food amid the debris of other animals, he is graced with organs which inhale it insensibly in the vital atmosphere, just as here our lungs nourish themselves without our being aware of it. There, instead of laboriously studying the science of the world, man assimilates the marvels of creation and its universal laws by means of the most delicate senses and a more perfect understanding. There, golden bonds of love unite all members of humanity into an immense family; brother is no longer enslaved by brother, and neither the bloody rivalries to the glory of combat nor the discords of envy spread the venom of death. There, men worship the Creator without enclosing themselves under a sky of stone: nature is the temple and man is the priest. There, men communicate by thought, or perhaps a universal language replaces all of our national idioms. There, finally, man contemplates the splendid panorama of the infinite without a sight, and converges with the inhabitants of neighboring spheres by means of marvelous faculties.

Flammarion ends his study by moving beyond the solar to the stellar regions, noting the vast number of stars and even of Milky Ways beyond our own. This leads him to reflect on the immensity of the heavens, the power of God to produce such immensity, and the comparatively insignificant size of "that grain of dust called Earth" (91).

ALFRED RUSSEL WALLACE (1823–1913)

In the final forty years of the nineteenth century, a number of authors cited evolution by natural selection in support of extraterrestrials. Among these was Harvard's John Fiske (1842–1901), who in 1874 published his *Outlines of Cosmic Philosophy Based on the Doctrine of Evolution.* Another was Robert S. Ball (1840–1913), who stated that when in college he read Darwin's *Origin of Species* and became an "instantaneous convert," which helps explain the evolutionary approach that Ball, who became professor of astronomy at Cambridge University and director of its observatory, took in writing more than a dozen books explaining astronomy to the public.[37] Despite such figures, one should not assume that scientists who have accepted

37. On Fiske and Ball, see Crowe, *Extraterrestrial Life Debate*, 465 and 460–61, respectively. See also Steven J. Dick and James E. Strick, *The Living Universe: NASA and the Development of Astrobiology* (New Brunswick, N.J.: Rutgers University Press, 2004), 11–12.

evolution by natural selection have invariably been advocates of the existence of extraterrestrial intelligent life. For example, in the latter half of the twentieth century experts on evolution as prominent as George Gaylord Simpson, Ernst Mayr, and Simon Conway Morris challenged the existence of extraterrestrial intelligent life.[38] Moreover, as the next selection indicates, A. R. Wallace adopted a negative position in the extraterrestrial life debate.

Near the end of the nineteenth century, Wallace, already in his late seventies, published a book, *The Wonderful Century* (1898), surveying the progress of science in the nineteenth century. The review of astronomy that he undertook in preparing this book led him to publish in 1903 a controversial article titled "Man's Place in the Universe," which later in 1903 he expanded into a long book titled *Man's Place in the Universe: A Study of the Results of Scientific Research in Relation to the Unity or Plurality of Worlds.* In his paper, Wallace shocked his contemporaries by claiming that recent scientific research, especially in astronomy, pointed to the probability that humanity's place in the universe "is special and probably unique" and that "the supreme end and purpose of this vast universe was the production and development of the living soul in the perishable body of man."[39] In the concluding portion of his essay, Wallace asserted:

> The three startling facts—that we *are* in the centre of a cluster of suns, and that that cluster is situated not only precisely in the plane of the Galaxy, but also centrally in that plane, can hardly now be looked upon as chance coincidences without any significance in relation to the culminating fact that the planet so situated has developed humanity.[40]

Although during this period there was substantial astronomical evidence to support Wallace's first two claims, his paper and book created a major controversy. He attempted to strengthen his argument by adding to his book in September 1904 an important appendix, which presented an argument from "organic evolution." His

---

38. See, for example, George Gaylord Simpson, "The Nonprevalence of Humanoids," *Science* 143 (Feb. 21, 1964): 769–75; Ernst Mayr, "The Probability of Extraterrestrial Intelligence," in *Extraterrestrials: Science and Alien Intelligence*, ed. Edward Regis, 23–30 (Cambridge: Cambridge University Press, 1985); and Simon Conway Morris, *Life's Solution: Inevitable Humans in a Lonely Universe* (Cambridge: Cambridge University Press, 2003), esp. chap. 5.

39. Alfred R. Wallace, "Man's Place in the Universe," *Fortnightly Review* 73 (March 1, 1903): 395–411, at 396.

40. Wallace, "Man's Place," 411.

enthusiasm for this argument led him to describe it as being "as strong as, or, as I am inclined to think, even stronger than the purely physical argument set forth in my first edition."[41]

Alfred R. Wallace, *Man's Place in the Universe: A Study of the Results of Scientific Research in Relation to the Unity or Plurality of Worlds*, 4th ed. (London: Chapman and Hall, 1904), 326–36.

APPENDIX

AN ADDITIONAL ARGUMENT DEPENDENT ON THE THEORY OF EVOLUTION

Those among my critics who have expressed adverse opinions, usually agree that my proofs of the absence of human life in the other planets of our system are very cogent if not quite conclusive, but declare that they cannot accept my view that the unknown planets that may exist around other suns are also without intelligent inhabitants. They give no reasons for this view other than the enormous number of suns that appear to be as favourably situated as our own, and the probability that many of them have planets as suitable as our earth for the development of human life. Several of them consider it absurd, or almost ludicrous, to suppose that man, or some being equally well organised and intelligent, has not been developed many times over in many of the worlds which they assume must exist. But not one of those who thus argue gives any indication of having carefully weighed the evidence as to the number of very complex and antecedently improbable conditions which are absolutely essential for the development of the higher forms of organic life from the elements that exist upon the earth or are known to exist in the universe. Neither

41. Alfred R. Wallace, *Man's Place in the Universe: A Study of the Results of Scientific Research in Relation to the Unity or Plurality of Worlds*, 4th ed. (London: Chapman and Hall, 1904), vi.

does any one of them take account of the enormous rate at which improbability increases with each additional condition which is itself improbable. Now the numerous conditions and sub-conditions essential to the development of the higher organic life, which must all exist simultaneously and which must all have continued to exist for enormous periods of time, are each of them improbable in various degrees, as clearly shown by the fact that the great majority of them are not present in the degree required in any other planet of the solar system. The actual degree of improbability of each of these conditions cannot be determined except vaguely, but any one who will carefully consider and weigh my very imperfect exposition of them will, I think, admit that they are usually very considerable, the chances against each of them being in some cases perhaps ten to one, in others ten thousand to one or even more. But if we take the whole of the simultaneous conditions requisite as only fifty in number, and the chances against each occurring simultaneously with the rest to be only ten to one, an estimate which seems to me absurdly low because many of them are quantitative within narrow limits, then, the chances against the simultaneous occurrence of the whole fifty would be a million raised to the eighth power ($1,000,000^8$), or a million multiplied by a million eight times successively to 1. These figures are suggested merely to give some indication to the general reader of the way in which the chances against any event happening more than once mount up to unimaginable numbers when the event is a highly complex one. But, to make this still clearer, let us take a concrete example.

The chances are certainly very large against two persons, strangers to each other, finding on getting into conversation in a railway carriage that they were born on the same day in the same year. Here are only two coincidences, and the chances may be roughly calculated. But if we add successively other coincidences, each not very improbable in itself, we shall soon arrive at a series which will be held to be quite impossible. Let it be supposed, for instance, that the two persons meet accidentally on a particular day named by a third person beforehand; that they both have the same, not common,

Christian and surname; that their wives were also born on the same day; that they, too, had the same name though not related or known to each other; then let us add a final set of coincidences, that not *two* only, but *ten* persons with similar identities met together fortuitously on a day named beforehand in one compartment of a railway carriage; that *their* ten wives were found to be similarly identical in name and age; that the ten couples had each the same number of children born on the same days as those of the others, and each child having the same names as those of the corresponding children, then the series of coincidences becomes so overwhelming, that though their occurrence is theoretically possible, their actual existence is so wildly improbable that hardly any sane person would believe that they ever had occurred or ever would occur.

I will now point out the additional argument derived from the teachings of evolution, which when compounded with those set forth in the preceding pages produces a degree of improbability not unlike that above suggested, though, in my opinion, even less likely to have happened through the action of the known laws of nature.

The extraordinary and endless modifications of form and structure in organisms which often live under similar conditions, together with the great variety of species which inhabit very restricted areas, are themselves a proof of the rigidity of natural selection and the severity of the struggle for existence, leading to a close adaptation of every living form to certain portions of its ever changing environment. It is therefore held by almost all those biologists who are well acquainted with animals or plants in a state of nature, that no species has ever arisen independently in different places or at different times from the convergent modification of distinct ancestors. The fact that the environment as a whole, inorganic or organic, is never identical in any two distinct areas, or in the same area at any two distinct epochs — epochs comparable with those that we know have been required for the modification of species — is sufficient proof that an identical specific evolution cannot take place a second time. Some biologists have indeed held that the same species may, occasionally, have had two independent origins, but they have never grappled with the difficulties of such a supposition, and their views have not been adopted by those best acquainted with the ever-fluctuating conditions under which organic forms have been evolved.

But if the independent development of any one species, more than once, is almost inconceivable, what are the probabilities that not one only, but a whole series, starting from the first dawn of life and extending in an unbroken succession till it culminated in the human organism, has been developed many times over? The course of development of organic life is almost universally admitted to be best represented by a gigantic forest tree, the shoots, leaves, and flowers *now* produced corresponding to the *living* species, while all the branches, twigs, leaves, and flowers which have been produced year by year since the first shoot appeared above the ground correspond generally with the *extinct* species since life first originated. Of course the *diversity* of actual species is not represented, but merely their enormous multitude and the ever branching and diverging succession in which they have appeared. Fully to appreciate this suggestive analogy the last fifteen pages of the fourth chapter of Darwin's *Origin of Species* should be carefully read.

Now if we could have watched such a tree during its whole life from the first shoot that appeared above the ground, we should have seen in its early stages many twigs and branches eaten off by herbivorous animals, many buds destroyed by insects or pecked out by birds. And as it grew and became a lofty tree, storms and tempests would often tear off a branch, other branches would die for want of light and air, arboreal mammals would devour the young twigs as well as the fruits and flowers, and hundreds of species of birds and reptiles, insects and molluscs, would feed upon its tenderest growing buds and leaves. All this corresponds with the way in which whole groups of species have become extinct through many and varied causes, just as a tree has been modified in its growth and the number and position of its branches by the various agencies just enumerated. In both cases there has been a continuous struggle for existence.

Any being who could have watched the whole course of development of life upon the earth, would have been able to follow out from beginning to end the exact succession of forms, from the primaeval living germ through a long series of lowly marine organisms into the primitive vertebrate, supposed to be allied to the very low type of fish, the Amphioxus. Thence it would have progressed into a type which was the progenitor of both reptiles and mammals, and thence again into some lowly mammal, and through some type allied to the lemurs to the generalised ape which became modified, in one direction into the existing four types of the anthropoid apes, and in another direction through a long series of unknown extinct forms, into man.

The point to which I wish to call special attention is this: that if it is true that each species has arisen from one parental species and one only, then the whole line of descent from any living species (and therefore from man) back to the earliest form of life, has been fixed and immutable; so that if any *one* of the thousands or millions of successive species in the line of descent had become extinct *before* it had been modified into the next species in the line of descent (or, which is the same thing, if it had been differently modified owing to some different change in the environment from what actually occurred), then this particular species which constitutes the last link in that particular line of descent—and this also applies to man—would never have come into existence.

The ultimate development of man has, therefore, roughly speaking, depended on something like a million distinct modifications, each of a special type and dependent on some precedent changes in the organic and inorganic environments, or on both. The chances against such an enormously long series of definite modifications having occurred twice over, even in the same planet but in different isolated portions of it, as in the eastern and western hemispheres of the earth had they been completely separated from each other, are almost infinite, when we know how easily the balance of nature can be disturbed, as in those cases when man purposely or accidentally introduced pigs, rabbits, cats, or weeds into new countries. Even if it be said that those biologists may be right who deny the necessary unity of origin of every species, that would not make much difference, for it is never claimed that all species have had a multiple origin, but only that such an origin is not impossible, and that in rare cases it has actually occurred.

But if such long-continued identity of the whole course of evolution is hardly conceivable on different parts of the *same* planet, where all the great and essential conditions are equally fulfilled, how infinitely improbable it becomes that such an identity should have arisen, and have been maintained during millions of ages, on other planets of other suns, where the whole series of fundamental conditions which I have shown to be essential for *any* high development of life, though they might in rare cases approximate those of the earth, could certainly never have been quite identical. And without absolute identity to the smallest details, any identity of development, resulting after millions of ages in the *same* forms of the higher animals, is manifestly impossible.

Of course it may be said that a creature with a mind and spiritual nature equal to that of man might have been developed in a very different form. To discuss this question fully would need another volume, but I may briefly state why it seems quite inadmissible. In the first place, man differs from all other animals in the range and speciality of his mental nature even more than in his physical structure. It is generally admitted that his mental development has been rendered possible by a combination of three factors: the erect posture and free hand, the specialised vocal organisation rendering articulate speech possible, and the exceptional development of the brain. The combination of these has made man the superior of all other animals, has given him the power to modify nature, to create the arts, and to develop to a marvellous and apparently inexhaustible extent his intellectual, moral, and idealistic faculties, enabling him to extend his vision to the remotest limits of the material universe, to discover its laws, and to penetrate ever deeper and deeper into the hidden secrets of nature.

No other animal types make the slightest approach to any of these high faculties or show any indication of the possibility of their development. In very many directions they have reached a limit of organic perfection beyond which there is no apparent scope for further advancement. Such perfect types we see in the dog, the horse, the cat-tribe, the deer and the antelopes, the elephants, the beaver, and the greater apes; while many others have become extinct because they were so highly specialised as to be incapable of adaptation to new conditions. All these are probably about equal in their mental faculties, and there is no indication that any of them are or have been progressing towards man's elevation, or that such progression, either physically or mentally, is possible. The mere assertion, therefore, that a being possessing man's intellectual and moral nature combined with a very different animal form, might have been developed, is wholly valueless. We have no evidence for it, while the fact that no other animal than man *has* developed his special faculties even to a lower degree, is strong evidence against it.

If all aspects of this problem are considered, it will be seen that the improbability of an *organic* development on any other planet resulting in some animal type that could possibly lead to such a very definite and peculiar mental and moral nature as that of man, is far greater than that of the simultaneous occurrence there of the numerous *physical* conditions which we have found to be essential for the existence and development of any of the higher forms of life. But the two improbabilities have to be compounded, that is, to be multiplied by each other. If the physical or cosmical improbabilities as set forth in the body of this volume are somewhere about a million to one, then the evolutionary improbabilities now urged cannot be considered to be less than perhaps a hundred millions to one; and the total chances against the evolution of man, or an equivalent moral and intellectual being, in any other planet, through the known laws of evolution, will be represented by a hundred millions of millions to one. This argument, I feel sure, will appeal to all biological students of evolution, who have not, so far as I know, taken any part in the discussion aroused by my work; while even the general reader will be able to see that the chances against the independent origin of man in other worlds, as explained in the first edition of this volume, are enormously increased by the additional and totally distinct series of improbabilities here set forth

I submit, therefore that the improbabilities of the independent development of man, even in one other world—and far more in thousands or millions of worlds, as usually supposed—are now shown to be so great as to approach very closely, if not quite to attain, the actually impossible. Of course this whole argument applies only to those who believe that the entire material universe, inclusive of man himself, is the product of the immutable laws and forces of nature, and most of my critics appear to adopt this view—the view of pure science, or, as some prefer to call it, the monistic view.

But to those who believe that the universe is the product of mind, that it shows proofs of design, and that man is the designed outcome of it, and who yet urge that other worlds in unknown numbers have also been designed to produce man, and have actually produced him—to these I reply, that such a view assumes a knowledge of the Creator's purpose and mode of action which we do not possess; that we have no guide to His purposes but the facts we actually know; that we *do* know that here, on our earth, man is the culmination of one line of evolution, not of many, and that the presumption, therefore, is, that no line of evolution in other worlds under other conditions could produce him.

But, further, we have no reason to suppose that the purpose of man's existence in the universe requires him to exist in infinite numbers, or that the number of human beings that have been or will be produced here is insufficient. On a moderate estimate of the antiquity of man, there must already have been produced many trillions of millions of human souls, and these millions may yet be enormously increased by such a course of development of the stellar universe as may keep up the needful supply of light and heat for a few more millions of centuries—a thing far more probable, far easier to conceive, and far more in accordance with the working of natural law than the independent development of a whole series of almost identical animal forms up to man, in many other worlds.

Whether we look at this great problem from the agnostic or religious point of view; whether we study it as scientific monists or as philosophical spiritualists, our only safe guide is to be found in the facts and the laws of nature as we know them, and in the conclusions to be logically derived from an unbiassed application of those laws to the question at issue. Such an application I have attempted to make in the present volume.

THIRTEEN

THE LATE NINETEENTH CENTURY

SOME POETIC AND RELIGIOUS WRITINGS

John Tyndall (1820–1893)

Tyndall's scientific writings extended to numerous areas and well equipped him to serve as superintendent of the Royal Institution, a leading London scientific center. Tyndall was also a popular writer. One historian commented concerning Tyndall's famous presidential address at the 1874 Belfast meeting of the British Association for the Advancement of Science that among Victorian battles regarding science and religion, "Probably no single incident in the conflict of religion and science raised so much furor."[1] The published draft of his "Additional Remarks on Miracles" (1867) contains the statement:

> To other miracles similar remarks apply. Transferring our thoughts from this little sand-grain of an earth to the immeasurable heavens, where countless

1. Frank M. Turner, "The Victorian Conflict between Science and Religion: A Professional Dimension," *Isis* 69 (1978): 373.

worlds with freights of life probably revolve unseen, the very suns which warm them being barely visible across abysmal space; reflecting that beyond these sparks of solar fire, suns innumerable may burn, whose light can never stir the optic nerve at all; and bringing these reflections face to face with the idea of the Builder and Sustainer of it all showing Himself in a burning bush, exhibiting His hinder parts, or behaving in other familiar ways ascribed to Him in Jewish Scripture, the Incongruity must appear.[2]

Then in his 1874 Presidential "Belfast Address" to the British Association for the Advancement of Science, Tyndall commented:

The impregnable position of science may be described in a few words. We claim, and we shall wrest from theology, the entire domain of cosmological theory. (2:210)

[Should there not be a] temptation to close to some extent . . . with Bruno, when he declares that Matter is not "that mere empty capacity which philosophers have pictured her to be, but the universal mother who brings forth all things. . . ?" Believing, as I do, in the continuity of nature, . . . I cross the boundary of the experimental evidence, and discern in . . . Matter . . . the promise and potency of all terrestrial life. (2:203–4)

Frederic Harrison (1831–1923)

Educated at King's College, London, and at Oxford, Harrison became a professor of jurisprudence and international law. He was also the leader of the positivist movement in England in the late nineteenth century, authoring many books in support of the positivist perspective.

With a geocentric astronomy . . . , the anthropomorphic Creator, the celestial resurrection, and the Divine Atonement, were natural and homogeneous ideas. . . . But with a science where this planet shrinks into an unconsidered atom . . . , the Augustan Theology goes overboard.[3]

---

2. John Tyndall, *Fragments of Science* (New York: P. F. Collier, 1901), 2:41–42.

3. As quoted without reference by Richard Holt Hutton in *Criticisms on Contemporary Thought and Thinkers* (London: Macmillan, 1894), 1:288.

Coventry Patmore (1823–1896)

While working at the British Museum, Patmore published much poetry, including religious poetry. A friend of Ruskin and of Tennyson, he converted to Catholicism in 1864. One example of his poetry is his "The Two Deserts" (1866):

Not greatly moved with awe am I
To learn that we may spy
Five thousand firmaments beyond our own.
The best that's known
Of the heavenly bodies does them credit small.
View'd close, the Moon's fair ball
Is of ill objects worst,
A corpse in Night's highway, naked, fire-scarr'd, accurst;
And now they tell
That the Sun is plainly seen to boil and burst
Too horribly for hell.
So, judging from these two,
As we must do,
The Universe, outside our living Earth,
Was all conceiv'd in the Creator's mirth.
Forecasting at the time Man's spirit deep,
To make dirt cheap.
Put by the Telescope!
Better without it man may see,
Stretch'd awful in the hush'd midnight,
The ghost of his eternity.
Give me the nobler glass that swells to the eye
The things which near us lie,
Till Science rapturously hails,
In the minutest water-drop
A torment of innumerable tails
These at the least do live.
But rather give
A mind not much to pry
Beyond our royal-fair estate
Betwixt these deserts blank of small and great.
Wonder and beauty our own courtiers are,
Pleasing to catch our gaze,

And out of obvious ways
Ne'er wandering far.[4]

Robert Browning (1812–1889)

A prolific poet who is buried in Westminster Abbey, Robert Browning is known also as the husband of Elizabeth Barrett Browning, whose poetic gifts probably surpassed those of her husband. In his *The Ring and the Book* (1868–69), Browning put the following words in the mouths of a seventeenth-century pope:

I it is who have been appointed here
To represent Thee, in my turn, on earth,
Just as, if new philosophy know aught,
This one earth, out of all the multitude
Of peopled worlds, as stars are now supposed,—
Was chosen, and no sun-star of the swarm,
For stage and scene of Thy transcendent act
Beside which even the creation fades
Into a puny exercise of power.
Book X, lines 1332–41[5]

William Kingdon Clifford (1845–1879)

Educated at King's College, London, and at Trinity College, Cambridge, Clifford became professor of mathematics at University College, London. Clifford was known not only for his mathematical but also for his philosophical and popular writings. In his "The Influence upon Morality of a Decline in Religious Belief," he stated:

A little field-mouse, which busies itself in the hedge, and does not mind my company, is more to me than the longest ichthyosaurus that ever lived, even if he lived a thousand years. When we look at a starry sky, the spectacle whose awfulness Kant compared with that of the moral sense, does it help out our poetic emotion to reflect that these specks are really very very big, and very very hot, and very very far away? Their heat and their bigness oppress us; we

4. Coventry Patmore, "The Two Deserts," in *The Poems of Coventry Patmore*, ed. Frederick Page (London: Oxford University Press, 1949), 381–82.

5. Robert Browning, *The Ring and the Book*, in *The Works of Robert Browning* (New York: Barnes and Noble, 1966 repr. of 1912 orig.), 6:199.

should like them to be taken still farther away, the great blazing lumps. But when we think of the unseen planets that surround them, of the wonders of life, of reason, of love that may dwell therein, then indeed there is something sublime in the sight. Fitness and kinship: these are the truly great things for us, not force and massiveness and length of days.[6]

Gerard Manley Hopkins (1844–1889)

Educated at Oxford, Hopkins converted to Catholicism in 1866 and entered the Jesuit order. He taught Greek at Dublin University. Although he wrote much poetry, he published none of it during his lifetime. Hopkins is now widely considered the most gifted Catholic poet who wrote in English during the nineteenth century. The sermon from which the following words are drawn was probably written around 1881.

As besides the actual world there is an infinity of possible worlds, differing in all degrees of difference from what now is down to the having nothing in common with it but virgin matter, each of which possible worlds, and this actual one are like so many 'cleaves' or exposed faces of some pomegranate (or other fruit) cut in all directions across: so there is an infinity of possible strains of action and choice for each possible self in these worlds and the sum of these strains would be also like a pomegranate in the round, which God sees whole but of which we see at best only one cleave.[7]

Thomas Hardy (1840–1928)

One of the most highly regarded of all English novelists, Hardy in his *Two on a Tower* (1882) treated a love affair between a handsome young astronomer and a wealthy local woman, setting this romance against the background of the bleak universe that had been sketched in the writings of such authors as R. A. Proctor.[8]

This slightly-built romance was the outcome of a wish to set the emotional history of two infinitesimal lives against the stupendous background of the stel-

6. William Kingdon Clifford, *Lectures and Essays* (London: Macmillan, 1886), 390.

7. From *Sermons and Writings of Gerard Manley Hopkins*, ed. Christopher Devlin (London: Oxford University Press, 1959), 151.

8. The following passages from Hardy's *Two on a Tower* cite page numbers from a 1975 printing (London: Macmillan).

lar universe, and to impart to the readers the sentiment that of these contrasting magnitudes the smaller might be greater to them as men. (29)

She looked around over the magnificent stretch of sky that their high position unfolded. "O, thousands,—hundreds of thousands," she said absently.

"No, there are only about three thousand. Now, how many do you think are brought within sight by the help of a powerful telescope?"

"I won't guess."

"Twenty millions. So that, whatever the stars were made for, they were not made to please our eyes. It is just the same in everything; nothing is made for man." (56)

"The imaginary picture of the sky as the concavity of a dome whose base extends from the horizon of our earth is grand, simply grand, and I wish I had never got beyond looking at it in that way. But the actual sky is a horror."

"A new view of our old friends, the stars," she said, smiling up at them.

"But such an obviously true one!" said the young man. "You would hardly think, at first, that horrid monsters lie up there waiting to be discovered by any moderately penetrating mind—monsters to which those of the oceans bear no sort of comparison."

"What monsters may they be?"

"Impersonal monsters, namely, Immensities. Until a person has thought out the stars and their interspaces, he has hardly learnt that there are things much more terrible than monsters of shape, namely monsters of magnitude without known shape. Such monsters are the voids and waste places of the sky. Look, for instance, at those pieces of darkness in the Milky Way. . . . In these our sight plunges quite beyond any twinkler we have yet visited. Those are deep wells for the human mind to let itself down into, leave alone the human body!" (57)

"And to add a new weirdness to what the sky possesses in its size and formlessness, there is involved the quality of decay. For all the wonder of these everlasting stars, eternal spheres. and what not, they are not everlasting, they are not eternal! they burn out like candles. You see that dying one in the body of the Greater Bear? Two centuries ago it was as bright as the others. . . . Imagine them all extinguished, and your mind feeling its way through a heaven of total darkness, occasionally striking against the black, invisible cinders of those stars. . . . If you are cheerful, and wish to remain so, leave the study of astronomy alone. Of all the sciences, it alone deserves the character of the terrible." (58)

Hardy invoked the same sort of universe when, in 1902, he published his poem "God-Forgotten":

I towered far, and lo! I stood within
The presence of the Lord Most High,
Sent thither by the sons of Earth, to win
Some answer to their cry.

—"The Earth, sayest thou? The Human race?
By Me created? Sad its lot?
Nay: I have no remembrance of such place:
Such world I fashioned not.",—

—"O Lord forgive me when I say
Thou spak'st the word that mad'st it all."
"The Earth of men—let me bethink me. . . .
Yea! I dimly do recall.

"Some tiny sphere I framed long back
(Mid millions of such shapes of mine)
So named . . . it perished surely—not a wrack
Remaining, or a sign?

"It lost my interest from the first,
My aims therefor succeeding ill;
Haply it died of doing as it durst?"—
"Lord, it existeth still."

"Dark then its life! For not a cry
Of aught it bears do I now hear;
Of its own act the threads were snapt whereby
Its plaints had reached mine ear.

"It used to ask for gifts of good,
Till came its severance, self-entailed
When sudden silence on that side ensued,
And has till now prevailed.

"All other orbs have kept in touch;
Their voicings reach me speedily:
Thy people took upon them overmuch
In sundering them from me!

"And it is strange—though sad enough—
Earth's race should think that one whose call
Frames, daily, shining spheres of flawless stuff
Must heed their tainted ball! . . .

"But sayest thou 'tis by pangs distraught,
And strife, and silent suffering?—
Deep grieved am I that injury should be wrought
Even on so poor a thing!

"Thou shouldst have learnt that *Not to Mend*
For Me could mean but *Not to Know*:
Hence Messengers! and straightway put an end
To what men undergo.". . .

Homing at dawn, I thought to see
One of the Messengers standing by.
—Oh, childish thought! . . . Yet often it comes to me
When trouble hovers nigh.[9]

Aubrey de Vere (1814–1902)

Educated at Trinity College, Dublin, de Vere became a well-known poet. He converted to Catholicism in 1851. His poem "The Death of Copernicus" consists of reflections he attributed to Copernicus on his death bed as he ponders the effects that his heliocentric theory might have:

Was Earth too small to be of God created?
Why then too small to be redeemed?
. . . .

9. Thomas Hardy, "God-Forgotten," in *Poems of the Past and Present*, 2nd. ed (New York: Harper & Brothers, 1902), 99–102.

They that know not of a God
How know they that the stars have habitants?
'Tis Faith and Hope that spread delighted hands
To such belief; no formal proof attests it.
Concede them peopled; can the sophist prove
Their habitants are fallen? That too admitted,
Who told him that redeeming foot divine
Ne'er trod those spheres? That fresh assumption granted
What then? Is not the Universe a whole?
Does not the sunbeam herald from the sun
Gladden the violet's bosom? Moons uplift
The tides: remotest stars lead home the lost:
Judæa was one country, one alone:
Not less Who died there died for all. The Cross
Brought help to vanished nations: Time opposed
No bar to Love: why then should Space oppose one?
We know not what Time is, nor what is Space;—
Why dream that bonds like theirs can bind the Unbounded?
If Earth be small, likelier it seems that Love
Compassionate most and condescending most
To Sorrow's nadir depths, should choose that Earth
For Love's chief triumph, missioning thence her gift
Even to the utmost zenith![10]

Alfred Lord Tennyson (1809–1892)

Cambridge educated, Tennyson is typically ranked as the most productive poet of nineteenth-century England. Tennyson's strong interest in astronomy found expression in many of his writings.

From his "Despair" (1885):

III

And the suns of the limitless universe sparkled and shone in the sky,
Flashing with fires as of God, but we knew that their light was a lie—
Bright as with deathless hope—but however they sparkled and shone,

10. Aubrey de Vere, "The Death of Copernicus," *Contemporary Review* 56 (1889): 421–30, at 424.

The dark little worlds running round them were worlds of woe like our own—
No soul in the heaven above, no soul on the earth below,
A fiery scroll written over with lamentation and woe.[11]

From "Locksley Hall Sixty Years After" (1886):

Warless? war will die out late then. Will it ever? late or soon?
Can it, till this outworn earth be dead as yon dead world the moon?

Dead the new astronomy calls her.—On this day and at this hour,
In this gap between the sandhills, whence you see the Locksley tower,

Here we met, our latest meeting—Amy—sixty years ago—
She and I—the moon was falling green thro' a rosy glow,

Just above the gateway tower, and even where you see her now—
Here we stood and clapst each other, swore the seeming-deathless vow.—

Dead, but how her living glory lights the hall, the dune, the grass!
Yet the moonlight is the sunlight, and the sun himself will pass.

Venus near her! smiling downward at this earthlier earth of ours,
Closer on the sun, perhaps a world of never fading flowers.

Hesper, whom the poet call'd Bringer home of all good things—
All good things may move in Hesper, perfect peoples, perfect kings.

Hesper—Venus—were we native to that splendor or in Mars,
We should see the globe we groan in, fairest of their evening stars.

Could we dream of wars and carnage, craft and madness, lust and spite,
Roaring London, raving Paris, in that point of peaceful light?

Might we not in glancing heavenward on a star so silver-fair,
Yearn, and clasp the hands and murmur, "Would to God that we were there"?

---

11. *The Poems and Plays of Alfred Lord Tennyson* (New York: Modern Library, 1938), 807. His other poems in this section have also been taken from this volume.

Forward, backward, backward, forward, in the immeasurable sea,
Sway'd by vaster ebbs and flows than can be known to you or me.

All the suns—are these but symbols of innumerable man,
Man or Mind that sees a shadow of the planner or the plan?

Is there evil but on earth? or pain in every peopled sphere?
Well, be grateful for the sounding watchword 'Evolution' here.

Evolution ever climbing after some ideal good,
And Reversion ever dragging Evolution in the mud.

What are men that He should heed us? cried the king of sacred song;
Insects of an hour, that hourly work their brother insect wrong,

While the silent heavens roll and suns along their fiery way,
All their planets whirling round them, flash a million miles a day.

Many an aeon moulded earth before her highest man, was born,
Many an aeon too may pass while earth is manless and forlorn,

Earth so huge, and yet so bounded—pools of salt, and plots of land—
Shallow skin of green and azure—chains of mountain, grains of sand!

Only That which made us meant us to be mightier by and by,
Set the sphere of all the boundless heavens within the human eye

Sent the shadow of Himself, the boundless, thro' the human soul;
Boundless inward in the atom, boundless outward in the Whole.

lines 171–212

From his "Vastness" (1889):

I

Many a hearth upon our dark globe sighs after many a vanish'd face,
Many a planet by many a sun may roll with the dust of a vanish'd race.

II

Raving politics, never at rest—as this poor earth's pale history runs,—
What is it all but a trouble of ants in the gleam of a million million of suns?

Finally, two statements attributed to Tennyson and published by his nephew's wife, Agnes Grace Weld, after Tennyson's death in 1892 are relevant:

> When I think of the immensity of the universe, I am filled with the sense of my own utter insignificance, and am ready to exclaim with David: "What is man that Thou art mindful of him!" The freedom of the human will and the starry heavens are the two greatest marvels that come under our observation, and when I think of all the mighty worlds around us, to which ours is but a speck, I feel what poor little worms we are, and ask myself, What is greatness? I do not like a word as design to be applied to the Creator of all these worlds, it makes Him seem a mere artificer. A certain amount of anthropomorphism must, however, necessarily enter into our conception of God, because, though there may be infinitely higher being than ourselves in the worlds beyond ours, yet to our conception man is the highest form of being.
>
> We shall have much to learn in a future world, and I think we shall all be children to begin with when we get to heaven, whatever our age when we die, and shall grow on there from childhood to the prime of life, at which we shall remain forever. My idea of heaven is to be engaged in perpetual ministry to souls in this and other worlds.[12]

Alice Meynell (1847–1922)

Poet, essayist, and critic, Meynell was, like Patmore, de Vere, and Hopkins, a Catholic. Her "Christ in the Universe,"[13] a selection from which follows, has been reprinted in many volumes on extraterrestrial life issues.

> With this ambiguous earth
> His dealings have been told us. These abide:
> The signal to a maid, the human birth,
> The lesson, and the young Man crucified.
>
> But not a star of all
> The innumerable host of stars has heard

12. Agnes Grace Weld, "Talks with Tennyson," *Contemporary Review* 63 (1893): 394–97, at 395 and 397.

13. "Christ in the Universe" was published in 1913 or possibly earlier. The poem can be found in *The Poems of Alice Meynell* (New York: C. Scribner's Sons, 1923), 92.

How He administered this terrestrial ball.
Our race have kept their Lord's entrusted Word.

Of His earth-visiting feet
None knows the secret, cherished, perilous,
The terrible, shamefast, frightened, whispered, sweet
Heart-shattering secret of His way with us.

No planet knows that this
Our wayside planet, carrying land and wave,
Love and life multiplied, and pain and bliss,
Bears, as chief treasure, one forsaken grave.

Nor, in our little day,
May His devices with the heavens be guessed,
His pilgrimage to tread the Milky Way,
Or His bestowals there be manifest.

But, in the eternities,
Doubtless we shall compare together, hear
A million alien Gospels, in what guise
He trod the Pleiades, the Lyre, the Bear.

O be prepared, my soul!
To read the inconceivable, to scan
The million forms of God those stars unroll
When, in our turn, we show to them a Man.

FYODOR DOSTOEVSKY (1821–1881)

The degree to which ideas of extraterrestrial intelligent life permeated nineteenth-century culture can be illustrated in many ways, not least from the writings of its leading literary figures. In a century of great novelists, few if any attained the international acclaim accorded the great Russian author Dostoevsky. His *Notes from the Underground* (1864) ranks among the most powerful and probing short stories ever composed, whereas his four great novels—*Crime and Punishment* (1866), *The Idiot* (1868), *The Possessed* (1871), and especially *The Brothers Karamazov* (1879)—continue to be seen as creations of extraordinary importance. It might seem unlikely that ideas of extraterrestrial life would find a place in the writings of such an author, but the contrary is the case. For example, in a climactic scene in *The Brothers Karamazov*, Alyosha, the saintly youngest of the three Karamazov brothers, undergoes a transforming experience after the death of his spiritual master, Fr. Zossima.

> What was he [Alyosha] weeping over?
>
> Oh! in his rapture he was weeping even over those stars, which were shining to him from the abyss of space, and "he was not ashamed of that ecstasy." There seemed to be threads from all those innumerable worlds of God, li[n]king his soul to them, and it was trembling all over "in contact with other worlds." He longed to forgive every one and for everything, and to beg forgiveness. Oh, not for himself, but for all men, for all and for everything. "And others are praying for me too," echoed again in his soul. But with every instant he felt clearly, and as it were tangibly, that something firm and unshakable as that vault of heaven had entered in to his soul. It was as though some idea has seized the sovereignty of his mind—and it was for all his life and for ever and ever. He had fallen on the earth a weak boy, and he rose up a resolute champion, and he knew and felt it suddenly at the very moment of his ecstasy. And never, never, all his life long, could Alyosha forget this minute.
>
> "Some one visited my soul in that hour," he used to say afterwards, with implicit faith in his words.
>
> Within three days he had left the monastery in accordance with the words of his elder, who had bidden him "sojourn in the world."[14]

Passages appear elsewhere in Dostoevsky's writings, but extraterrestrials appear most prominently in his last short story, "The Dream of a Ridiculous Man" (1877), which in its heavy use of introspection can be seen as comparable to his *Notes from the Underground.*

Fyodor Dostoevsky, "The Dream of a Ridiculous Man," in his *An Honest Thief and Other Stories*, trans. Constance Garnett (New York: Macmillan, 1919), 307, 314–15, 315–18, 319–26.

I am a ridiculous person. Now they call me a madman. That would be a promotion if it were not that I remain as ridiculous

14. Fyodor Dostoevsky, *The Brothers Karamazov*, trans. Constance Garnett (New York: Modern Library, 1950), 436–37.

in their eyes as before. But now I do not resent it, they are all dear to me now, even when they laugh at me—and, indeed, it is just then that they are particularly dear to me. I could join in their laughter—not exactly at myself, but through affection for them, if I did not feel so sad as I look at them. Sad because they do not know the truth and I do know it. Oh, how hard it is to be the only one who knows the truth! But they won't understand that. No, they won't understand it.

After some further discussion of his ridiculousness, the narrator recounts events that began on the night of November 3, when, as he was walking home one evening, he contemplated committing suicide. Sighting a particularly bright star moved him to decide to take his life that night. Shortly thereafter, a young girl asked his help, which he refused. Arriving home, he sat in a chair and prepared to take his life by placing his revolver beside him. Then he began to think of the girl and was puzzled by the twinge of compassion he felt for her even as he prepared to end his own life. Moreover, he began to think about two questions: "if I had lived before on the moon or on Mars and there had committed the most disgraceful and dishonourable action and had there been put to such shame and ignominy as one can only conceive and realise in dreams and nightmares, and if, afterwards finding myself on earth, I were able to retain the memory of what I had done on the other planet and at the same time knew that I should never, under any circumstance, return there, then looking from the earth to the moon—*should I care or not?* Should I feel shame for that action or not?" Pondering these questions, he fell asleep and began to dream. He recalls (as part of the dream) firing the revolver at his heart, being placed in a coffin, and being buried. Although dead, the narrator recounts his thoughts and experiences.

. . . behold my grave suddenly was rent asunder, that is, I don't know whether it was opened or dug up, but I was caught up by some dark and unknown being and we found ourselves in space. I suddenly regained my sight. It was the dead of night, and never, never had there been such darkness. We were flying through space far away from the earth. I did not question the being who was taking me; I was proud and waited. I assured myself that I was not afraid, and was thrilled with ecstasy at the thought that I was not afraid. I do not know how long we were flying, I cannot imagine; it happened as it always does in dreams when you skip over space and time, and the laws of thought and existence, and only pause upon the points for which the heart yearns. I remember that I suddenly saw in the darkness a star. "Is that Sirius?" I asked impulsively, though I had not meant to ask questions.

"No, that is the star you saw between the clouds when you were coming home," the being who was carrying me replied.

After mentioning some further exchanges with the being transporting him, he returns to his story.

We were flying through dark, unknown space. I had for some time lost sight of the constellations familiar to my eyes. I knew that there were stars in the heavenly spaces the light of which took thousands or millions of years to reach the earth. Perhaps we were already flying through those spaces. I expected something with a terrible anguish that tortured my heart. And suddenly I was thrilled by a familiar feeling that stirred me to the depths: I suddenly caught sight of our sun! I knew that it could not be *our* sun, that gave life to *our* earth, and that we were an infinite distance from our sun, but for some reason I knew in my whole being that it was a sun exactly like ours, a duplicate of it. A sweet, thrilling feeling resounded with ecstasy in my heart: the kindred power of the same light which had given me light stirred an echo in my heart and awakened it, and I had a sensation of life, the old life of the past for the first time since I had been in the grave.

"But if that is the sun, if that is exactly the same as our sun," I cried, "where is the earth?"

And my companion pointed to a star twinkling in the distance with an emerald light. We were flying straight towards it.

"And are such repetitions possible in the universe? Can that be the law of Nature? . . . And if that is an earth there, can it be just the same earth as ours . . . just the same, as poor, as unhappy, but precious and beloved for ever, arousing in the most ungrateful of her children the same poignant love for her that we feel for our earth?" I cried out, shaken by irresistible, ecstatic love for the old familiar earth which I had left.

The image of the poor child whom I had repulsed flashed through my mind.

"You shall see it all," answered my companion, and there was a note of sorrow in his voice.

But we were rapidly approaching the planet. It was growing before my eyes; I could already distinguish the ocean, the outline of Europe; and suddenly a feeling of a great and holy jealousy glowed in my heart.

"How can it be repeated and what for? I love and can love only that earth which I have left, stained with my blood, when, in my ingratitude, I quenched my life with a bullet in my heart. But I have never, never ceased to love that earth, and perhaps on the very night I parted from it I loved it more than ever. Is there suffering upon this new earth? On our earth we can only love with suffering and through suffering. We cannot love otherwise, and we know of no other sort of love. I want suffering in order to love. I long, I thirst, this very instant, to kiss with tears the earth that I have left, and I don't want, I won't accept life on any other!"

But my companion had already left me. I suddenly, quite without noticing how, found myself on this other earth, in the bright light of a sunny day, fair as paradise. I believe I was standing on one of the islands that make up on our globe the Greek archipelago, or on the coast of the mainland facing that archipelago. Oh, everything was exactly as it is with us, only everything seemed to have a festive radiance, the splendour of some great, holy triumph attained at last. The caressing sea, green as emerald, splashed softly upon the shore and kissed it with manifest, almost conscious love. The tall, lovely trees stood in all the glory of their blossom, and their innumerable leaves greeted me, I am certain, with their soft, caressing rustle and seemed to articulate words of love. The grass glowed with bright and fragrant flowers. Birds were flying in flocks in the air, and perched fearlessly on my shoulders and arms and joyfully struck me with their darling, fluttering wings. And at last I saw and knew the people of this happy land. That came to me of themselves, they surrounded me, kissed me. The children of the sun, the children of their sun—oh, how beautiful they were! Never had I seen on our own earth such beauty in mankind. Only perhaps in our children, in their earliest years, one might find, some remote faint reflection of this beauty. The eyes of these happy people shone with a clear brightness. Their faces were radiant with the light of reason and fullness of a serenity that comes of perfect understanding, but those faces were gay; in their words and voices there was a note of childlike joy. Oh, from the first moment, from the first glance at them, I understood it all! It was the earth untarnished by the Fall; on it lived people who had not sinned. They lived just in such a paradise as that in which, according to all the legends of mankind, our first parents lived before they sinned; the only difference was that all this earth was the same paradise. These people, laughing joyfully, thronged round me and caressed me; they took me home with them, and each of them tried to reassure me. Oh, they asked me no questions, but they seemed, I fancied, to know everything without asking, and they wanted to make haste to smoothe away the signs of suffering from my face.

And do you know what? Well, granted that it was only a dream, yet the sensation of the love of those innocent and beautiful people has remained with me for ever, and I feel as though their love is still flowing out to me from over there. I have seen them myself, have known them and been convinced; I loved them, I suffered for them afterwards. Oh, I understood at once even at the time that in many things I could not understand them at all; as an up-to-date Russian progressive and contemptible Petersburger, it struck me as inexplicable that, knowing so much, they had, for instance, no science like ours. But I soon realised that their knowledge was gained and fostered by intuitions different from those of us on earth, and that their aspirations, too, were quite different. They desired nothing and were at peace; they did not aspire to knowledge of life as we aspire to understand it, because their lives were full. But their knowledge was higher and deeper than ours; for our science seeks to explain what life is, aspires to understand it in order to teach others how to love, while they without science knew how to live; and that I understood, but I could not understand their knowledge. They showed me their trees, and I could not understand the intense love with which they looked at them; it was as though they were talking with creatures like themselves. And perhaps I shall not be mistaken if I say that they conversed with them. Yes, they had found their language, and I am convinced that the trees understood them. They looked at all Nature like that—at the animals who lived in peace with them and did not attack them, but loved them, conquered by their love. They pointed to the stars and told me something about them which I could not understand, but I am convinced that they were somehow in touch with the stars, not only in thought, but by some living channel. Oh, these people did not persist in trying to make me understand them, they loved me without that, but I knew that they would never understand me, and so I hardly spoke to them about our earth. I only kissed in their presence the earth on which they lived and mutely worshipped them themselves. And they saw that and let me worship them without being abashed at my adoration, for they themselves loved much. They were not unhappy on my account when at times I kissed their feet with tears, joyfully conscious of the love with which they would respond to mine. At times I asked myself with wonder how it was they were able never to offend a creature like me, and never once to arouse a feeling of jealousy or envy in me? Often I wondered how it could be that, boastful and untruthful as I was, I never talked to them of what I knew—of which, of course, they had no notion—that I was never tempted to do so by a desire to astonish or even to benefit them.

Two paragraphs are omitted at this point in which the narrator elaborates on the joyfulness, goodness of heart, cooperation, and freedom from fear of death manifest in the extraterrestrials he has encountered. Their planet is presented as a paradise free of the evils so characteristic of Earth. He also discusses the pleasure the inhabitants take in song.

Some of their songs, solemn and rapturous, I scarcely understood at all. Though I understood the words I could never fathom their full significance. It remained, as it were, beyond the grasp of my mind, yet my heart unconsciously absorbed it more and more. I often told them that I had had a presentiment of it long before, that this joy and glory had come to me on our earth in the form of a yearning melancholy that at times approached insufferable sorrow; that I had had a foreknowledge of them all and of their glory in the dreams of my heart and the visions of my mind; that often on our earth I could not look at the setting sun without tears . . . that in my hatred for the men of our earth there was always a yearning anguish: why could I not hate them without loving them? why could I not help forgiving them? and in my love for them there was a yearning grief: why could I not love them without hating them? They listened to me, and I saw they could not conceive what I was saying, but I did not regret that I had spoken to them of it: I knew that they understood the intensity of my yearning anguish over those whom I had left. But when they looked at me with their sweet eyes full of love, when I felt that in their presence my heart, too, became as innocent and just as theirs, the feeling of the fullness of life took my breath away, and I worshipped them in silence.

Oh, everyone laughs in my face now, and assures me that one cannot dream of such details as I am telling now, that I only dreamed or felt one sensation that arose in my heart in delirium and made up the details myself when I woke up. And when I told them that perhaps it really was so, my God, how they shouted with laughter in my face, and what mirth I caused! Oh, yes, of course I was overcome by the mere sensation of my dream, and that was all that was preserved in my cruelly wounded heart; but the actual forms and images of my dream, that is, the very ones I really saw at the very time of my dream, were filled with such harmony, were so lovely and enchanting and were so actual, that on awakening I was, of course, incapable of clothing them in our poor language, so that they were bound to become blurred in my mind; and so perhaps I really was forced afterwards to make up the details, and so of course to distort them in my passionate desire to convey some at least of them as quickly as I could. But on the other hand, how can I help believing that it was all true? It was perhaps a thousand times brighter, happier and more joyful than I describe it. Granted that I dreamed it, yet it must have been real. You know, I will tell you a secret: perhaps it was not a dream at all! For then something happened so awful, something so horribly true, that it could not have been imagined in a dream. My heart may have originated the dream, but would my heart alone have been capable of originating the awful event which happened to me afterwards? How could I alone have invented it or imagined it in my dream? Could my petty heart and fickle, trivial mind have risen to such a revelation of truth? Oh, judge for yourselves: hitherto I have concealed it, but now I will tell the truth. The fact is that I . . . corrupted them all!

Yes, yes, it ended in my corrupting them all! How it could come to pass I do not know, but I remember it clearly. The dream embraced thousands of years and left in me only a sense of the whole. I only know that I was the cause of their sin and downfall. Like a vile trichina, like a germ of the plague infecting whole kingdoms, so I contaminated all this earth, so happy and sinless before my coming. They learnt to lie, grew fond of lying, and discovered the charm of falsehood. Oh, at first perhaps it began innocently, with a jest, coquetry, with amorous play, perhaps indeed with a germ, but that germ of falsity made its way into their hearts and pleased them. Then sensuality was soon begotten, sensuality begot jealousy, jealousy—cruelty. . . . Oh, I don't know, I don't remember; but soon, very soon the first blood was shed. They marvelled and were horrified, and began to be split up and divided. They formed into unions, but it was against one another. Reproaches, upbraidings followed. They came to know shame, and shame brought them to virtue. The conception of honour sprang up, and every union began waving its flags. They began torturing animals, and the animals withdrew from them into the forests and became hostile to them. They began to struggle for separation, for isolation, for individuality, for mine and thine. They began to talk in different languages. They became acquainted with sorrow and loved sorrow; they thirsted for suffering, and said that truth could only be attained through suffering. Then science appeared. As they became wicked they began talking of brotherhood and humanitarianism, and understood those ideas. As they became criminal, they invented justice and drew up whole legal codes in order to observe it, and to ensure their being kept, set up a guillotine. They hardly remembered what they had lost, in fact refused to believe that they had ever been happy and innocent. They even laughed at the possibility of this happiness in the past, and called it a dream. They could not even imagine it in definite form and shape, but, strange and wonderful to relate, though they lost all faith in their past happiness and called it a legend, they so longed to be happy and innocent once more that they succumbed to this desire like children, made an idol of it, set up temples and worshipped their own idea, their own desire; though at the same time they fully believed that it was unattainable and could not be realised, yet they bowed down to it and adored it with tears! Nevertheless, if it could have happened that they had returned to the innocent and happy condition which they had lost, and if someone had shown it to them again and had asked them whether they wanted to go back to it, they would certainly have refused. They answered me:

"We may be deceitful, wicked and unjust, we know it and weep over it, we grieve over it; we torment and punish ourselves more perhaps than that merciful Judge Who will judge us and whose Name we know not. But we have science, and by the means of it we shall find the truth and we shall arrive at it consciously. Knowledge is higher than feeling, the consciousness of life is higher than life. Science will give us wisdom, wisdom will reveal the laws, and the knowledge of the laws of happiness is higher than happiness."

That is what they said, and after saying such things everyone began to love himself better than anyone else, and indeed they could not do otherwise. All became so jealous of the rights of their own personality that they did their very utmost to curtail and destroy them in others, and made that the chief thing in their lives. Slavery followed, even voluntary slavery;

the weak eagerly submitted to the strong, on condition that the latter aided them to subdue the still weaker. Then there were saints who came to these people, weeping, and talked to them of their pride, of their loss of harmony and due proportion, of their loss of shame. They were laughed at or pelted with stones. Holy blood was shed on the threshold of the temples. Then there arose men who began to think how to bring all people together again, so that everybody, while still loving himself best of all, might not interfere with others, and all might live together in something like a harmonious society. Regular wars sprang up over this idea. All the combatants at the same time firmly believed that science, wisdom and the instinct of self-preservation would force men at last to unite into a harmonious and rational society; and so, meanwhile, to hasten matters, 'the wise' endeavoured to exterminate as rapidly as possible all who were 'not wise' and did not understand their idea, that the latter might not hinder its triumph. But the instinct of self-preservation grew rapidly weaker; there arose men, haughty and sensual, who demanded all or nothing. In order to obtain everything they resorted to crime, and if they did not succeed—to suicide. There arose religions with a cult of non-existence and self-destruction for the sake of the everlasting peace of annihilation. At last these people grew weary of their meaningless toil, and signs of suffering came into their faces, and then they proclaimed that suffering was a beauty, for in suffering alone was there meaning. They glorified suffering in their songs. I moved about among them, wringing my hands and weeping over them, but I loved them perhaps more than in old days when there was no suffering in their faces and when they were innocent and so lovely. I loved the earth they had polluted even more than when it had been a paradise, if only because sorrow had come to it. Alas! I always loved sorrow and tribulation, but only for myself, for myself; but I wept over them, pitying them. I stretched out my hands to them in despair, blaming, cursing and despising myself. I told them that all this was my doing, mine alone; that it was I who had brought them corruption, contamination and falsity. I besought them to crucify me, I taught them how to make a cross.

I could not kill myself, I had not the strength, but I wanted to suffer at their hands. I yearned for suffering, I longed that my blood should be drained to the last drop in these agonies. But they only laughed at me, and began at last to look upon me as crazy. They justified me, they declared that they had only got what they wanted themselves, and that all that now was could not have been otherwise. At last they declared to me that I was becoming dangerous and that they should lock me up in a madhouse if I did not hold my tongue. Then such grief took possession of my soul that my heart was wrung, and I felt as though I were dying; and then . . . then I awoke.

It was morning, that is, it was not yet daylight, but about six o'clock. I woke up in the same arm-chair; my candle had burnt out; everyone was asleep in the captain's room, and there was a stillness all round, rare in our flat. First of all I leapt up in great amazement: nothing like this had ever happened to me before, not even in the most trivial detail; I had never, for instance, fallen asleep like this in my arm-chair. While I was standing and coming to myself I suddenly caught sight of my revolver lying loaded, ready—but instantly I thrust it away! Oh, now, life, life! I lifted up my hands and called upon eternal truth, not with words, but with tears; ecstasy, immeasurable ecstasy flooded my soul. Yes, life and spreading the good tidings! Oh, I at that moment resolved to spread the tidings, and resolved it, of course, for my whole life. I go to spread the tidings, I want to spread the tidings—of what? Of the truth, for I have seen it, have seen it with my own eyes, have seen it in all its glory.

And since then I have been preaching! Moreover I love all those who laugh at me more than any of the rest. Why that is so I do not know and cannot explain, but so be it. I am told that I am vague and confused, and if I am vague and confused now, what shall I be later on? It is true indeed: I am vague and confused, and perhaps as time goes on I shall be more so. And of course I shall make many blunders before I find out how to preach, that is, find out what words to say, what things to do, for it is a very difficult task. I see all that as clear as daylight, but, listen, who does not make mistakes?

And yet, you know, all are making for the same goal, all are striving in the same direction anyway, from the sage to the lowest robber, only by different roads. It is an old truth, but this is what is new: I cannot go far wrong. For I have seen the truth; I have seen and I know that people can be beautiful and happy without losing the power of living on earth. I will not and cannot believe that evil is the normal condition of mankind. And it is just this faith of mine that they laugh at. But how can I help believing it? I have seen the truth—it is not as though I had invented it with my mind, I have seen it, seen it, and *the living image* of it has filled my soul for ever. I have seen it in such full perfection that I cannot believe that it is impossible for people to have it. And so how can I go wrong? I shall make some slips no doubt, and shall perhaps talk in second-hand language, but not for long: the living image of what I saw will always be with me and will always correct and guide me. Oh, I am full of courage and freshness, and I will go on and on if it were for a thousand years! Do you know, at first I meant to conceal the fact that I corrupted them, but that was a mistake—that was my first mistake! But truth whispered to me that I was *lying*, and preserved me and corrected me. But how establish paradise—I don't know, because I do not know how to put it into words. After my dream I lost command of words. All the chief words, anyway, the most necessary ones. But never mind, I shall go and I shall keep talking, I won't leave off, for anyway I have seen it with my own eyes, though I cannot describe what I saw. But the scoffers do not understand that. It was a dream, they say, delirium, hallucination. Oh! As though that meant so much! And they are so proud! A dream! What is a dream? And is not our life a dream? I will say more. Suppose that this paradise will never come to pass (that I understand), yet I shall go on preaching it. And yet how simple it is: in one day, *in one hour* everything could be arranged at once! The chief thing is to love others like yourself, that's the chief thing, and that's everything; nothing else is wanted—you will find out at once how to arrange it all. And yet it's an old truth which has been told and retold a billion times—but it has not formed part of our lives! The consciousness of life is higher than life, the knowledge of the laws of happiness is higher than happiness—that is what one must contend against. And I shall. If only everyone wants it, it can be arranged at once.

And I tracked down that little girl . . . and I shall go on and on!

MARK TWAIN (1835–1910)

At various points in the life of the famous American author and humorist Mark Twain, astronomy and particularly ideas of extraterrestrial life influenced his thought and appeared in his writings. An early example of this occurred around 1858, when Twain read Thomas Paine's *Age of Reason*. The degree to which it moved him is suggested by Twain's remark: "I . . . read it with fear and hesitation, but marveling at its fearlessness and wonderful power."[15] It seems probable that the "wonderful power" that Twain attributed to Paine's volume was what led Twain on a number of occasions in the period around 1870 to attack Christianity on the basis of Paine's astronomical objection to it.

Twain launched one such attack in 1870 when, in discussing an astronomy book, he revealed to his wife his difficulties with Christianity.

> How insignificant we are, with our pigmy little world!—an atom glinting with uncounted myriads of other atom worlds . . . & yet prating complacently of our speck as the Great World, & regarding the other specks as pretty trifles made to steer our schooners by & inspire the reveries of "puppy" lovers. Did Christ live 33 years in each of the millions & millions of worlds that hold their majestic courses above our heads? Or was *our* small globe the favored one of all?[16]

Twain's hostility to Christianity ultimately took a far more public form in a novella he began to compose in the late 1860s, twice revised in the 1870s, but, realizing its controversial nature and heeding the sensitivities of his devout wife, withheld from publication until 1907; this is his *Extract from Captain Stormfield's Visit to Heaven*.

15. As quoted in Minnie M. Brashear, *Mark Twain: Son of Missouri* (Chapel Hill: University of North Carolina Press, 1934), 245.

16. As quoted in Dixon Wecter, ed., *The Love Letters of Mark Twain* (New York: Harper, 1949), 133.

The story opens with Stormfield, who has died, travelling through space and engaging in a race with a comet. Twain puts part of this adventure to use as a way of suggesting humanity's insignificance in the cosmos.

Mark Twain, *Extract from Captain Stormfield's Visit to Heaven* (New York: Harper & Brothers, 1909), 9–12, 15–20, 22–24, and 90–92.

Well, sir, I gained and gained, little by little, till at last I went skimming sweetly by the magnificent old conflagration's nose. By this time the captain of the comet had been rousted out, and he stood there in the red glare for'ard, by the mate, in his shirtsleeves and slippers, his hair all rats' nests and one suspender hanging, and how sick those two men did look! I just simply couldn't help putting my thumb to my nose as I glided away and singing out:

"Ta-ta! ta-ta! Any word to send to your family?"

Peters, it was a mistake. Yes, sir, I've often regretted that—it was a mistake. You see, the captain had given up the race, but that remark was too tedious for him—he couldn't stand it. He turned to the mate, and says he—

"Have we got brimstone enough of our own to make the trip?"

"Yes, sir."

"Sure?"

"Yes, sir—more than enough."

"How much have we got in cargo for Satan?"

"Eighteen hundred thousand billion quintillions of kazarks."

"Very well, then, let his boarders freeze till the next comet comes. Lighten ship! Lively, now, lively, men! Heave the whole cargo overboard!"

Peters, look me in the eye, and be calm. I found out, over there, that a kazark is exactly the bulk of a *hundred and sixty-nine worlds like ours!* They hove all that load overboard. When it fell it wiped out a considerable raft of stars just as clean as if they'd been candles and somebody blowed them out. As for the race, that was at an end. The minute she was lightened the comet swung along by me the same as if I was anchored. The captain stood on the stern, by the after-davits, and put his thumb to his nose and sung out—

"Ta-ta! ta-ta! Maybe *you've* got some message to send your friends in the Everlasting Tropics!"

Then he hove up his other suspender and started for'ard, and inside of three-quarters of an hour his craft was only a pale torch again in the distance. Yes, it was a mistake, Peters—that remark of mine. I don't reckon I'll ever get over being sorry about it. I'd 'a' beat the bully of the firmament if I'd kept my mouth shut.

Stormfield's travels take him to beautiful gates, which he realizes are gates to heaven.

I was pointed straight for one of these gates, and a-coming like a house afire. Now I noticed that the skies were black with millions of people, pointed for those gates. What a roar they made, rushing through the air! The ground was as thick as ants with people, too—billions of them, I judge.

I lit. I drifted up to a gate with a swarm of people, and when it was my turn the head clerk says, in a business-like way—

"Well, quick! Where are you from?"

"San Francisco," says I.

"San Fran—*what?*" says he.

"San Francisco."

He scratched his head and looked puzzled, then he says—

"Is it a planet?"

By George, Peters, think of it! "*Planet?*" says I; "it's a city. And moreover, it's one of the biggest and finest and—"

"There, there!" says he, "no time here for conversation. We don't deal in cities here. Where are you from in a *general* way?"

"Oh," I says, "I beg your pardon. Put me down for California."

I had him *again,* Peters! He puzzled a second, then he says, sharp and irritable—

"I don't know any such planet—is it a constellation?"

"Oh my goodness!" says I. "Constellation, says you? No—it's a State."

"Man, we don't deal in states here. *Will* you tell me where you are from *in general—at large,* don't you understand?"

"Oh, now I get your idea," I says. "I'm from America,—the United States of America."

Peters, do you know I had him *again?* If I hadn't I'm a clam! His face was as blank as a target after a militia shooting match. He turned to an under clerk and says—

"Where is America? *What* is America?"

The under clerk answered up prompt and says—

"There ain't any such orb."

"*Orb?*" says I. "Why, what are you talking about, young man? It ain't an orb; it's a country; it's a continent. Columbus discovered it; I reckon you've heard of *him,* anyway. America—why, sir, America—"

"Silence!" says the head clerk. "Once and for all where—are—you—*from?*"

"Well," says I, "I don't know anything more to say—unless I lump things, and just say I'm from the world."

"Ah," says he, brightening up, "now that's something like! *What* world?"

Peters, he had *me,* that time. I looked at him, puzzled, he looked at me, worried. Then he burst out—

"Come, come, what world?"

Says I, "Why, *the* world, of course."

"*The* world!" he says. "H'm! there's billions of them! . . . Next!"

That meant for me to stand aside. I done so, and a sky-blue man with seven heads and only one leg hopped into my place. I took a walk. It just occurred to me, then, that all the myriads I had seen swarming to that gate, up to this time, were just like that creature. I tried to run across somebody I was acquainted with, but they were out of acquaintances of mine just then. So I thought the thing over and finally sidled back there pretty meek and feeling rather stumped, as you might say.

"Well?" said the head clerk.

"Well, sir," I says, pretty humble, "I don't seem to make out which world it is I'm from. But you may know it from this—it's the one the Saviour saved."

He bent his head at the Name. Then he says, gently—

"The worlds He has saved are like to the gates of heaven in number—none can count them. What astronomical system is your world in?—perhaps that may assist."

Stormfield then mentions the planets in the solar system, the mention of Jupiter catching the clerk's interest.

"What system is Jupiter in?" [head clerk speaking to under clerk]

"I don't remember, sir, but I think there is such a planet in one of the little new systems away out in one of the thinly worlded corners of the universe. I will see."

He got a balloon and sailed up and up and up, in front of a map that was as big as Rhode Island. He went on up till he was out of sight, and by and by he came down and got something to eat and went up again. To cut a long story short, he kept on doing this for a day or two, and finally he came down and said he thought he had found that solar system, but it might be fly specks. So he got a microscope and went back. It turned out better than he feared. He had rousted out our system, sure enough. He got me to describe our planet and its distance from the sun, and then he says to his chief—

"Oh, I know the one he means, now, sir. It is on the map. It is called the Wart."

Says I to myself, "Young man, it wouldn't be wholesome for you to go down *there* and call it the Wart."

It having been determined which section of heaven he is from, Stormfield gains admission and proceeds to the area assigned to earthlings. Here he meets an old acquaintance, Sandy, who, having died earlier than Stormfield, is able to supply the captain with information on various aspects of heaven. In particular, Sandy informs Captain Stormfield that

". . . There were some prophets and patriarchs there that ours ain't a circumstance to, for rank and illustriousness and all that. Some were from Jupiter and other worlds in our own system, but the most celebrated were three poets, Saa, Bo and Soof, from great planets in three different and very remote systems. These three names are common and familiar in every nook and corner of heaven, clear from one end of it to the other—fully as well known as the eighty Supreme Archangels, in fact—whereas our Moses, and Adam, and the rest, have not been heard of outside of our world's little corner of heaven, except by a few very learned men scattered here and there—and they always spell their names wrong, and get the performances of one mixed up with the doings of another, and they almost always locate them simply *in our solar system,* and think that is enough without going into the details such as naming the particular world they are from. It is like a learned Hindoo saying Longfellow lives in the United States—as if he lived all over the United States, and as if the country was so small you couldn't throw a brick there without hitting him. Between you and me, it does gravel me, the cool way people from those worlds outside our system snub our little world, and even our system. Of course we think a good deal of Jupiter, because our world is only a potato in it, for size; but then there are worlds in other systems that Jupiter isn't even a mustard-seed to—like the planet Goobra, for instance, which you couldn't squeeze inside the orbit of Halley's comet without straining the rivets. Tourists from Goobra (I mean parties that have lived and died there—natives) come here, now and then, and inquire about our world, and when they find out it is so little that a streak of lightning can flash clear around it in the eighth of a second, they have to lean up against something to laugh. Then they screw a glass in their eye and go to examining *us,* as if we were a curious kind of foreign bug, or something of that sort. One of them asked me how long our day was; and when I told him it was twelve hours long, as a general thing, he asked me if people where I was from considered it worth while to get up and wash for such a day as that. That is the way with those Goobra people—

> they can't seem to let a chance go by to throw it in your face that their day is three hundred and twenty-two of our years long. . . ."

Scholars, investigating Twain's manuscripts in the decades after his death, have come to realize that in his later years he became more bitterly hostile to Christian convictions. One indication of this is Twain's last long composition before his death, his "Letters from the Earth," which he wrote in 1909 but which was first published only in 1962. In that work, Twain discussed the statement in Genesis that God made the universe in six days. Regarding the Earth, Twain remarks that God "constructed it in five days–and then? It took him only one day to make twenty million suns and eighty million planets!"[17] Paine would have been pleased.

17. Mark Twain, "Letters from the Earth" in his *What is Man? and Other Philosophical Writings*, ed. Paul Baender (Berkeley: University of California Press, 1973), 413. I am indebted to Professor Thomas Werge of the University of Notre Dame for calling this essay to my attention.

FOURTEEN

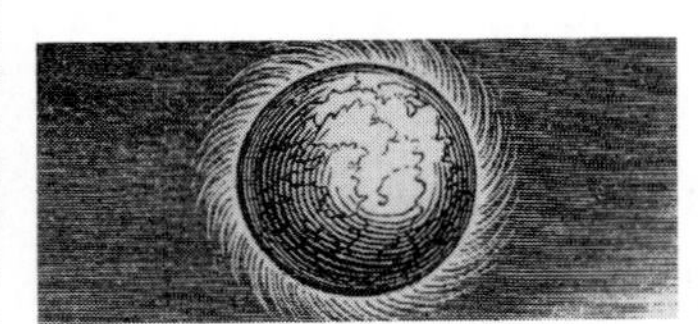

THE CONTROVERSY OVER THE CANALS OF MARS

GIOVANNI SCHIAPARELLI (1835–1910)

In the last quarter of the nineteenth century, a dramatic situation was developing. Astronomers had by then come to believe that few if any of the other planets in our solar system can, at least at present, support intelligent life. A possible exception was Mars, which is nearly the size of our Earth and moves in an orbit not greatly distant from the Earth's orbit. Mars thus became the last, best hope of those who argued that, in our system, intelligent life extends beyond our Earth. Between 1877 and 1913, dozens of books, hundreds of telescopes, thousands of articles, and millions of people focused on whether intelligent beings, possibly desperately struggling to survive, conceivably seeking to signal us, roam the surface of Mars.

This sensational controversy about Mars began in 1877 when Giovanni Schiaparelli, a prominent Italian astronomer, attained a result that to some constituted compelling evidence of intelligent life on Mars. This was Schiaparelli's detection in that year of the so-called canals of Mars. In other words, he reported observing fine lines crisscrossing the Martian surface. He labeled these lines "canali," which in Italian can mean either channel or canal, the important difference being that humans build canals whereas channels are formed by nature. As word of his reported

Fig. 12. Giovanni Schiaparelli

observations spread throughout Britain and the United States, the tendency was to call the objects "canals." Some respected astronomers questioned Schiaparelli's observations, whereas others supported them and in some cases reported success in observing the canals. Credulity was further tested when in 1879 Schiaparelli reported that he had seen some of the canals "geminate" or double. This typically short term appearance was both hard to understand and also to confirm. By 1890, over a dozen astronomers had reported success in sighting the canals and a few managed even to observe the elusive geminations. If Schiaparelli's observations were difficult to accept, their potential significance and his reputation and restraint in interpreting their meaning made astronomers hesitant to dismiss them.

Probably the most widely read of all Schiaparelli's presentations of his Martian investigations was a study that he published in 1893 and that within a year appeared in English translation in a leading American journal, *Astronomy and Astrophysics.* The selection that follows is from that paper.

Giovanni Schiaparelli, "The Planet Mars," trans. W. H. Pickering, *Astronomy and Astrophysics* 13 (1894): 635–39, 717–23.

Many of the first astronomers who studied Mars with the telescope, had noted on the outline of its disc two brilliant white spots of rounded form and of variable size. In process of time it was observed that whilst the ordinary spots upon Mars were displaced rapidly in consequence of its daily rotation, changing in a few hours both their position and their perspective, that the two white spots remained sensibly motionless at their posts. It was concluded rightly from this, that they must occupy the poles of rotation of the planet, or at least must be found very near to them. Consequently they were given the name of polar caps or spots. And not without reason is it conjectured, that these represent upon Mars that immense mass of snow and ice, which still to-day prevents navigators from reaching the poles of the Earth. We are led to this conclusion not only by the analogy of aspect and of place, but also by another important observation.

* * * * * * * * * * * * *

As things stand, it is manifest, that if the above mentioned white polar spots of Mars represent snow and ice, they should continue to decrease in size with the approach of summer in those places, and increase during the winter. Now this very fact is observed in the most evident manner. In the second half of the year 1892 the southern polar cap was in full view; during that interval, and especially in the months of July and August, its rapid diminution from week to week was very evident, even to those observing with common telescopes. This snow, (for we may well call it so,) which in the beginning reached as far as latitude 70°, and formed a cap of over 2000 kilometers (1200 miles) in diameter, progressively diminished, so that two or three months later little more of it remained than an area of perhaps 300 kilometers, (180 miles) at the most, and still less was seen later in the last days of 1892. In these months the southern hemisphere of Mars had its summer; the summer solstice occuring [*sic*] upon October 13. Correspondingly the mass of snow surrounding the northern pole should have increased; but this fact was not observable, since that pole was situated in the hemisphere of Mars which was opposite to that facing the Earth. The melting of the northern snow was seen in its turn in the years 1882, 1884 and 1886.

• • •

The mass of the northern snow-cap of Mars is on the other hand centered almost exactly upon its pole. It is located in a region of yellow color, which we are accustomed to consider as representing the continent of the planet. From this arises a singular phenomenon which has no analogy upon the Earth. At the melting of the snows, accumulated at that pole during the long night of ten months and more, the liquid mass produced in that operation is diffused around the circumference of the snowy region, converting a large zone of surrounding land into a temporary sea, and filling all the lower regions. This produces a gigantic inundation, which has led some observers to suppose the existence of another ocean in those parts, but which does not really exist in that place, at least as a permanent sea. We see then, (the last opportunity was in 1884), the white spot of the snow surrounded by a dark zone, which follows its perimeter in its progressive diminution, upon a circumference ever more and more narrow. The outer part of this zone branches out into dark lines, which occupy all the surrounding region, and seem to be distributary canals, by which the liquid mass may return to its natural position. This produces in these regions very extensive lakes, such as that designated upon the map by the name of *Lacus Hyperboreus;* the neighboring interior sea called *Mare Acidalium* becomes more black, and more conspicuous. And it is to be remembered as a very probable thing, that the flowing of this melted snow is the cause which determines principally the hydrographic state of the planet, and the variations that are periodically observed in its aspect.

• • •

As has been stated, the polar snows of Mars prove in an incontrovertable [*sic*] manner, that this planet, like the Earth, is surrounded by an atmosphere capable of transporting vapor from one place to another. These snows are in fact precipitations of vapor, condensed by the cold, and carried with it successively. How carried with it, if not by atmospheric movement? The existence of an atmosphere charged with vapor has been confirmed also by spectroscopic observations, principally those of Vogel; according to which this atmosphere must be of a composition differing little from our own, and above all *very rich in aqueous vapor.* This is a fact of the highest importance, because from it we can rightly affirm with much probability, that to water, and to no other liquid is due the seas of Mars and its polar snows. When this conclusion is assured beyond all doubt, another one may be derived from it, of not less importance,—that the temperature of the Arean climate, notwithstanding the greater distance of that planet from the Sun, is of the same order as the temperature of the terrestrial one. Because, if it were true, as has been supposed by some investigators, that the temperature of Mars was on the average very low, (from 50° to 60° below zero!) it would not be

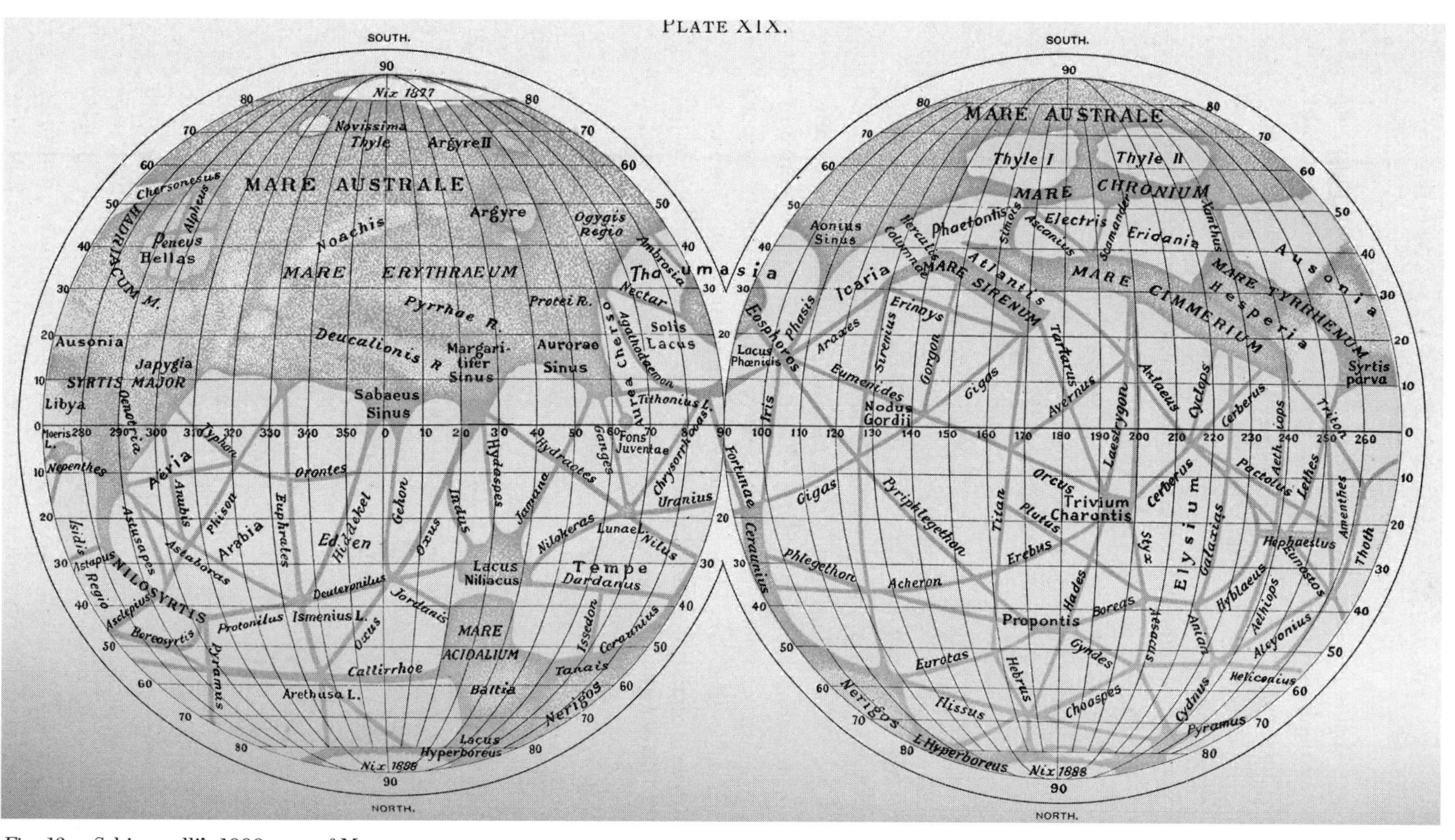

Fig. 13. Schiaparelli's 1888 map of Mars

possible for water vapor to be an important element in the atmosphere of that planet, nor could water be an important factor in its physical changes; but would give place to carbonic acid, or to some other liquid whose freezing point was much lower.

The elements of the meteorology of Mars seem then to have a close analogy to those of the Earth. But there are not lacking, as might be expected, causes of dissimilarity. From circumstances of the smallest moment, nature brings forth an infinite variety in its operations. Of the greatest influence must be the different arrangement of the seas and the continents upon Mars, and upon the Earth, regarding which, a glance at the map will say more than would be possible in many words. We have already emphasized the fact of the extraordinary periodical flood, which at every revolution of Mars inundates the northern polar region at the melting of the snow. Let us now add that this inundation is spread out to a great distance by means of a network of canals, perhaps constituting the principal mechanism (if not the only one) by which water (and with its organic life) may be diffused over the arid surface of the planet. Because on Mars it rains very rarely, *or perhaps even, it does not rain at all.*

• • •

All the vast extent of the continents is furrowed upon every side by a network of numerous lines or fine stripes of a more or less pronounced dark color, whose aspect is very variable. These traverse the planet for long distances in regular lines, that do not at all resemble the winding courses of our streams. Some of the shorter ones do not reach 500 kilometers (300 miles), others on the other hand extend for many thousands, occupying a quarter or sometimes even a third of a circumference of the planet. Some of these are very easy to see, especially that one which is near the extreme left-hand limit of our map, and is designated by the name Nilosyrtis. Others in turn are extremely difficult, and resemble the finest thread of spider's web drawn across the disc. They are subject also to great variations in their breadth, which may reach 200 or even 300 kilometers (120 to 180 miles) for the Nilosyrtis, whilst some are scarcely 30 kilometers (18 miles) broad.

These lines or stripes are the famous canals of Mars, of which so much has been said. As far as we have been able to observe them hitherto, they are certainly fixed configurations upon the planet. The Nilosyrtis has been seen in that place for nearly one hundred years, and some of the others for at least thirty years. Their length and arrangement are constant, or vary only between very narrow limits. Each of them always begins and ends between the same regions. But their appearance and their degree of visibility vary greatly, for all of them, from one opposition to another, and even from one week to another, and these variations do not take place simultaneously and according to the same laws for all, but in most cases happen apparently capriciously, or at least according to laws not sufficiently simple for us to be able to unravel. Often one or more become indistinct, or even wholly invisible, whilst others in their vicinity increase to the point of becoming conspicuous even in telescopes of moderate power. The first of our maps shows all those that have been seen in a long series of observations. This does not at all correspond to the appearance of Mars at any given period, because generally only a few are visible at once.

• • •

Of the remainder, that the lines called canals are truly great furrows or depressions in the surface of the planet, destined for the passage of the liquid mass, and constituting for it a true hydrographic system, is demonstrated by the phenomena which are observed during the melting of the northern snows. We have already remarked that at the time of melting they appeared surrounded by a dark zone, forming a species of temporary sea. At that time the canals of the surrounding region become blacker and wider, increasing to the point of converting, at a certain time, all of the yellow region comprised between the edge of the snow and the parallel of 60° north latitude, into numerous islands of small extent. Such a state of things does not cease, until the snow, reduced to its minimum area, ceases to melt.

Then the breadth of the canals diminishes, the temporary sea disappears, and the yellow region again returns to its former area. The different phases of these vast phenomena are renewed at each return of the seasons, and we have been able to observe them in all their particulars very easily during the Oppositions of 1882, 1884 and 1886, when the planet presented its northern pole to terrestrial spectators. The most natural and the most simple interpretation is that to which we have referred, of a great inundation produced by the melting of the snows,—it is entirely logical, and is sustained by evident analogy with terrestrial phenomena. We conclude therefore that the canals are such in fact, and not only in name. The network formed by these was probably determined in its origin in the geological state of the planet, and has come to be slowly elaborated in the course of centuries. It is not necessary to suppose them the work of intelligent beings, and notwithstanding the almost geometrical appearance of all of their system, we are now inclined to believe them to be produced by the evolution of the planet, just as on the Earth we have the English Channel and the Chanel [*sic*] of Mozambique.

• • •

But the most surprising phenomenon pertaining to the canals of Mars is their gemination, which seems to be produced principally in the months which precede, and in those which follow the great northern inundation, at about the times of the equinoxes. In consequence of a rapid process, which certainly lasts at most a few days, or even perhaps only a few hours, and of which it has not yet been possible to determine the particulars with certainty, a given canal changes its appearance, and is found transformed through all its length, into two lines or uniform stripes, more or less parallel to one another, and which run straight and equal with the exact geometrical precision of the two rails of a railroad. But this exact course is the only point of resemblance with the rails, because in dimensions there is no comparison possible, as it is easy to imagine. The two lines follow very nearly the direction of the original canal, and end in the place where it ended.

. . .

The observation of the geminations is one of the greatest difficulty, and can only be made by an eye well practiced in such work, added to a telescope of accurate construction, and of great power. This explains why it is that it was not seen before 1882. In the ten years that have transpired since that time, it has been seen and described at eight or ten observatories. Nevertheless, some still deny that these phenomena are real, and tax with illusion (or even imposture) those who declare that they have observed it.

Their singular aspect, and their being drawn with absolute geometrical precision, as if they were the work of rule or compass, has led some to see in them the work of intelligent beings, inhabitants of the planet. I am very careful not to combat this supposition, which includes nothing impossible. (To mi guarderò bene dal combattere questa supposizione, la quale nulla include d'impossibile). But it will be noticed that in any case the gemination cannot be a work of permanent character, it being certain that in a given instance it may change its appearance and dimensions from one season to another. If we should assume such a work, a certain variabil[i]ty would not be excluded from it[,] for example, extensive agricultural labor and irrigation upon a large scale. Let us add further that the intervention of intelligent beings might explain the geometrical appearance of the gemination, but it is not at all necessary for such a purpose.

. . .

Having regard then to the principle that in the explanation of natural phenomena it is universally agreed to begin with the simplest suppositions, the first hypotheses on the nature and cause of the geminations have for the most part put in operation only the laws of inorganic nature. Thus, the gemination is supposed to be due either to the effects of light in the atmosphere of Mars, or to optical illusions produced by vapors in various manners, or to glacial phenomena of a perpetual winter, to which it is known all the planets will be condemned, or to double cracks in its surface, or to single cracks of which the images are doubled by the effect of smoke issuing in long lines and blown laterally by the wind. The examination of these ingenious suppositions leads us to conclude that none of them seem to correspond entirely with the observed facts, either in whole or in part. Some of these hypotheses would not have been proposed, had their authors been able to examine the geminations with their own eyes. Since some of these may ask me directly,—Can you suggest anything better? I must reply candidly, No.

It would be far more easy if we were willing to introduce the forces pertaining to organic nature. Here the field of plausible supposition is immense, being capable of making an infinite number of combinations capable of satisfying the appearances even with the smallest and simplest means. Changes of vegetation over a vast area, and the production of animals, also very small, but in enormous multitudes, may well be rendered visible at such a distance. An observer placed in the Moon would be able to see such an appearance at the times in which agricultural operations are carried out upon one vast plain,—the seed time and the gathering of the harvest. In such a manner also would the flowers of the plants of the great steppes of Europe and Asia be rendered visible at the distance of Mars,—by a variety of coloring. A similar system of operations produced in that planet may thus certainly be rendered visible to us. But how difficult for the Lunarians and the Areans [Martians] to be able to imagine the true causes of such changes of appearance, without having first at least some superficial knowledge of terrestrial nature! So also for us, who know so little of the physical state of Mars, and nothing of its organic world, the great liberty of possible supposition renders arbitrary all explanations of this sort, and constitutes the gravest obstacle to the acquisition of well founded notions. All that we may hope is that with time the uncertainty of the problem will gradually diminish, demonstrating, if not what the geminations are, at least what they can

> not be. We may also confide a little in what Galileo called "the courtesy of Nature," thanks to which, sometime from an unexpected source, a ray of light will illuminate an investigation at first believed inaccessible to our speculations, and of which we have a beautiful example in celestial chemistry. Let us therefore hope and study.

One feature of this paper that Schiaparelli's fellow astronomers found extremely impressive is what it does not contain: excessive speculation about Martian life. Schiaparelli was, at least in one sense, laying out his observations, but proceeding very cautiously in interpreting them. This has, in fact, been the dominant image of Schiaparelli in the English-speaking world. A somewhat different view of him can be derived from a paper that Schiaparelli published in 1895 in Italian. Parts of this paper appeared in 1896 in a German translation and in 1898 in a French translation, but it was never published in English. In that paper, Schiaparelli began by endorsing the idea that the canals constitute the Martian irrigation system and by repeating his claim, disputed even by some of his supporters, that oceans and lakes can be seen on Mars. He also advocated the reasonable claim that what is seen when the canals are observed is not the canals themselves but the surrounding vegetation, perhaps lining the sides of a shallow valley. Then, setting aside his cautiousness, he proceeded to offer an explanation of the "mysterious geminations." Suggesting that the idea that these "may be due to intelligent beings ought not be rejected as an absurdity,"[1] he speculated that Martian engineers may have built dikes at various levels along the slopes of the shallow valleys through which the waterways run. When the spring inundations begin, the "Minister of Agriculture orders the opening of the most elevated sluices and fills the upper canals with water. . . . The irrigation then spreads to the two (lower) lateral-zones . . . , the valley changes color in these two lateral zones, and the terrestrial astronomer perceives a gemination."[2] Gradually the engineers release the water into the lower portions of the valley, fertilizing the lowest region of the valley and producing a single "canal" appearance. Having gone this far, he did not hesitate to go further:

1. Giovanni Schiaparelli, "La vie sur la planète Mars," *Société astronomique de France bulletin* 12 (1898): 423–29, at 427.

2. Schiaparelli, "La vie," 428.

> The institution of a collective socialism ought indeed result from a parallel community of interests and of a universal solidarity among the citizens, a veritable phalanstery which can be considered a paradise of societies. One may also imagine a great federation of humanity in which each valley constitutes an independent state. The interests of all are not distinguished from the other; the mathematical sciences, meteorology, physics, hydrography, and the art of construction are certainly developed to a high degree of perfection; international conflicts and wars are unknown; all the intellectual efforts which, among the insane inhabitants of a neighboring world are consumed in mutually destroying each other, are [on Mars] unanimously directed against the common enemy, the difficulty which penurious nature opposes at each step.[3]

That Schiaparelli himself recognized he had in this paper gone rather beyond what strict scientific methodology dictated is suggested by his concluding remark in the paper: "I now leave to the reader the need to continue these considerations, and, as for myself, I descend from the hippogryph."[4] Moreover, across the top of a copy he sent the astronomer Camille Flammarion, Schiaparelli wrote: "Semel in anno licet insanire," which means "Once a year it is permissible to act like a madman."[5]

In the late 1890s, Schiaparelli began to recognize that his eyesight was failing, at least for the precision work of astronomical observation. He lived on until 1910, holding fast to the accuracy of his claims for canals. Shortly before his death he wrote to the astronomer Antoniadi: "The polygonations and geminations of which you show such horror . . . are an established fact against which it is useless to protest."[6]

PERCIVAL LOWELL (1855–1916)

A major event in the huge controversy over the question of canals on Mars occurred in 1893. In that year, Percival Lowell, a wealthy Harvard educated businessman and

3. Schiaparelli, "La vie," 429.

4. Schiaparelli, "La vie," 429. A hippogryph is a mythical flying beast, which is part horse and part eagle.

5. As quoted by Flammarion in an editorial comment to Schiaparelli, "La vie," 429.

6. As quoted in E. M. Antoniadi, *La planète Mars* (Paris: Hermann, 1930), 31.

Fig. 14. Percival Lowell (Courtesy of Lowell Observatory Archives)

orientalist, returned from Japan to announce his plan to enter the canal controversy. Lowell almost immediately became founder, funder, director, publicist, as well as chief observer and theoretician for a new observatory, the Lowell Observatory, which he erected in Flagstaff, Arizona, and equipped with a splendid 24-inch aperture refracting telescope. Seen by some as brilliant and charming but described by Harvard President C. W. Eliot as "intensely egoistic and unreasonable,"[7] Lowell in an address in Boston announced his intention to have the "main object" of his observatory be "an investigation into the conditions of life in other worlds" and urged that "there is strong reason to believe that we are on the eve of pretty definite discovery in the matter."[8]

By 1894, Lowell had begun observing, aided by two astronomers, A. E. Douglass and W. H. Pickering, whom he had borrowed from Harvard Observatory. In 1895, Lowell published his widely read book titled *Mars* and blitzed various journals and magazines with papers about the planet. In his *Mars*, he reported observing 183 canals or rather canals with vegetation running along their sides, and also announced the discovery of Martian "oases," which typically occur at the intersections of canals. He presented Mars as an old, rather dry planet with snowy polar caps and Martians as super-engineers desperately trying to irrigate their dying planet. Lowell's engaging style and exciting claims made his book both extremely popular and highly controversial. W. W. Campbell, an astronomer who reviewed Lowell's *Mars* for the journal *Science*, accused Lowell of taking "the popular side of the most popular scientific question afloat. The world at large is anxious for the discovery of intelligent life on Mars, and every advocate gets an instant and large audience."[9] Campbell also argued that Lowell possessed neither adequate observations nor mastery of the Mars literature. After quoting from Lowell's 1894 Boston address, Campbell charged: "Mr. Lowell went direct from the lecture hall to his observatory in Arizona, and how well his observations established his pre-observational views is told in this book." And Campbell lamented that this was the practice of an author who "has written vigorously . . . of the dangers of bias [from] preconceived notions."[10]

7. As quoted in Bessie Zaban Jones and Lyle Gifford Boyd, *The Harvard College Observatory: The First Four Directorships, 1839–1919* (Cambridge, Mass.: Harvard University Press, 1971), 473.

8. Percival Lowell, "The Lowell Observatory," *Boston Commonwealth*, May 26, 1894, 3.

9. W. W. Campbell, "[Review of] *Mars* By Percival Lowell," *Science*, n.s. 4 (August 21, 1896), 232.

10. Campbell "[Review]," 232.

Percival Lowell, *Mars* (Boston and New York: Houghton Mifflin, 1895), 201–12.

CONCLUSION

To review, now, the chain of reasoning by which we have been led to regard it probable that upon the surface of Mars we see the effects of local intelligence. We find, in the first place, that the broad physical conditions of the planet are not antagonistic to some form of life; secondly, that there is an apparent dearth of water upon the planet's surface, and therefore, if beings of sufficient intelligence inhabited it, they would have to resort to irrigation to support life; thirdly, that there turns out to be a network of markings covering the disk precisely counterparting what a system of irrigation would look like; and, lastly, that there is a set of spots placed where we should expect to find the lands thus artificially fertilized, and behaving as such constructed oases should. All this, of course, may be a set of coincidences, signifying nothing; but the probability points the other way. As to details of explanation, any we may adopt will undoubtedly be found, on closer acquaintance, to vary from the actual Martian state of things; for any Martian life must differ markedly from our own.

The fundamental fact in the matter is the dearth of water. If we keep this in mind, we shall see that many of the objections that spontaneously arise answer themselves. The supposed herculean task of constructing such canals disappears at once; for, if the canals be dug for irrigation purposes, it is evident that what we see, and call by ellipsis the canal, is not really the canal at all, but the strip of fertilized land bordering it,—the thread of water in the midst of it, the canal itself, being far too small to be perceptible. In the case of an irrigation canal seen at a distance, it is always the strip of verdure, not the canal, that is visible, as we see in looking from afar upon irrigated country on the Earth.

We may, perhaps, in conclusion, consider for a moment how different in its details existence on Mars must be from existence on the Earth. One point out of many bearing on the subject, the simplest and most certain of all, is the effect of mere size of habitat upon the size of the inhabitant; for geometrical conditions alone are most potent factors in the problem of life. Volume and mass determine the force of gravity upon the surface of a planet, and this is more far-reaching in its effects than might at first be thought. Gravity on the surface of Mars is only a little more than one third what it is on the surface of the Earth. This would work in two ways to very different conditions of existence from those to which we are accustomed. To begin with, three times as much work, as for example, in digging a canal, could be done by the same expenditure of muscular force. If we were transported to Mars, we should be pleasingly surprised to find all our manual labor suddenly lightened threefold. But, indirectly, there might result a yet greater gain to our capabilities; for if Nature chose she could afford there to build her inhabitants on three times the scale she does on Earth without their ever finding it out except by interplanetary comparison. Let us see how.

As we all know, a large man is more unwieldy than a small one. An elephant refuses to hop like a flea; not because he considers the act undignified, but simply because he cannot bring it about. If we could, we should all jump straight across the street, instead of painfully paddling through the mud. Our inability to do so depends upon the size of the Earth, not upon what it at first seems to depend, on the size of the street.

To see this, let us consider the very simplest case, that of standing erect. To this every-day feat opposes itself the weight of the body simply, a thing of three dimensions, height, breadth, and thickness, while the ability to accomplish it resides in the cross-section of the muscles of the knee, a thing of only two dimensions, breadth and thickness. Consequently, a person half as large again as another has about twice the supporting capacity of that other, but about three times as much to support. Standing therefore tires him out more quickly. If his size were to go on increasing, he would at last reach a stature at which he would no longer be able to stand at all, but would have to lie down. You shall see the same effect in quite inanimate objects. Take two cylinders of paraffine wax, one made into an ordinary candle, the other into a gigantic facsimile of one, and then stand both upon their bases. To the small one nothing happens. The big one, however, begins to settle, the base actually made viscous by the pressure of the weight above.

Now apply this principle to a possible inhabitant of Mars, and suppose him to be constructed three times as large as a human being in every dimension. If he were on Earth, he would weigh twenty-seven times as much, but on the surface of Mars, since gravity there is only about one third of what it is here, he would weigh but nine times as much. The cross-section of his muscles would be nine times as great. Therefore the ratio of his supporting power to the weight he must support would be the same as ours. Consequently, he would be able to stand with as little fatigue as we. Now consider the work he might be able to do. His muscles, having length, breadth, and thickness, would all be twenty-seven times as effective as ours. He would prove twenty-seven times as strong as we, and could accomplish twenty-seven times as much. But he would further work upon what required, owing to decreased gravity, but one third the effort to overcome. His effective force, therefore, would be eighty-one times as great as man's, whether in digging canals or in other bodily occupation. As gravity on the surface of Mars is really a little more than one third that at the surface of the Earth, the true ratio is not eighty-one, but about fifty; that is, a Martian would be, physically, fifty-fold more efficient than man.

As the reader will observe, there is nothing problematical about this deduction whatever. It expresses an abstract ratio of physical capabilities which must exist between the two planets, quite irrespective of whether there be denizens on either, or how other conditions may further affect their forms. As the reader must also note, the deduction refers to the possibility, not to the probability, of such giants; the calculation being introduced simply to show how different from us any Martians may be, not how different they are.

It must also be remembered that the question of their size has nothing to do with the question of their existence.

The arguments for their presence are quite apart from any consideration of avoirdupois. No Herculean labors need to be accounted for; and, if they did, brain is far more potent to the task than brawn.

Something more we may deduce about the characteristics of possible Martians, dependent upon Mars itself, a result of the age of the world they would live in.

A planet may in a very real sense be said to have life of its own, of which what we call life may or may not be a subsequent detail. It is born, has its fiery youth, sobers into middle age, and just before this happens brings forth, if it be going to do so at all, the creatures on its surface which are, in a sense, its offspring. The speed with which it runs through its gamut of change prior to production depends upon its size; for the smaller the body the quicker it cools, and with it loss of heat means beginning of life for its offspring. It cools quicker because, as we saw in a previous chapter, it has relatively less inside for its outside, and it is through its outside that its inside cools. After it has thus become capable of bearing life, the Sun quickens that life and supports it for we know not how long. But its duration is measured at the most by the Sun's life. Now, inasmuch as time and space are not, as some philosophers have from their too mundane standpoint supposed, forms of our intellect, but essential attributes of the universe, the time taken by any process affects the character of the process itself, as does also the size of the body undergoing it. The changes brought about in a large planet by its cooling are not, therefore, the same as those brought about in a small one. Physically, chemically, and, to our present end, organically, the two results are quite diverse. So different, indeed, are they that unless the planet have at least a certain size it will never produce what we call life, meaning our particular chain of changes or closely allied forms of it, at all. As we saw in the case of atmosphere, it will lack even the premise to such conclusion.

Whatever the particular planet's line of development, however, in its own line, it proceeds to greater and greater degrees of evolution, till the process stops, dependent, probably, upon the Sun. The point of development attained is, as regards its capabilities, measured by the planet's own age, since the one follows upon the other.

Now, in the special case of Mars, we have before us the spectacle of a world relatively well on in years, a world much older than the Earth. To so much about his age Mars bears evidence on his face. He shows unmistakable signs of being old. Advancing planetary years have left their mark legible there. His continents are all smoothed down; his oceans have all dried up. *Teres atque rotundus*, he is a steady-going body now. If once he had a chaotic youth, it has long since passed away. Although called after the most turbulent of the gods, he is at the present time, whatever he may have been once, one of the most peaceable of the heavenly host. His name is a sad misnomer; indeed, the ancients seem to have been singularly unfortunate in their choice of planetary cognomens. With Mars so peaceful, Jupiter so young, and Venus bashfully draped in cloud, the planet's names accord but ill with their temperaments.

Mars being thus old himself, we know that evolution on his surface must be similarly advanced. This only informs us of its condition relative to the planet's capabilities. Of its actual state our data are not definite enough to furnish much deduction. But from the fact that our own development has been comparatively a recent thing, and that a long time would be needed to bring even Mars to his present geological condition, we may judge any life he may support to be not only relatively, but really older than our own.

From the little we can see, such appears to be the case. The evidence of handicraft, if such it be, points to a highly intelligent mind behind it. Irrigation, unscientifically conducted, would not give us such truly wonderful mathematical fitness in the several parts to the whole as we there behold. A mind of no mean order would seem to have presided over the system we see,—a mind certainly of considerably more comprehensiveness than that which presides over the various departments of our own public works. Party politics, at all events, have had no part in them; for the system is planet wide. Quite possibly, such Martian folk are possessed of inventions of which we have not dreamed, and with them electrophones and kinetoscopes are things of a bygone past, preserved with veneration in museums as relics of the clumsy contrivances of the simple childhood of the race. Certainly what we see hints at the existence of beings who are in advance of, not behind us, in the journey of life.

Startling as the outcome of these observations may appear at first, in truth there is nothing startling about it whatever. Such possibility has been quite on the cards ever since the existence of Mars itself was recognized by the Chaldean shepherds, or whoever the still more primeval astronomers may have been. Its strangeness is a purely subjective phenomenon, arising from the instinctive reluctance of man to admit the possibility of peers. Such would be comic were it not the inevitable consequence of the constitution of the universe. To be shy of anything resembling himself is part and parcel of man's own individuality. Like the savage who fears nothing so much as a strange man, like Crusoe who grows pale at the sight of footprints not his own, the civilized thinker instinctively turns from the thought of mind other than the one he himself knows. To admit into his conception of the cosmos other finite minds as factors has in it something of the weird. Any hypothesis to explain the facts, no matter how improbable or even palpably absurd it be, is better than this. Snow-caps of solid carbonic acid gas, a planet cracked in a positively monomaniacal manner, meteors ploughing tracks across its surface with such mathematical precision that they must have been educated to the performance, and so forth and so on, in hypotheses each more astounding than its predecessor, commend themselves to man, if only by such means he may escape the admission of anything approaching his kind. Surely all this is puerile, and should as speedily as possible be outgrown. It is simply an instinct like any other, the projection of the instinct of self-preservation. We ought, therefore, to rise above it, and, where probability points to other things, boldly accept the fact provisionally, as we should the presence of oxygen, or iron, or anything else. Let us not cheat ourselves with words. Conservatism sounds finely, and covers any amount of ignorance and fear.

We must be just as careful not to run to the other extreme, and draw deductions of purely local outgrowth. To talk of Martian beings is not to mean Martian men. Just as the probabilities point to the one, so do they point away from the other. Even on this Earth man is of the nature of an accident. He is the survival of by no means the highest physical organism. He is not even a high form of mammal. Mind has been his making. For aught we can see, some lizard or batrachian might just as well have popped into his place early in the race, and been now the dominant creature of this Earth. Under different physical conditions, he would have been certain to do so. Amid the surroundings that exist on Mars, surroundings so different from our own, we may be practically sure other organisms have been evolved of which we have no cognizance. What manner of beings they may be we lack the data even to conceive.

For answers to such problems we must look to the future. That Mars seems to be inhabited is not the last, but the first word on the subject. More important than the mere fact of the existence of living beings there, is the question of what they may be like. Whether we ourselves shall live to learn this cannot, of course, be foretold. One thing, however, we can do, and that speedily: look at things from a standpoint raised above our local point of view; free our minds at least from the shackles that of necessity tether our bodies; recognize the possibility of others in the same light that we do the certainty of ourselves. That we are the sum and substance of the capabilities of the cosmos is something so preposterous as to be exquisitely comic. We pride ourselves upon being men of the world, forgetting that this is but objectionable singularity, unless we are, in some wise, men of more worlds than one. For, after all, we are but a link in a chain. Man is merely this earth's highest production up to date. That he in any sense gauges the possibilities of the universe is humorous. He does not, as we can easily foresee, even gauge those of this planet. He has been steadily bettering from an immemorial past, and will apparently continue to improve through an incalculable future. Still less does he gauge the universe about him. He merely typifies in an imperfect way what is going on elsewhere, and what, to a mathematical certainty, is in some corners of the cosmos indefinitely excelled.

If astronomy teaches anything, it teaches that man is but a detail in the evolution of the universe, and that resemblant though diverse details are inevitably to be expected in the host of orbs around him. He learns that, though he will probably never find his double anywhere, he is destined to discover any number of cousins scattered through space.

EDWARD WALTER MAUNDER (1851–1928)

In the many retellings of the story of the Martian canals, Giovanni Schiaparelli and Percival Lowell usually receive primacy of place. Important as the canals of Mars were in the history of astronomy, it seems significant and interesting to ask: who destroyed the canals of Mars? Who removed this basis for claims for life on Mars? Strong evidence points toward crediting this achievement to E. W. Maunder and E. M. Antoniadi, whose contributions are represented in this and the subsequent selection.

Maunder, after studies at King's College in London, worked for over four decades as a spectroscopist on the staff of England's Royal Greenwich Observatory. Although his earliest critique of claims for life on Mars dates from 1882, one of his most important papers in this area appeared in 1894. A slightly abridged version of this paper is provided below. In examining this paper, it is important to recognize that its primary goal is not to challenge *theoretical* claims for life on Mars put forward by Schiaparelli, Lowell, and others, but rather to dispute their *observational* claims for the canals. This paper by no means succeeded in fully undermining the canals. Rather it should be seen as the first major paper, followed by many more published by Maunder, Antoniadi, and others, that by around 1913 had convinced the astronomical community that the canal observations were illusory. The discrediting of the observational basis for the canals had the additional effect that it removed the empirical foundation for the claims of Schiaparelli and Lowell for intelligent life on the outer planet nearest to the Earth.

Fig. 15. Edward W. Maunder

E. Walter Maunder, "The Canals of Mars," *Knowledge* 17 (Nov. 1, 1894): 249–52.

Seventeen years ago our knowledge of Mars appeared to be in a very satisfactory state. The principal markings had been often and long observed, and had been found to be permanent. The Kaiser Sea, the Oculus, the Maraldi Sea, had been observed by Hooke and Cassini, Herschel and Schroeter, Beer and Mädler, Dawes, Lockyer, Knobel and Green. The inference, from the annual waxing and waning of the white polar caps, that cloud, snow, and rain were features of the meteorology of the planet, had been confirmed by the testimony of the spectroscope, and that the white caps themselves were composed of ice or snow was a natural conclusion. It followed necessarily that there must be water on the surface of the planet, and the dark spots were considered as seas, leaving the brighter districts to be regarded as land. Mars was, in short, a smaller copy of our own world. "The analogy between Mars and the earth" was pronounced to be "by far the greatest in the whole solar system."

Schiaparelli's discovery of the "canals"—to use a term which, however misleading it has been, has now been too strongly sanctioned by custom to be easily changed—was the beginning of a new epoch in Martian observation, and its chief and most patent result has been to disturb our old conceptions of the analogy between Mars and the earth, and consequently to unsettle our notions of the physical condition of the planet. The reaction has gone so far that Prof. Schaeberle has reversed the old identification of the dark spots as seas, and claimed that the bright districts are to be thus regarded. The grounds for this change of view have not been generally accepted, and more is to be said for an objection raised by Prof. W. H. Pickering, that just as snow is not the only substance to give a bright white reflection—hoar-frost or cloud would serve as well—so water is not the only surface which would appear dark; forests and prairies would appear as at least sombre districts.

The "canals," however, have been the great element of disturbance. The report of their discovery was received in 1877 with very considerable mistrust. But time has been on the side of the patient and keen-eyed Italian astronomer. Each succeeding opposition has seen more of his "canals" identified by other astronomers; each construction of a new and more powerful telescope, each enlistment of a fresh earnest and skilled observer, has meant a further confirmation of his work. So that now there is a bulk of evidence in favour, not merely of some of Schiaparelli's "canals," but of his canal system as a whole, by no means lightly to be set aside.

The positive evidence has not, however, destroyed or overcome the negative. Both still hold the field, and both must be considered. The principal points against the actual existence of the "canals" as represented by Schiaparelli may be summarized as follows.

1. They are extremely narrow, approaching the limits of visibility, even when Mars is at its closest approach. Thus a fine drawing reproduced in Flammarion's great monograph on "Mars," p. 658, shows "canals" of breadths not exceeding 0.04" or 0.05" of arc. The two branches of a *doubled* "canal" in one part of the sketch are distant from each other only 0.25". Nor is this all. In his account of his 1888 observations (maximum diameter of the planet, 15.4"), he states that in a set of canals on Mädler Continent he could make out, not merely the two banks of the canal, but also "very small undulations in their two banks, which could be distinguished from each other."

2. The distinctness of some of those objects, though so narrow, does not seem impaired by distance. Thus, in the opposition of 1877, the Indus was *better* seen by its discoverer when the planet had receded a great distance from opposition, and was very well seen when the planet was only 5.7" in diameter, and the canal 0.2" in breadth.

3. The great divergency between the descriptions of different observers. To take the one (and all important) feature of breadth. In the opposition of 1890, when Schiaparelli was observing "canals" down to a breadth of 0.05", Holden and

Keeler, at the Lick Observatory, always saw the canals as dark, broad, somewhat diffused bands. We may judge what is meant by "broad bands," for the record goes on:—"In bad vision they were drawn in that way by Schaeberle also. Under good conditions, however, the latter observer described them as narrow lines, a second of arc or so in width." A second of arc corresponded at this time to a minimum of 6° of a Martian great circle; that is to say, to about the breadth of Herschel Strait, and very nearly to that of the Mare Sirenum, the narrower end of the Maraldi Sea, or to nearly double the breadth of the Nasmyth Inlet, or the point of the Kaiser Sea, markings seen even by the earliest observers. If Schaeberle could describe markings of such dimensions—markings twenty times as broad as Schiaparelli's narrowest canals—as "narrow lines," what must have been the breadth of the "broad bands" of his two colleagues? It is clear that the phenomena observed by the Lick observers were quite of a different order to those recorded by Schiaparelli.

4. The greatness and suddenness of the changes remarked in the "canal" system. The "gemination" or doubling of the "canals" has been remarked "to take place in a relatively short space of time, and by a rapid metamorphosis. . . . Sometimes the metamorphosis has been completed in the interval of twenty four hours between two consecutive observations. So far as the observer could judge, the phenomenon took place simultaneously along the entire length of the canal doubled." Schiaparelli himself draws attention to the strange and rapid changes taking place on the planet, and remarks: "Evidently the planet has fixed geographical details similar to those of the earth, with gulfs, canals, &c., on an irregular plan. There comes a certain moment and all this disappears, to give place to those grotesque polygons and geminations, which clearly represent approximately the former state; but it is a coarse, and, I might say, almost a ridiculous mask."

5. The "canals," when near the edge of the disc, are apt to be represented as much straighter than they could possibly be.

6. To these difficulties may be added that at the very time when to some observers the canal system was most developed,

others have sometimes only been able to perceive the usual markings of the planet in their customary configuration.

How are we to explain these curious discrepancies?

First of all, many of these differences are to be explained by difference in "seeing" power, including in that term not mere atmospheric conditions, but instrumental and personal differences, such as the aperture of the telescope, its defining power, the magnification employed, and the keenness of sight and artistic skill of the observer.

Maunder at this point illustrated his position by citing various cases in which astronomers reported quite different observations of identical regions of Mars.

I ventured to lay stress, in my paper on "The Tenuity of the Sun's Surroundings" in the March number of *Knowledge*, on the fact, which we easily overlook, that "the smallest portion of the sun's surface visible by us as a separate entity, even as a mathematical point, is yet really a widely extended area." The same truth applies in its degree to the planet Mars. We have no right to assume, and yet we do habitually assume, that our telescopes reveal to us the ultimate structure of the surface of the planet.

An illustration of this point was afforded me some time ago, when a question arose as to the limit of visibility to the naked eye of sunspots. I was astonished to find that a group of spots had been recorded as seen directly, when their total area was much less than that of many well defined circular spots that had entirely escaped scrutiny. A few experiments convinced me, however, that the observation was perfectly correct, and that it was often possible to see a straggling group of small unimportant spots, when a single spot of considerably greater total area would be invisible.

I then tried how small an object could be detected without optical assistance, the objects being always black marks (Indian ink) on white glazed paper, illuminated by dull diffused daylight.

The limit of my vision for a circular dot ranged from a diameter of 60" to 86" of arc. One of 20" was quite invisible;

of 40", distinctly seen. This was decidedly smaller than I had anticipated. But the limit for a straight line, to my surprise, was as low as 7" or 8"; 12" was easy and conspicuous. More than this, a pair of lines, each only 4" in breadth, and the pair separated by say 20", was visible as a faint single line; two lines, even of only 8", meeting at a very acute angle, were visible after their separation had diminished below about 25". In each case the object was unmistakably *discerned,* and appeared as a line or dot; it was not, of course, *defined* so as to be seen in its true form.

Further, a chain of dots, each of 20", irregularly disposed along a straight line, the average interval between any two dots being three times the diameter of a dot, was easily seen as a continuous straight line, whilst a double chain of yet smaller dots, each 4" in diameter, and the two chains some 40" apart, was visible as a very faint line.

The theoretical limit of visibility has been given as 40" or a little greater, a limit with which the above observations are really in tolerable accord; for when the angular diameter of the object fell much below 40" it was seen, not as a minute defined black dot, but as a grey diffused spot of about 40" or 45" in diameter. It would seem, then, that the smallest perceptible area is about 40", but that there be within any such given area a sufficiency of dark markings, however individually minute, to turn the white to a decided grey, then that area will be visible as a grey spot. Two lines or a number of dots, easily visible as one object when close together, can readily be made invisible by a greater angular separation.

It seems to me that these rough experiments have a decided bearing upon the "canal system" and the supposed changes of Mars. It seems a violent hypothesis to call in inundations extending over many thousands of square miles to account for merely temporary changes, for sooner or later the old districts take on the old configurations, more especially since, as I have already shown, the meteorology of Mars must necessarily be a languid one. Indeed, it may happen that whilst several independent observers have recorded a change, others equally skilled have seen the planet as before.

But if what we see is not the ultimate structure of the planet's surface, if, especially in the half-tone regions like the De la Rue Ocean, we have an intermingling of minute areas of dark and light—be they water and land, forest and bare rock, prairie and sandy desert, or what you will—it is easy to see how enormous changes may apparently occur in a very little time. What we actually see is a greyish spot, contrasted with dark spots but little darker than it is, and with bright spots but little brighter. What is required, then, in the observer is not so much keenness of vision to detect minuteness of detail as power to appreciate delicate differences of tone. And the formation or dissipation of thin cirrus cloud above such a half-tone district will readily make it indistinguishable from one of the "continents," or from one of the "seas," as the case may be; cloud on Mars necessarily taking the form of our lightest and highest terrestrial clouds, rather than that of the densest and heaviest. When the difference of tone in two contiguous markings is but small, but a little defect in the transparency and steadiness of our own atmosphere will be sufficient to render them indistinguishable.

It is easy to see, if these causes are the principal reasons for the apparent changes on the planet, that different markings will have two or three different forms under which they present themselves, but will not pass through an indefinite number of changes. This is actually the case in more than one locality. To take perhaps the best authenticated case, the marking to which reference has already been made, Herschel II strait is sometimes closed at its western end, and sometimes open, its southern shore being then a cigar-shaped island. . . .

The rough little experiments to which I have alluded may, I think, throw some light on the "canal system." It must not, of course, be imagined that a power of 100 on Mars when 20" in diameter will show it with equal distinctness to the moon as seen without telescopic assistance nor, if a line 8" in breadth be visible to the naked eye, will a power of 400 show a line 0.02" in breadth on Mars, even in the steadiest air. To begin with, all the contrasts on Mars are subdued. Then, the gain by increasing the power of the eyepiece is always less than the numerical ratio of the magnification, till a point is reached when it vanishes, either on account of optical limitations, defects of the instrument, or atmospheric conditions. But a narrow dark line *can* be seen when its breadth is far less than the diameter of the smallest visible dot. Further, a line of detached dots will produce the impression of a continuous line, if the dots be too small or too close together for separate vision. There are some intimations that this may be the next phase of the "canal" question, Mr. Gale, of Paddington, New South Wales, having broken up one "canal" into a chain of "lakes" on a night of superb definition, Mars being near the zenith, and Professor W. H. Pickering, at Arequipa, having under equally favourable circumstances detected a vast number of small "lakes" in the general structure of the "canal system."

If this be so, if the canals are, generally speaking, beyond the limit of distinct definite vision, but producing the impression of lines, it will be readily understood how it is that different observers differ so widely as to their breadth and as to their character, whether diffused or sharp. If they are too narrow to be *definedly* seen, and yet broad and dark enough to make their presence felt, this is precisely what would ensue. The apparent breadth of the "canal" would vary with the definition each observer enjoyed, and with his personal idiosyncrasies. Indeed, one man might see the veritable "canal" itself as a hard, sharp, well-defined line—if that be its actual character—whilst to another, less fortunate in his climate, his telescope, or his sight, it would only be a diffused shade. . . .

[In conclusion,] I should like once again to emphasize what I feel to be an important point, not only with regard to the sun and to Mars, but to all the planets. We cannot assume that what we are able to discern is really the ultimate structure of the body we are examining.

Gradually after Maunder's 1894 suggestion that the canals may be an illusion, other astronomers began to adopt similar positions. One such astronomer was an Italian, Vincenzo Cerulli, who in 1898, working apparently independently of

Maunder, published the first of a number of writings in which he argued that the canals are an illusion resulting from the eye linking together detail below the limits of visibility. More important than Cerulli was a Greek astronomer living in Paris named Eugène Antoniadi, who although at first a supporter of the canal observations gradually began to shift to Maunder's position, and by around 1903 had become Maunder's chief ally.[11]

An important development in the history of the illusion theory of the canals occurred in 1902 when an Englishman, B. W. Lane, published a paper recounting experiments of a quite elementary nature in which persons shown a small sketch of Mars (without canals) drew lines (canals) connecting oppositely situated projecting points.

Then in 1903, Maunder published a number of papers recounting more refined experiments in which schoolboys, shown at a distance a canalless map of Mars (see fig. 16), drew canals. A somewhat abbreviated version of one of these papers by Maunder is provided below. An important aspect of this period is that when at a 1903 meeting of the Royal Astronomical Society Maunder presented an account of his illusion theory and the experiments he had carried out in support of it, the leading American astronomer from that period, Simon Newcomb, happened to be present. Although Newcomb had earlier reported observations of canals, he came forth at that meeting as a supporter of Maunder. This is significant because in 1907 Newcomb published his "Optical and Psychological Principles Involved in the Interpretation of the So-Called Canals of Mars," in which he recounted experiments he had conducted similar to those he had heard Maunder discuss in 1903. The conclusions that Newcomb drew from his experiments were that the mind is no passive processor of visual stimuli and, in particular, that the canals are an illusion. Maunder's position had thus won the support of the most influential astronomer in the United States.

E. Walter Maunder, "The Canals of Mars," *Knowledge* 26 (November 1903): 249–51.

> Just a year ago a very interesting paper under the above heading by Mr. B. W. Lane appeared in *Knowledge*; in which he described a series of experiments tending to show that "the mere shape of the oceans of Mars is sufficient to give rise to

11. For more information on Antoniadi, see the next section.

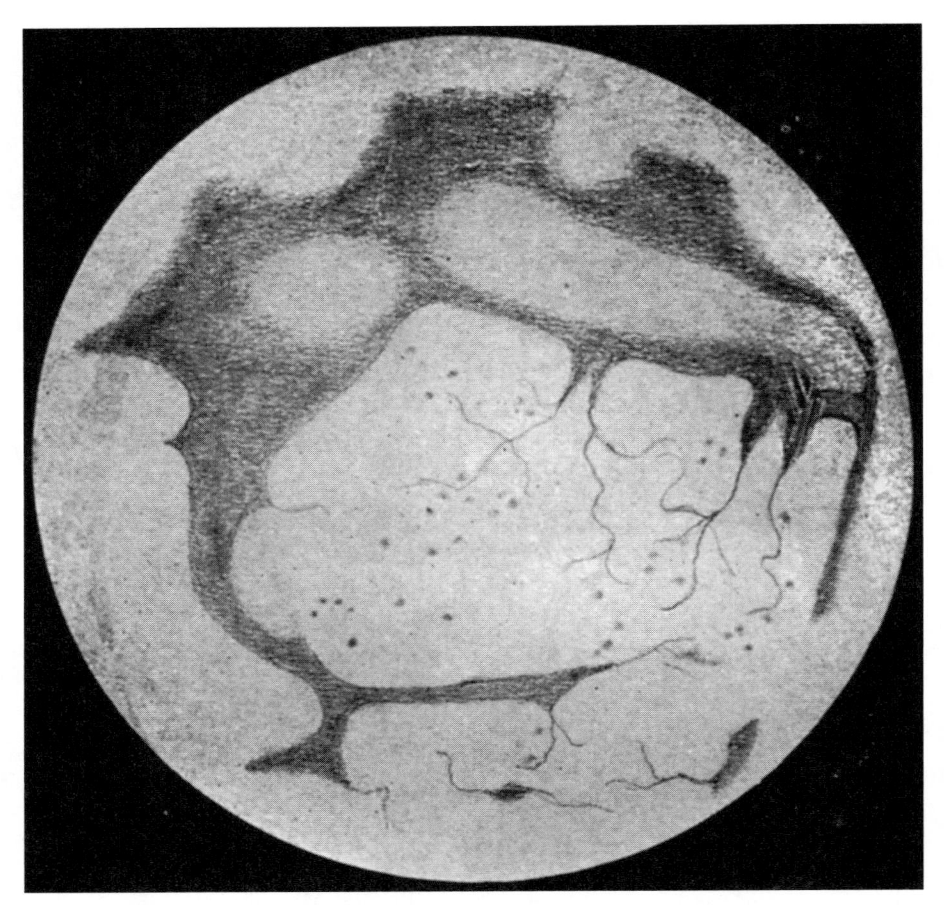

Fig. 16. Diagram used by Maunder

the appearance of the complicated system discovered by Schiaparelli."[12] Mr. Lane's paper led to Mr. J. E. Evans, headmaster of the Royal Hospital School, Greenwich, and myself, undertaking together a series of experiments with boys in that school, from which some interesting results have been obtained. At the time when Mr. Lane's paper was published only a small proportion of these experiments had been carried

12. *Knowledge*, 1902, November, p. 250.

out, and I was not therefore able to speak as definitely as I now can as to the conclusions to be drawn from them. These experiments were supplemented by others, made chiefly by Mrs. Maunder and myself, on the impressions produced by lines and dots when near the limit of vision, which further illustrate the subject.

The experiments at the Royal Hospital School were made in the following manner. A class of about twenty boys, from twelve to fourteen years of age, were seated in four or five rows at different distances from a carefully-lighted diagram, which they were told to copy. The diagram was reproduced from some published drawing of Mars, but in nearly every experiment the canals were omitted. For the most part any boy was used in only one experiment; but a few were set to draw the same diagram twice, the second time at a different distance from the first. The diagram was generally about six inches in diameter, and the distances of the boys from the diagram ranged from fifteen to forty feet, except in two experiments where the range extended up to sixty feet.

At this point, Maunder described the test diagrams he used. He employed three types of circular diagrams, some containing only "oases," that is, dark dots, some showing only "canals," and some showing irregular lines. In all cases, the pattern of arrangement was similar to that in a diagram of the Martian surface published by Percival Lowell.

The general result was striking. In several of these experiments nearly all the boys drew "canals" on their copies, though there were none on the original from which they were copying. And these "canals" were not placed at random; they were just in the very places where canals are seen in the charts of Schiaparelli and Lowell. The boys agreed on the whole rather better in reproducing the position and direction of canals like Phison, Hiddekel, Euphrates, and Arnon, which were not on the original before them, than in reproducing the position and outlines of large and prominent markings like the Syrtis Major and Sinus Sabaeus, which were very distinctly on the original. . . .

In a section omitted at this point, Maunder disputed some instances where B. W. Lane's experimental results differed from Maunder's findings.

Whence then did the "canals" come which were drawn by the boys of the Hospital School?

One cause was the prolongation of dark indentations invading the brighter regions. The Gehon and Hiddekel, for example, on the boys' drawings were clearly partly due to the two arms of Dawes' Forked Bay. This was my first suggestion as to the probable true nature of Schiaparelli's discovery, a suggestion made as long ago as 1882. But so far as our experiments go I do not think that many of the canals can be explained in this way. The Forked Bay and the two lakes Ismenius Lacus and Siloe Fons would be enough to some of our boys to show them Hiddekel and Gehon, but not to the majority.

A more fruitful source of the "canals" was the introduction of regions slightly darker or slightly brighter than their surroundings. Meroe Island figured as an example in the first category, Elysium as one in the second, in two different experiments. And no one could wish for straighter and sharper "canals" than were drawn by a good proportion of the boys to express these regions. A few put in shading, fewer still put in shading and drew also a dark canal as its boundary. There can, I think, be no reasonable doubt that Mr. Green, when in 1879 he suggested that the borders of faint shadings might have given rise to some of the canals, pointed out a cause not only equal to producing such an effect, but one that does so in reality.

Another cause which proved decidedly effective was the tendency of the eye to join together two small spots, where these were not too small to be separately seen, by a wholly subjective line. But the cause which was the most effective within the limits of our experiments with the Hospital School boys was the way in which the eye summed up together minute irregular markings, each too small to be separately perceived as straight streaks. An examination of the drawings with reference to the placing of the boys rendered this very evident. At fifteen or seventeen feet the boys were near enough to detect some, if not all, of the minuter details as separate entities, and hence drew few canals. At thirty-five or thirty-eight feet these details were for the most part too small to produce any effect, even in the gross, and therefore the boys here also drew few if any canals. But between these two extremes nearly every boy drew canals. Halfway between them every boy without exception saw "canals." For here, whilst the details were each one separately invisible, they were yet capable of creating in the sum a distinct impression.

And, seen at this distance, it was very striking to see how the eye, as it were "took the average" of all irregularities. Two irregular wavy lines drawn so as to cross each other, took form as a beautiful "oasis" with four straight canals radiating from it. Lines as meandering as the Thames or Trent with their tributaries on a map of England, straightened out as rigidly as the Phison or Euphrates on Lowell's chart. If Mr. Lane's effects were never secured with an entirely blank "continent," it was astonishing to see how effectively three or four dots, absolutely invisible at the distances occupied, would suffice to make those effects plain. Nor was it at all necessary that the dots should be put in a straight line where the canal ought to run. They might wander from it a good deal on either side, if only the mean line between them ran in the right direction.

It may be objected that very little has been gained if we recognise that the canals are either the edges of half-tone districts, or the summation of very minute details. The general distribution of the true markings on the planet must approximate to that shown on the charts of Schiaparelli and Lowell, and the details if not straight lines in their ultimate conceivable resolution are at least straight lines to the eye.

But the gain is really great. For so long as we conceive of that elaborate reticulation as being a true feature of the actual surface of the planet, we can hardly escape from Mr. Lowell's induction. Lines so straight, so formal, so uniform in width, so regular in their intersections, so symmetrical, with dark spots so inevitably marking their intersections, must be accounted, as he accounts them, artificial; the handiwork of intelligent beings. But if actual details of perfectly irregular and unsymmetrical character, details having no sign of artificiality about them, can present exactly the appearance, and make just the impression which the network of the canal system does, the argument for the existence of inhabitants on Mars has vanished. We are freed, too, from the necessity of considering such bizarre theories as would make out the planet to have been scored into its present form by grazing meteorites, or to have assumed it through crystallization. To have been set free from the grotesque in observation is to have been freed also from the grotesque in speculation.

This service I think the drawings of the Hospital School boys have effectually rendered to us. They have shown that perfectly unbiased observers will see and draw the Schiaparellian canals when the actual markings presented to them are as little regular and artificial as any which our own earth might present to an outside spectator.

I do not think we need claim more for these experiments than this; a complete explanation of the canal system would probably involve other factors than those indicated above. Thus a most able letter by M. Antoniadi, which appeared in the *English Mechanic* for July 31st, 1903, sums up the canal-impression under five heads

(*a*) Entirely physiological markings, like those seen by Mr. Lowell on other planets, by myself on Mars, and by Mr. Lane on his artificial discs.
(*b*) Subjective lines, generated by the topographical details.
(*c*) Edges of physiological half-tones, begotten by contrast.
(*d*) Edges of objective half-tones, arising from the same reason. And
(*e*) Incontestably real canals, which, were we to see Mars better, would resolve themselves into groups of knotted or unevenly shaded areas.

This classification cannot be much bettered with our present knowledge. I should myself give the first place to *b* as the most fruitful source of canals, and the next to *d*, these two classes being those with which our Greenwich experiments were concerned. The two physiological classes, *a* and *c*, were not included in our work. But I am inclined to think that they are intimately combined with the others in producing the canal system as we know it; and that its geometrical appearance is largely due to them. In the last class *c*, I should myself prefer to withhold the name "canal" from features which, like Nasmyth Inlet and Huggins Inlet, were well known long before the canal system as such had its commencement. I well remember that my first disposition to be sceptical of Schiaparelli's discoveries arose from the fact that he drew so many canals which I could not see at all as being equally distinct with Nasmyth Inlet which I had seen conspicuously. That may have been a bit of personal bias, but I think it is clear that whilst the canal system as a whole must he considered as subjective only, some markings which have been included in it have been too thoroughly well defined to be anything but real. At the present moment the most interesting feature with regard to the entire discussion is the tendency of the ablest and most favourably circumstanced observers to see the chief canals no longer as straight uniform lines, but as close sequences of spots, as if an approach had been made to their complete resolution. It is now a quarter of a century since the canals as such were first introduced to us. It may well be that the next quarter of a century may see such an advance that ere its close we may have detected the actual component details of what we now consider as straight lines, and that therefore the canals — as such — may have entirely disappeared.

EUGÈNE M. ANTONIADI (1870–1944)

E. W. Maunder and E. M. Antoniadi rank as the most creative and effective proponents of the claim that the canals of Mars are illusory. If Maunder was the first person to develop this theory, Antoniadi was the more active in championing this position. Born in Constantinople in 1870 to Greek parents, Antoniadi exhibited early interests in astronomy, especially Mars; in fact, by the early 1890s he was submitting reports of his observations to a French astronomical journal, *L'Astronomie*, edited by Camille Flammarion, and to the British Astronomical Society's *Journal* and *Memoirs*. During that period, he observed Mars from Flammarion's observatory in Juvisy, which is near Paris. From there he had by 1894 observed as many as

Fig. 17. Eugène M. Antoniadi (Courtesy SPL/Royal Astronomical Society)

42 canals. His astronomical skills and interests led to his becoming in 1896 the director of the British Astronomical Association Mars Section, a post he continued to fill for about two decades. In that position, he received and coordinated reports from the Association's membership of observations on Mars, drawing these results into a substantial report published after each Martian opposition. Antoniadi's first report was very supportive of the canals, but gradually, very possibly under the influence of Maunder, Antoniadi began to grow more skeptical of reports claiming sightings of numerous canals.

By the time of his report (given below) on the 1909 opposition, Antoniadi had concluded that essentially all the canals are illusions. One of the chief arguments he made was to report that with the 26-inch aperture Meudon telescope located on a mountain top in France, he had been able to break the canals down into fine detail. What Antoniadi was claiming can be seen from the accompanying drawings (Fig. 18) taken from a book by Antoniadi on Mars.[13] The drawing shows a particular region of Mars. The drawing on the left shows the region as drawn by Schiaparelli. On the right is Antoniadi's drawing, indicating how he had seen Mars with the Meudon telescope. He discussed this observation of Mars in the report given below, in which he also praised Maunder's 1894 and 1903 papers, referring to the earlier document as a "masterly interpretation." Antoniadi's paper, it should be noted, is one of many that he published during this period arguing the illusion theory. Maunder also returned to the fray, in which they had been joined by various other astronomers, including Simon Newcomb, A. E. Douglass, and Vincenzo Cerulli. Their cause gained additional support from publications by W. W. Campbell, an American spectroscopist, whose precise studies showing the very small quantity of water vapor in the Martian atmosphere supported the opponents of the canals much more than their advocates.

The level of success resulting from these efforts is suggested by the fact that in 1910 J. Comas Solá, a Spanish astronomer, asserted regarding the 1909 Martian opposition: "The marvelous legend of the canals of Mars has disappeared with this opposition."[14] Moreover, in 1913 Antoniadi added: "Ponderous volumes will still be written to record the discovery of new canals. But the astronomer of the future will sneer at these wonders; and the canal fallacy, after retarding the progress of astronomy for a third of a century, is doomed to be relegated into the myths of the past."[15]

13. E. M. Antoniadi, *La planète Mars* (Paris: Hermann, 1930).

14. J. Comas Solá, "Quelques considérations sur la planète Mars," *Société astronomique de France bulletin* 24 (1910): 36–37.

15. E. M. Antoniadi, "Considerations on the Physical Appearance of the Planet Mars," *Popular Astronomy* 21 (1913): 424.

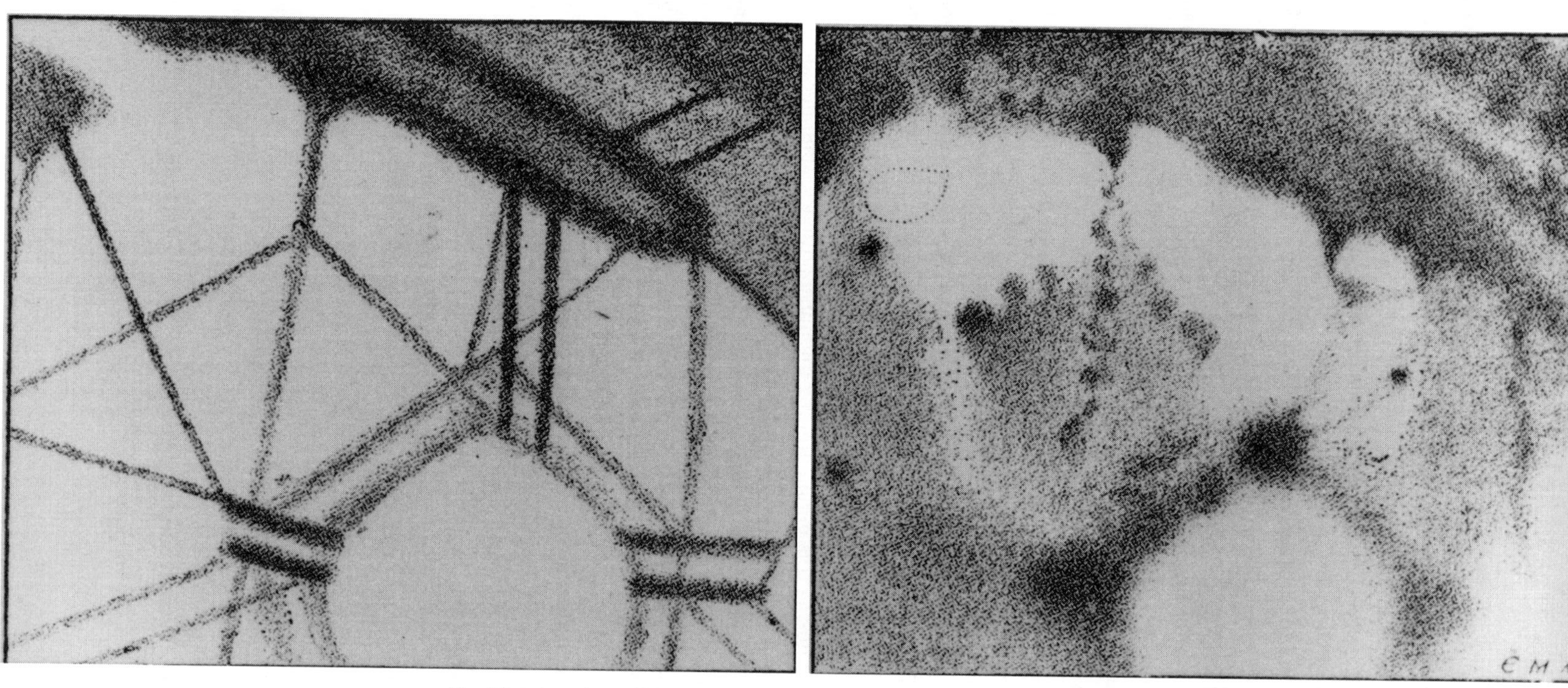

Fig. 18. A region of Mars as drawn by Schiaparelli (left) and Antoniadi (right)

E. M. Antoniadi, "Fifth Interim Report for 1909, Dealing with the Fact Revealed by Observation That Prof. Schiaparelli's 'Canal' Network Is the Optical Product of the Irregular Minor Details Diversifying the Martian Surface," *Journal of the British Astronomical Association* 20 (1909): 136–41.

In the third Interim Report of this Section[16] attention was called to the illusive character of the geometrical network of Mars, and to the fact that, of the various "canal" systems known, it was really that of Prof. Schiaparelli which could stand the severe test of a large aperture. Since then it was possible to continue the observations of these markings with the following additional results:—

Antœus.—Seen as a group of detached, uneven, and irregular dusky spots.

Cerberus.—Resolved into a knotted band.

Læstrygon.—Observed as a congeries of irregular, disjointed masses.

Sirenius.—Seen swelling out into a huge shading to N.

Tartarus.—As the irregular S. W. edge of shaded *Titan.*

Titan.—Emerging out of the N. W. front of *Mare Sirenum,* not out of *Titanum Sinus,* and expanding into a vast shading, running down as far as *Trivium Charontis.*

All these objects were held steadily under good seeing.

As "canals" having no objective basis should be mentioned the *Brontes,* seen on November 9 as a narrow, sharply defined black line, when the boiling rendered the form of the planet almost triangular.

Fifty "canals" having some real basis were seen at Meudon. Of these, 28 per cent were resolved into disconnected knots of diffuse shadings; 20 per cent appeared as more or less dark bands; 16 per cent were edges of faint shadings; another 16 per cent seemed to be broad and diffused streaks; 8 per cent were seen as irregular lines; 6 per cent swelled out into

16. [*British Astronomical Association*] *Journal,* 25–28.

vast shadings; and another 6 per cent had the form of irregular isolated "lakes."

From his observations (1909, September 20 to November 27) the Director [E. M. Antoniadi] was led to deduce the following general inferences:—

I. The true appearance of the planet Mars is a natural one, and comparable to that of the Earth and of the Moon;

II. Under good seeing there is no trace whatever of a geometrical network; and

III. The "continental regions" of the planet are variegated with innumerable dusky spots of very irregular outline and intensity, whose sporadic groupings give rise, in small telescopes, to the "canal" system of Schiaparelli.

No doubt we have never seen a single genuine canal on Mars; nor should we see any from Phobos, the nearest satellite to the planet, on a disc of 42°.

On October 6 and on November 9 it was permitted the Director to witness what he considers to be an elementary view of the true structure of the Martian deserts. The image was slightly tremulous on each date, when, suddenly, definition becoming perfect, a wonderful sight presented itself for a dozen seconds on both occasions. The soil of the planet then appeared covered with a vast number of dark knots and chequered fields, diversified with the faintest imaginable dusky areas, and marbled with irregular, undulating filaments, the representation of which was evidently beyond the powers of any artist. There was nothing geometrical in all this, nothing artificial, the whole appearance having something overwhelmingly natural about it.

In a section omitted at this point, Antoniadi discusses observations made by three prominent astronomers, E. E. Barnard, A. S. Williams, and Gaston Millochau, which observations support Antoniadi's claim that the canals are an illusion due to the eye integrating detail on the Martian surface in such a way as to give the appearance of canals.

The task of the experienced observer, using a very powerful telescope, is an easy one, and must fall short of the labour of the analyst, as seldom requiring the higher powers of the understanding. It is, therefore, a duty to honour here the man who, by a masterly interpretation of facts, saw clearly into the "canal" deadlock at a time when everything was darkness to all. In a profound paper, written in 1894, Mr. E. Walter Maunder laid particular stress on the illegitimacy of a very common assumption: "We have no right to assume," he said, "and yet we do habitually assume, that our telescopes reveal to us the ultimate structure of the surface of the planet."[17] An examination of the results obtained in 1892 by our valued colleague, Mr. W. F. Gale, and by Prof. W. H. Pickering, then led Mr. Maunder to predict that the discovery of "lakes" would be "the next phase of the 'canal' question."[18] A few weeks later Mr. Maunder framed his theory that the "canals" were "only the summation of a complexity of detail."[19] Lastly, in 1903, he conceived and conducted, with the aid of Mr. Evans, the famous experiments with the boys of the Royal Hospital School, Greenwich, experiments which M. André has fitly characterised as "one of the most interesting object lessons imaginable."[20] The result of that inquiry was that, whilst "Mr. Green's suggestion that the 'canals' are boundaries of regions of differing shade or tone" received some support, yet "the most fruitful source of the canal-like impression" was "the tendency to join together minute dot-like markings,"[21] and that, in the majority of cases, the "canals" were "simply the integration by the eye of minute details."[22]

Two years later the so-called "canals" of Mars were photographed by the wonderful ability of Messrs. Lowell and Lampland, and, for some time, the optical theory seemed almost tottering. But, careless of hostile criticism, and confident in the fact that photography had simply revealed the

17. *Knowledge*, November 1894, p. 251.
18. *Ibid.*, p. 252.
19. *Ibid.*, March 1895, p. 58.
20. *Les Planètes et leur Origine*, Paris, 1909, p. 59.
21. *Monthly Notices*, Royal Astronomical Society, Vol. LXIII, p. 498.
22. *Ibid.*, p. 499.

objective substratum underlying the "canal" impression, Mr. Maunder persisted in the path of truth and reason; and his patience was finally rewarded by seeing the spiders' webs wiped off for ever on a single night.

In 1898 Dr. Cerulli, of Teramo (Italy), published the results of his observations of the planet during the 1896–97 apparition, made with an excellent 15 1/2-in. equatorial by Messrs. T. Cooke and Sons, Limited, of York.[23] The "canal" *Lethes* was resolved into knots on 1897, January 4, and the distinguished Italian observer noted that "from July 1896 to February 1897 the lines of Mars did not seem either to increase or to decrease in visibility. Some of the canals were sufficiently easy already in July, on a disc of 7''. In December, the diameter being 17'', they ought to have become magnificent; but, on the contrary, they preserved the same appearance as in July."[24] He also remarked that the "canals" are less visible when transiting the central meridian than when near the edges of the disc.[25] Such curious facts militated, according to Dr. Cerulli, against the reality of narrow lines on Mars.[26] He then concluded that "these lines are formed by the eye . . . which utilises . . . the dark elements which it finds along certain directions"; that "a large number of these elements forms a broad band"; and that "a smaller number of them gives rise to a narrow line."[27] Also, "the marvellous appearance of the lines in question has its origin, not in the reality of the thing, but in the inability of the present telescope to show faithfully such a reality."[28] Then, pushing the theory further, Dr. Cerulli discovered the remarkable fact that an opera glass reveals "canals" on the Moon;[29] while, in a recent letter, he showed that small photographs of our satellite, about 1 centimetre in diameter, if held at the proper distance, re-

23. *Marte nel* 1896–97, Collurania, 1898.
24. *Ibid.*, p. 115.
25. *Ibid.*, p. 116.
26. *Ibid.*, pp. 115–19.
27. *Ibid.*, p. 117.
28. *Ibid.*, p. 114.
29. *Ibid.*, pp. 119–21.

veal, after prolonged staring, single and double "canals" on the Moon, as on Mars. To these lunar discs he attached the fitting epigram: "*Stat vero apparens Mars sicut Lunula Lunæ.*" Lastly, in 1900, Dr. Cerulli published his results of the next apparition,[30] from which we extract the following pregnant considerations: "The darker, larger, and easier canals will present, in the best telescopes, . . . symptoms of resolution."[31] Then the "canals," upon "a close approach to Mars to the Earth will seem to lose rather than gain through the increase [in size] of the disc. New lines of shade will appear, but the primitive ones of the smaller discs will be despoiled, to a large extent, of the sensations of lateral dusky markings, and will seem to become narrower."[32] Finally, he laid stress on the fact that "the optical theory finds . . . its justification in the instability of the areoscopic sensations."[33] All these conclusions of Dr. Cerulli's are in perfect harmony with the Director's experience. They are most important, and if they were not as generally known as they ought to be, it is only to the language in which they were written that the fact should be attributed.

In 1899 Mr. Stanley Williams admirably expressed his belief that "if we could approach Mars to within a few miles, the appearance presented by these so-called 'canals' would be so changed that we should not recognise them at all."[34]

Eight years later Newcomb arrived at a conclusion almost identical with those of Mr. Maunder and Signor Cerulli.[35]

In 1909 M. André affirmed that the "narrow straight canals . . . are optical illusions."[36]

In a note published on 1909, September 28, by the Ἀθηναι of Athens, and dated 1909, September 21, the Director wrote that, according to his first observation of Mars at Meudon,

30. *Nuove Ossercazioni di Marte*, 1898–99, Collurania, 1900.
31. *Ibid.*, p. 104.
32. *Ibid.*, p. 104.
33. *Ibid.*, p. 106.
34. *The Observatory*, 1899, June.
35. *Astrophysical Journal*, July 1907.
36. *Les Planètes et leur Origine*, p. 71.

"narrow black canals . . . were not visible at all, although details much smaller . . . were quite plain in that giant telescope." He then proceeded as follows: "I conclude that those geometrical spider's webs . . . do not exist, as vanishing in the great instrument, which shows us this neighbour world much more resembling the Earth than was hitherto believed to be the case."[37]

Such a conclusion is in fair agreement with the opinion of Mr. W. F. Denning on the subject; for, in a letter dated 1909, September 30, that distinguished observer of planetary surfaces and discoverer of so many comets and meteoric radiants, said: "Thank you for your letter and . . . drawing of Mars [September 20]. . . . There are undoubtedly a few streaks on the planet, and I think that the canals cannot be eliminated altogether. . . . But the network of lines . . . is an illusion."

We shall not expatiate on the circumstances attending the fall of the geometrical network. Yet we cannot refrain from pointing out that it vanished when the planet was practically at its closest approaches to the Earth, high above the horizon, and scrutinized with the best instruments of our time. And the fact that *no straight lines could be held steadily when much more delicate detail was continually visible* constitutes a fatal objection to their crumbling existence.

It is thus by a disregard of the dangers of pressing too closely the evidence of our senses that some observers framed those startling theories about the planet. And, whilst Mars became the abode of superhuman channel diggers, reason was gravely insulted by the representation of canals defying perspective[38]—the most ingenious mode of Martian signalling.

---

37. This document, which was sent to the Council, was thus written 15 days, and published 8 days, *before* the well-known telegram of Prof. Frost, dated 1909, October 6: "Yerkes telescope too powerful for canals." It was important to mention this, as it would be preposterous to claim any independence in the end of the canal network *after* the publication of the telegram in question.

38. This applies to the drawings of a M. Jonckheere (*A.N.* [*Astronomische Nachrichten*], No. 4368).

The favourable apparition of 1909 is now over, carrying away with it many a useless tale. But the Director is glad to be at least in a position to do justice and honour to Prof. Schiaparelli's monumental work, and to affirm that, as far as it was possible to judge (the planet being partly veiled by cloud lately), wherever the distinguished Italian astronomer had drawn a streak there *was* a group of irregular shadings on the surface of Mars.

1909, December 23. E. M. Antoniadi,
Director of the Section.

CONCLUSION

The question whether or not extraterrestrial intelligent beings exist remained unanswered in 1915—and remains unanswered today. Or perhaps a more accurate statement would be that over the last twenty-five hundred years, humans have supplied an abundance of answers to this question, but a paucity of direct evidence in support of those answers. If the materials presented in this volume are looked at broadly, it emerges that already in the premodern period extraterrestrials had made their entrance into terrestrial thought. Some authors such as Aristotle made efforts to dislodge them, whereas others such as the materialist Lucretius and the Christian Cusanus fought for their acceptance.

If it be asked—who from modern times must ultimately be held responsible for the invasion of the extraterrestrials and for the controversy this has caused?—the surprising answer is a conservative Polish canon writing from Frauenberg Cathedral. It was certainly not the intention of Copernicus to invite extraterrestrials into his heliocentric system; in fact, he did nothing more than open the door an inch, but this was enough to initiate the invasion. He conveniently exited our planet before the invasion, but it was nonetheless his highly mathematical volume, which was read only by experts and understood by only a portion of them, that provided the incentive for their immigration. To put the point differently, the celibate canon of the cathedral in Frauenberg acted in a manner that has left him open to the charge that he is the father, or at least the grandfather, of Darth Vader, ET, ALF, Mork, and the whole tribe of extraterrestrials whom we all know so well.

Copernicus's heliocentric system brought a new order to the planetary system but greatly disrupted the traditional cosmos. It pried the stars loose from the starry vault, scattered them throughout a nearly or actually infinite cosmos, and turned them into suns; as Edward Young put it, "One sun by day, by night ten thousand shine."[1] And the Copernican system transported humans from seeing their location as the center of the universe to viewing their home as a second-rate planet circling one of billions of stars composing our galaxy, which is but one of the billions of galaxies in our vast universe.

One of the first fully to realize the severity of this problem created by the Copernican universe was Blaise Pascal, whose partial resolution of the problem was to stress that although a knowledge of the infinite universe may humble humanity, such knowledge also indicates the greatness of humanity as composed of thinking beings who, despite their minuscule size and remote location, have managed to measure this vast universe and determine their location in it. Pascal stated:

> Man is but a reed, the most feeble thing in nature; but he is a thinking reed. The entire universe need not arm itself to crush him. A vapour, a drop of water suffices to kill him. But, if the universe were to crush him, man would still be more noble than that which killed him, because he knows that he dies and the advantage which the universe has over him; the universe knows nothing of this.
>
> All our dignity consists, then, in thought. By it we must elevate ourselves, and not by space and time which we cannot fill. Let us endeavour, then, to think well; this is the principle of morality.[2]

The invasion of our solar system by the extraterrestrials did not take place over night. Indeed, we have seen that it was only in the late seventeenth and in the eighteenth centuries that the invasion largely succeeded. By about 1800, the extraterrestrials had made more inroads on the human intellect than at any time before—or perhaps since. Essentially every body in our solar system, the Sun and Moon, the planets and their satellites, and comets, were territories that in the eyes of many, the extraterrestrials had occupied. And their presence was applauded by devotees of natural theology, by poets, preachers, and scientists.

---

1. Edward Young, *Night Thoughts*, ed. George Gilfillan (Edinburgh: J. Nichol, 1853), IX, line 748.

2. Blaise Pascal, *Thoughts*, trans. W. F. Trotter, Harvard Classics edition (New York: P. F. Collier and Son, 1938), #347 (Brunschvicg) = #200 (Lafuma).

The same point can be put somewhat differently. Recall that Carl Becker in a famous book on the Enlightenment, his *Heavenly City of the Eighteenth-Century Philosophers*, suggested that the Enlightenment "*philosophes* demolished the Heavenly City of St. Augustine only to rebuild it with more up-to-date materials."[3] What the readings in this volume suggest is that Becker's famous metaphor can be borrowed to say that various Enlightenment authors created an array of celestial cities and populated them with extraterrestrials.

Gradually, however, in the first half of the nineteenth century, a counter-offensive commenced. Among the authors anthologized in this volume, Richard Adams Locke can lay claim to having launched an attack on at least the Lunarians. Early in the 1850s, a polymathic priest, formerly a friend of the extraterrestrials, turned on them and secretly planned and launched from the Master's Lodge of Trinity College, Cambridge, a more ambitious offensive. Although many, indeed most of Whewell's contemporaries were scandalized by Whewell's efforts to expel the extraterrestrials, gradually some began to recognize that his claims were not without merit. Richard Proctor's conversion, never more than partial but always fully publicized, as well as the progress of astronomical observation and of the scientific spirit put the extraterrestrials if not into rout at least into retreat.

In 1877, the extraterrestrials countered these maneuvers by a new offensive launched from Mars, the planet of war and the extraterrestrials' sole seemingly secure fortress in our solar system. Vast quantities of energy were poured into rescuing the Martians, who were the last best hope for the possibility of our having a companion planet in our system. Nonetheless, by 1915, the fluids flowing through the Martian canals had evaporated, the canals structures had crumbled, and the Martians expelled from their planet, which was left a nearly barren waste. What resulted was confirmation of the largely deserted solar system of Tennyson's "Vastness" and Hardy's "God-Forgotten." Moreover, this solar system is very nearly identical to the solar system that, a century later, you and I inhabit. Put differently, Whewell, Proctor, Maunder, and Antoniadi were the first scientists to see the bleak if beautiful solar system of the early twenty-first century. And part of the significance of this is that our solar system is our best guide to the possibilities for systems surrounding other stars.

Pre-1915 astronomers did not confine their offensive to the extraterrestrials of the solar system. Extraterrestrials that earlier had been placed on comets or on planets orbiting double or variable stars, and in such objects as the Orion nebula also suffered attacks from which they have not recovered.

---

3. Carl Becker, *The Heavenly City of the Eighteenth-Century Philosophers* (New Haven: Yale University Press, 1932), 31.

Of course, the extraterrestrial life debate did not end in 1915. During the twentieth century the universe expanded at a remarkable rate, became filled with other galaxies, and, as we began to confirm only in 1995, other solar systems. Thus the locales available for extraterrestrials have increased, as has our knowledge of the universe and the power of our telescopic equipment. Were the authors anthologized in this volume to return to life in the twentieth-first century, they would find an area known for over twenty centuries as the plurality of worlds transformed in techniques and title. Moreover, astrobiology (the newest title) is pursued by thousands of scientists, enriched in tens of thousand of publications, and funded by many millions of dollars. Although the ultimate answer as yet eludes our efforts, many areas of the debate have seen astonishing progress.

APPENDIX

The History of Ideas of Extraterrestrial Life Research Program

The following short essay can be read as an invitation to engage in a developing research program, the focus of which is the history of ideas of extraterrestrial life. The first point to be made is that ideas of extraterrestrial life do have a history! Many persons assume that the question of extraterrestrial life arose only in the twentieth century. In fact, a number of historical studies[1] show that such ideas have a history extending from the ancient Greeks to the present.[2] Moreover, these ideas have been a subject of historical study for more than a century. Among those

1. Four major studies can be cited at this point as examples of this research program: Steven J. Dick, *Plurality of Worlds: The Origins of the Extraterrestrial Life Debate from Democritus to Kant* (Cambridge: Cambridge University Press, 1982); Karl S. Guthke, *The Last Frontier: Imagining Other Worlds from the Copernican Revolution to Modern Science Fiction* (Ithaca: Cornell University Press, 1990); Michael J. Crowe, *The Extraterrestrial Life Debate 1750–1900: The Idea of a Plurality of Worlds from Kant to Lowell* (Cambridge: Cambridge University Press, 1986); Steven J. Dick, *The Biological Universe: The Twentieth-Century Extraterrestrial Life Debate and the Limits of Science* (Cambridge: Cambridge University Press, 1996).

2. For a listing of ca. 130 books published on this subject before 1917, see Crowe, *Extraterrestrial Life Debate*, 646–57.

who in the period before 1980 published significant writings on this history were Camille Flammarion, Pierre Duhem, Arthur Lovejoy, Marjorie Hope Nicolson, and Stanley Jaki.[3]

The primary goal of this research program is to construct an accurate history of the ideas that humanity has had about extraterrestrial intelligent life (hereafter EIL). Included under humanity are both leading intellectuals in any area of learning and also members of the public. To be candid, nearly all the research in this area has focused not on the entire human family, but rather on authors in the Western tradition. The history of ideas of extraterrestrial life outside the western tradition awaits its historian.

The writings of scientists, especially astronomers, deserves special attention in the construction of this history. One reason for this is that their ideas about extraterrestrials have had a strong influence on others. Moreover, the issue of whether extraterrestrials exist is on the first level scientific in nature. Although scientists are a group of special importance, other groups of intellectuals deserve considerable attention. Many persons concerned about religion have taken a strong interest in this issue, which is loaded with implications for such questions as the nature of the divine and whether, as Christianity teaches, the divine became incarnate on this planet and undertook to redeem its inhabitants (and possibly intelligent beings on other planets). Philosophers have also been of much importance, partly because philosophy takes as one of its tasks dealing with large questions about life and the universe. Another area of philosophical relevance concerns the proper methodology for deciding issues regarding extraterrestrials. Literary figures, especially poets, have given much attention to ideas of extraterrestrial life and have thereby influenced popular beliefs regarding such beings. Learning how poets, for example, have reacted to ideas of EIL can teach us much about humanity. Science fiction writers have also had a major role in the extraterrestrial life debate. This is obvious, but what is less clear is whether the writings of such authors teach us more about what they can imagine than about what humans believe.

In pursing this program, a number of questions need to be kept in mind.

- To what extent and in what ways do astronomical ideas about EIL influence religious and literary authors? Do these authors give serious and informed consideration to what the scientists are saying?

---

3. For references to the relevant writings of these authors, see the bibliography at the end of this volume.

- Do religious or philosophical issues or concerns influence what scientists do and claim? How in general does belief about extraterrestrial life influence science? This may seem an unpromising question to ask, but numerous cases can be cited where beliefs about extraterrestrials have drawn people to astronomy or provided financial support for astronomers, astronomical writers, or observatories. Allied to this is the question of the role that ideas of extraterrestrial life have played in the popularization of astronomy and, more generally, its communication to the public.
- It has been found productive to investigate to what degree the EIL conceptions of an age, of a group, or of a person are responsibly linked with the scientific information available at the time. Put somewhat differently, the issue is whether the extraterrestrials of an era are more the creation of the science of that period or of its philosophical or religious notions. The widespread acceptance of such notions as the Great Chain of Being, the Copernican Principle, and the Principle of Plenitude in various periods suggests that broad conceptions of this sort have powerfully influenced our ideas of EIL. In this context, it can be asked: Do the historiographical methods appropriate for this area have greater affinity with those used for, say, the history of a technical area in astronomy, or, at the other extreme, for the history of pseudoscience?

Although historical investigations in this area are unlikely to lead to resolution of the question of whether extraterrestrials exist, this type of historical research can shed light on such issues as the methodology of the quest and on what constitutes responsible work in this area, which at times borders on the sensational and in which values are very close to the surface. It is also possible that some of the arguments—be they scientific, religious, philosophical, or whatever—that have been made in the past may still merit consideration. Moreover, although studying the history of ideas of extraterrestrial life may not shed light on such beings, it gives promise of telling humanity about itself. Just as inkblot tests are not about inkblots but rather tell us about their interpreters, just as a study of the paintings of saints may tell us little about those saints but much about the artists and the era in which they painted, so also learning about how humans have thought of extraterrestrials can be deeply revealing of the fears and hopes of persons from the past as well as the images they have of the universe, whether it is friendly or malevolent, ordered or chaotic, fundamentally spiritual or material.

The final point to be made is that the history of ideas of extraterrestrial life research program is a developing program. In the last twenty years, a dozen or so new histories of EIL issues have appeared, two of my favorites being Roger Hennessey's *Worlds without End: The Historical Search for Extraterrestrial Life* (1999), and on a more advanced level, George Basalla's *Civilized Life in the Universe: Scientists on Intelligent Extraterrestrials* (2005). John Moffitt, an art historian, has written a book filled with important insights, including many illuminating earlier materials: *Picturing Extraterrestrials: Alien Images in Modern Mass Culture* (2003). For an institutional study of NASA, which is very relevant to the EIL debate, see Steven J. Dick and James E. Strick, *The Living Universe: NASA and the Development of Astrobiology* (2004). The historical studies of UFO-related reports and claims is very extensive and sometimes of uneven quality. Jerome Clark's two volume encyclopedia is a good place to start: *The UFO Encyclopedia*, vol. 1, *UFOs in the 1980s* (1990), and vol. 2, *The Emergence of a Phenomenon: UFOs from the Beginning through 1959* (1992). New biographies of key figures in the EIL debate continually appear, for example, the biographies of Thomas Dick, Percival Lowell, and Carl Sagan.[4] The number of doctoral dissertations and master theses relevant to this research program is also increasing. A final important point is that educational materials are available for the offering of undergraduate courses in the history of EIL ideas or more generally on the historical, philosophical, theological, and social scientific study of issues and persons involved in the EIL debate. Courses of this nature have been offered for a number of years at the University of Notre Dame and at Truman State University.

4. See the bibliography at the end of this volume for more and fuller references.

SELECTED BIBLIOGRAPHY

No comprehensive bibliography on historical and humanistic aspects of the extraterrestrial life debate has yet been published. This bibliography, which concentrates on but is not limited to the period before 1915, is far from definitive. It is designed to provide helpful leads into the literature. It is divided into six sections:

1. Sources of Bibliography
2. Surveys of the Extraterrestrial Life Debate
3. Other Useful Publications Taking a Survey Approach
4. Some Especially Important Studies
5. Studies of Particular Individuals
6. Religious Aspects of the Debate: A Sample

1. SOURCES OF BIBLIOGRAPHY

Crowe, Michael J. "Bibliography of Books on the Question of a Plurality of Worlds Published before 1917." In *The Extraterrestrial Life Debate, 1750–1900: The Idea of a Plurality of Worlds from Kant to Lowell.* Cambridge: Cambridge University Press, 1986. New ed., Mineola, N.Y.: Dover Publications, 1999, 646–57. This bibliography lists about 140 items.

Eberhart, George M., ed. *UFOs and the Extraterrestrial Contact Movement: A Bibliography.* 2 vols. Metuchen, N.J.: Scarecrow Press, 1986. Although far from complete, this bibliography provides 15,631 references.

Kuiper, Thomas B. H., and Glen David Brin. "Resource Letter ETC-1: Extraterrestrial Civilization." *American Journal of Physics* 57 (1989): 12–18.

Rynin, N. A. *Interplanetary Flight and Communication*, 3 vols. in 3 parts. Translated from Russian. Jerusalem: Israel Program for Scientific Translations, 1971. Contains both a bibliography and extensive historical information.

Schatzberg, Walter, Ronald A. Waite, and Jonathan K. Johnson, eds. *The Relations of Literature and Science: An Annotated Bibliography of Scholarship, 1880–1980.* New York: Modern Language Association, 1987. According to the above work (viii): for the period after 1980, see *The Relations of Literature and Science: A Bibliography of Scholarship, 1980–81.* Worcester, Mass.: Clarke University Press, 1982. . . . *1981–82.* Worcester, Mass.: Clarke University Press, 1983. . . . *1982–83.* Worcester, Mass.: Clarke University Press, 1984. This is the single most useful source for bibliography on literary figures who were involved in the debate. Later issues appeared for a while in *Publications of the Society for Literature and Science* and currently appear in the journal *Configurations.*

2. SURVEYS OF THE EXTRATERRESTRIAL LIFE DEBATE

Basalla, George. *Civilized Life in the Universe: Scientists on Intelligent Extraterrestrials.* New York: Oxford University Press, 2005.

Benz, Ernst. *Kosmische Bruderschaft: Die Pluralität der Welten.* Freiburg: Aurum, 1978.

Crowe, Michael J. *The Extraterrestrial Life Debate, 1750–1900: The Idea of a Plurality of Worlds from Kant to Lowell.* Cambridge: Cambridge University Press, 1986; a two-page addendum on new materials appears in the 1988 paperback reprinting. A new preface was added for the 1999 Dover reprinting, which carries the abbreviated title: *The Extraterrestrial Life Debate, 1750–1900.* Mineola, N.Y.: Dover Publications, 1999.

Dick, Steven J. *Plurality of Worlds: The Origins of the Extraterrestrial Life Debate from Democritus to Kant.* Cambridge: Cambridge University Press, 1982. Contains a useful bibliography.

———. *The Biological Universe: The Twentieth-Century Extraterrestrial Life Debate and the Limits of Science.* Cambridge: Cambridge University Press, 1996.

———. *Life on Other Worlds: The 20th-Century Extraterrestrial Life Debate.* Cambridge: Cambridge University Press, 1998. An abridged and updated version of Dick's *Biological Universe.* Provides a brief but useful bibliography.

Fisher, David E., and Marshall Jon Fisher. *Strangers in the Night: A Brief History of Life on Other Worlds.* Washington, D.C.: Counterpoint, 1998.

Guthke, Karl S. *The Last Frontier: Imagining Other Worlds from the Copernican Revolution to Modern Science Fiction.* Ithaca, N.Y.: Cornell University Press, 1990. Originally appeared as *Der Mythos der Neuzeit: Das Thema der Mehrheit der Welten in der Literatur- und Geistesgeschichte von der kopernickanischen Wende bis zur Science Fiction.* Berne: Francke, 1983.

Hennessey, Roger. *Worlds without End: The Historic Search for Extraterrestrial Life.* Stroud: Tempus Publishing Ltd., 1999.

Lemonick, Michael. *Other Worlds: The Search for Life in the Universe.* New York: Simon and Schuster, 1998.

Moffitt, John F. *Picturing Extraterrestrials: Alien Images in Modern Mass Culture.* Amherst, N.Y.: Prometheus Books, 2003. The author, an art historian, draws effectively on his background to bring a perspective not usually available. He is also knowledgeable about claims regarding UFOs, alien abduction, and ideas of extraterrestrial life.

Shapiro, Robert. *Planetary Dreams: The Quest to Discover Life beyond Earth.* New York: John Wiley and Sons, 1999.

Smith, Cynthia Anne Miller. "Shadows of Things to Come: The Theological Implications of Intelligent Life on Other Worlds." M.A. thesis, Georgia State University, 2003. Available at http://www.romancatholic.org/th.html (last accessed June 2007.)

Vizgin, V. P. *Ideya Mnozhestvennosti Mirov.* In Russian; title in translation is *Plurality of Worlds Idea.* Moscow: Nauka Publishing House, 1988.

3. OTHER USEFUL PUBLICATIONS TAKING A SURVEY APPROACH

Bartholomew, Robert, and George Howard. *UFOs and Alien Contact: Two Centuries of Mystery.* Amherst, N.Y.: Prometheus Books, 1998. Shows that over the past two centuries, repeated sightings of nonexistent flying objects (e.g., airplanes and dirigibles before their invention) have been reported and uses social psychology to explain reported sightings of such objects and of UFOs.

Beck, Lewis White. "Extraterrestrial Intelligent Life." *Proceedings of the American Philosophical Association* 45 (1971–72): 5–21.

Bell, Trudy E. "The Grand Analogy: History of the Idea of Extraterrestrial Life." *Griffith Observer* 2 (August 1978): 2–16.

Chamberlin, R. V. "Life in Other Worlds." *Bulletin of the University of Utah, Biological Series* 1, no. 6 (1932): 1–32. A useful, early historical study.

Clark, Jerome. *The UFO Encyclopedia.* Vol. 1, *UFOs in the 1980s.* Detroit: Apogee Books, 1990. Vol. 2, *The Emergence of a Phenomenon: UFOs from the Beginning through 1959.* Detroit: Omnigraphics, 1992.

Crowe, Michael J. "A History of the Extraterrestrial Life Debate." *Zygon* 32 (June 1997): 147–62.

———. "Extraterrestrial Intelligence." In *History of Astronomy: An Encyclopedia,* ed. John Lankford, 207–9. New York: Garland, 1997.

———. "The Plurality of Worlds and Extraterrestrial Life." In *History of Science and Religion in the Western Tradition: An Encyclopedia,* ed. Gary Ferngren, 342–43. New York: Garland, 2000.

———. "Extraterrestrial Life and Christianity." In *Science, Religion, and Society: An Encyclopedia of History, Culture, and Controversy,* ed. Arri Eisen and Gary Laderman, 1:297–303. Armonk, N.Y.: M. E. Sharpe, 2007.

Dick , Steven J., and James E. Strick. *The Living Universe: NASA and the Development of Astrobiology.* New Brunswick, N.J.: Rutgers University Press, 2004.

Engdahl, Sylvia Louise. *The Planet-Girded Suns: Man's View of Other Solar Systems.* New York: Atheneum, 1974. Although written for teenagers, this book provides some fresh information on historical aspects of the extraterrestrial intelligent life debate.

Flammarion, N. C. *Les mondes imaginaires et les mondes rèels,* 20th ed. Paris: Didier (?), 1882. Especially useful for nineteenth-century French sources.

———. *La pluralité des mondes habités*, 29th ed. Paris: Didier (?), 1882. Especially useful for nineteenth-century French sources.

Guthke, Karl. "Are We Alone? The Idea of Extraterrestrial Intelligence in Literature and Philosophy from the Scientific Revolution to Modern Science." In *Trails in No-Man's-Land: Essays in Literary and Cultural History*, 152–71. Columbia, S.C.: Camden House, 1992.

Krafft, Fritz. "Wissenschaft und Weltbild (I): Die Wende von der Einheit zur Vielfalt; (II): Von der Einheit der Welt zur Vielfalt der Welten und des Menschen Stellung in ihnen." In *Naturwissenschaft und Theologie*, ed. Norbert A. Luyten, 53–78; 79–117. Düsseldorf: Patmos, 1981.

McColley, Grant. "The Seventeenth-century Doctrine of a Plurality of Worlds." *Annals of Science* 1 (1936): 385–430.

Mugler, Charles. *Deux thèmes de la cosmologie grecque: Devenir cyclique et pluralité des mondes*. Paris: Librairie C. Klincksieck, 1953.

Parker, Barry. *Alien Life: The Search for Extraterrestrials and Beyond*. New York: Plenum Publishing Company, 1998.

Tipler, Frank. "A Brief History of the Extraterrestrial Intelligence Concept." *Quarterly Journal of the Royal Astronomical Society* 22 (1981): 133–45.

Weber, Thomas P., ed. *Science and Fiction II: Leben auf anderen Sternen*. Frankfurt am Main: Fischer Taschenbuch, 2004. Contains ten essays on the history of extraterrestrial life ideas.

Zabilka, Ivan L. "Nineteenth Century British and American Perspectives on the Plurality of Worlds: A Consideration of Scientific and Christian Attitudes." Ph.D. dissertation, University of Kentucky, 1988.

4. SOME ESPECIALLY IMPORTANT STUDIES

Ashbrook, Joseph. *The Astronomical Scrapbook: Skywatchers, Pioneers, and Seekers in Astronomy*, ed. L. J. Robinson. Cambridge: Cambridge University Press, 1984. Scattered throughout the articles in this book is much valuable information on the history of extraterrestrial intelligent life ideas.

Baasner, Rainer. *Das Lob der Sternkunst: Astronomie in der deutschen Aufklärung*. Göttingen: Vandenhoeck & Ruprecht, 1987.

Beck, Daniel A. "Life on the Moon? A Short History of the Hansen Hypothesis." *Annals of Science* 41 (1984): 463–70.

Benz, Ernst. "Der kopernikanische Schock und seine theologische Auswirkung." In *The Variety of Worlds*, ed. Adolf Portmann and Rudolf Ritsema, Eranos Yearbook 1975. Leiden, 1977: 15–60. Chiefly on German authors of the eighteenth and nineteenth centuries and on their involvement with extraterrestrial life ideas.

Blumenberg, Hans. *The Genesis of the Copernican World*. Trans. Robert Wallace. Cambridge, Mass.: MIT Press, 1987.

Boss, Valentin. *Newton and Russia: The Early Influences, 1698–1796*. Cambridge, Mass.: Harvard University Press, 1972. Contains interesting material on the reception in Russia of the books by Fontenelle and Huygens.

Brooke, John H. "Natural Theology and the Plurality of Worlds: Observations on the Brewster-Whewell Debate." *Annals of Science* 34 (1977): 221–86.

Brownlee, Donald, and Peter Ward. *Rare Earth: Why Complex Life Is Uncommon in the Universe.* New York: Copernicus, 2000. Perhaps the most thorough and widely respected presentation of the idea that extraterrestrial intelligent life is quite rare.

Burger, William. *Perfect Planet, Clever Species: How Unique Are We?* Amherst, N.Y.: Prometheus Books, 2002.

Burke, John G. *Cosmic Debris: Meteorites in History.* Berkeley: University of California Press, 1986.

Burnett, J. "British Studies of Mars: 1877–1914." *British Astronomical Association Journal* 89 (1979): 136–43.

Cantril, Hadley. *The Invasion from Mars: A Study in the Psychology of Panic, with the Complete Script of the Famous Orson Welles Broadcast.* Princeton: Princeton University Press, 1940. For follow up on this, see Robert E. Bartholomew, "The Martian Panic Sixty Years Later: What Have We Learned?" *Skeptical Inquirer* 22, no. 6 (1998): 40–43.

Cornford, F. M. "Innumerable Worlds in the Pre-Socratic Philosophy." *Classical Quarterly* 28 (1964):1–16.

Cronin, Vincent. *The View from Planet Earth: Man Looks at the Cosmos.* New York: W. Morrow, 1981. Useful for interactions of astronomy with literature and religion.

Crowe, M. J. "New Light on the Moon Hoax." *Sky and Telescope* 62 (Nov. 1981): 428–29.

Curran, Douglas. *In Advance of the Landing: Folk Concepts of Outer Space.* New York: Abbeville Press, 1985.

Danielson, Dennis, ed. *The Book of the Cosmos: Imagining the Universe from Heraclitus to Hawking.* Cambridge, Mass.: Perseus Publishing, 2000. The editor is an English professor with extensive knowledge of the historical interactions of cosmology and culture. The eighty-five selections are made very judiciously.

Darling, David. *The Extraterrestrial Encyclopedia: An Alphabetical Reference to All Life in the Universe.* New York: Three Rivers Press, 2000.

———. *Life Everywhere: The Maverick Science of Astrobiology.* New York: Basic Books, 2001.

Dean, Jodi. *Aliens in America.* Ithaca, N.Y.: Cornell University Press, 1998.

Dick, Steven J. "The Origins of the Extraterrestrial Life Debate and Its Relation to the Scientific Revolution." *Journal of the History of Ideas* 41 (1980): 3–27.

———. "From the Physical World to the Biological Universe: Historical Developments Underlying SETI." In *Bioastronomy: The Search for Extraterrestrial Life—The Exploration Broadens,* proceedings of the Third International Symposium on Bioastronomy held in Val Cenis, Savoie, France, 18–23 June 1990, ed. J. Heidmann and M. J. Klein, 356–62. Berlin: Springer-Verlag, 1992.

———. "Other Worlds: The Cultural Significance of the Extraterrestrial Life Debate." *Leonardo: International Journal of the Contemporary Artist* 29 (1996): 133–37.

Dobbins, Thomas A., and William Sheehan. "The Canals of Mars Revisited." *Sky and Telescope* 107 (March 2004): 114–17.

Duhem, Pierre. *Etudes sur Léonard de Vinci,* vol. 2. Paris: A. Hermann, 1909, esp. 55–95 and 408–23.

———. *Le système du monde,* vol. 9. Paris: A. Hermann et Fils, 1958, esp. 363–430.

Eimer, Manfred. "Byron und der Kosmos." *Anglistische Forschungen* 34 (1912): 1–234.

Fairchild, Hoxie Neale. *Religious Trends in English Poetry*, 6 vols. New York: Columbia University Press, 1939–68.

Fara, Patricia. "Heavenly Bodies: Newtonianism, Natural Theology and the Plurality of Worlds Debate in the Eighteenth Century." *Journal for the History of Astronomy* 35 (2004): 143–60.

Feminò, Fabio. "La pluralità dei mondi nel pensiero debli anitichi." *L'Astronomia* 12 (Oct. 1990): 14–21.

Gardner, Martin. "Psychic Astronomy." In *The New Age: Notes of a Fringe-Watcher*, 252–63. Amherst, N.Y.: Prometheus Books, 1988. Discusses the claims made about extraterrestrials in a number of psychic authors, chiefly from the nineteenth century.

Genuth, Sara Schechner. "Devils' Hells and Astronomers' Heavens: Religion, Method, and Popular Culture in Speculations about Life on Comets." In *The Invention of Physical Science: Intersections of Mathematics, Theology and Natural Philosophy since the Seventeenth Century; Essays in Honor of Erwin N. Hiebert*, ed. M. J. Nye, J. L. Richards, and R. H. Stuewer, 3–26. Dordrecht: Kluwer, 1992.

Goldsmith, Donald. *The Quest for Extraterrestrial Life: A Book of Readings*. Mill Valley, Calif.: University Science Books, 1980. Among the fifty-eight selections in this anthology, nearly all are from the twentieth century.

Guthke, Karl S. "Die Mehrheit der Welten: Geistesgeschichtliche Perspektiven auf ein literarisches Thema im 18 Jahrhundert." *Zeitschrift für deutsche Philologie* 97 (1978): 481–512.

———. "The Idea of Extraterrestrial Intelligence." *Harvard Library Bulletin* 33 (1985): 196–210.

Haglund, Dick. "Svante Arrhenius och Panspermi-hypotesen [Svante Arrhenius and the Panspermic Hypothesis]." *Lychnos* (1967–68): 77–104.

Harley, Timothy. *Moon Lore*. Detroit: Singing Tree Press, 1969 reprinting of the 1885 original. Contains useful information on the history of lunar studies, including the question of lunar life.

Heeren, Fred. "Home Alone in the Universe." *First Things* 121 (March 2002): 38–46.

Heffernan, William. "The Singularity of Our Inhabited World: William Whewell and A. R. Wallace in Dissent." *Journal of the History of Ideas* 39 (1978): 81–100.

Hetherington, Norriss. "Amateur versus Professional: The British Astronomical Association and the Controversy over Canals on Mars." *British Astronomical Association Journal* 86 (1976): 303–8.

Hillegas, Mark. "Victorian Extraterrestrials." In *Worlds of Victorian Fiction*, ed. Jerome Buckley, 391–414. Cambridge, Mass.: Harvard University Press, 1975.

Jaki, Stanley. *Planets and Planetarians: A History of Theories of the Origin of Planetary Systems*. Edinburgh: Scottish Academic Press, 1978.

Kaiser, Christopher. "Extraterrestrial Life and Extraterrestrial Intelligence." *Reformed Review* 51 (1997–98): 77–91. Shows a number of difficulties in strong claims for the existence of extraterrestrials.

Kamminga, Harmke. "Life from Space: A History of Panspermia." *Vistas in Astronomy* 26 (1982): 67–86.

Knight, David. "Uniformity and Diversity of Nature in 17th Century Treatises on Plurality of Worlds." *Organon* 4 (1967): 61–68.

Kukla, André. "SETI: On the Prospects and Pursuitworthiness of the Search for Extraterrestrial Intelligence." *Studies in History and Philosophy of Science* 32 (2001): 31–67.

Lamb, David. *The Search for Extraterrestrial Intelligence: A Philosophical Inquiry.* New York: Routledge, 2001.

Lambert, Ladina. *Imagining the Unimaginable: The Poetics of Early Modern Astronomy.* Amsterdam: Rodopi, 2001.

Le Boeuffle, André. *Le ciel des Romains.* Paris: Boccard, 1989. See esp. chap. 10.

Leszl, Walter. "Infinito e pluralità dei mondi in alcuni autori greci." In *L'Infinito dei Greci e dei Romani.* Genova: Università di Genova, Facoltà di lettere, Dipartimento di archeoligia, filogia classica e loro tradizione, 1989. Publicazioni del D. Ar. FI. Cl. ET., nuova ser., n. 126: 49–85.

Ley, Willy. *Watchers of the Skies: An Informal History of Astronomy from Babylon to the Space Age.* New York: Viking Press, 1963. Contains much information on the history of extraterrestrial life ideas.

Locke, George. *Voyages in Space: A Bibliography of Interplanetary Fiction, 1801–1914.* London: Ferret Fantasy, 1975.

Lovejoy, Arthur. *The Great Chain of Being.* New York: Harper and Row, 1960. See esp. chap. 4.

Mash, Robert. "Big Numbers and Induction in the Case of Extraterrestrial Intelligence." *Philosophy of Science* 60 (1993): 204–22.

McColley, Grant. "The Theory of a Plurality of Worlds as a Factor in Milton's Attitude toward the Copernican Hypothesis." *Modern Language Notes* 47 (1932): 319–25.

———. "Nicholas Hill and the *Philosophia Epicurea.*" *Annals of Science* 4 (1939): 390–405.

———. "The Debt of Bishop John Wilkins to the *Apologia pro Galileo* of Tommaso Campanella." *Annals of Science* 4 (1939): 150–68.

McColley, Grant, and H. W. Miller. "Saint Bonaventure, Francis Mayron, William Vorilong and the Doctrine of a Plurality of Worlds." *Speculum* 12 (1937): 386–89.

McDannell, Colleen, and Bernhard Lang. *Heaven: A History.* New Haven: Yale University Press, 1988.

Meadows, A. J. *The High Firmament: A Survey of Astronomy in English Literature.* Leicester: Leicester University Press, 1969.

Michaud, Michael A. G. *Contact with Alien Civilizations: Our Hopes and Fears about Encountering Extraterrestrials.* New Springs, N.Y.: Copernicus Books, 2007.

Millhauser, Milton. "A Plurality of After-Worlds: Isaac Taylor and Alfred Tennyson." *Hartford Studies in Literature* 1 (1969): 37–49.

Munitz, Milton. "One Universe or Many?" In *Roots of Scientific Thought,* ed. P. P. Wiener and A. Noland, 593–617. New York: Basic Books, 1958.

Naeye, Robert. "OK, Where Are They?" *Astronomy* 24 (July 1996): 36–43. Presents evidence that we may be alone in the universe.

Nathan, Michel. *Le ciel des Fouriéristes: Habitants des étoiles et réincarnations de l'âme.* Lyon: Presses universitaires de Lyon, 1981.

Nicolson, Marjorie Hope. "A World in the Moon: A Study of the Changing Attitude toward the Moon in the Seventeenth and Eighteenth Centuries." *Smith College Studies in Modern Language* 17, no. 2 (1936).

———. *Voyages to the Moon.* New York: Macmillan Co., 1948.

———. *Science and Imagination.* Ithaca, N.Y.: Great Seal Books, 1956.

Numbers, Ronald. *Creation by Natural Law: Laplace's Nebular Hypothesis and American Thought.* Seattle: University of Washington Press, 1977.

Ordway, F. L. "The Legacy of Schiaparelli and Lowell." *Journal of the British Interplanetary Society* 39 (1986): 19–27.

Pohle, Joseph. *Die Sternenwelten und ihrer Bewohner,* 2nd ed. Köln: J. P. Bachem, 1899. Rich in historical information.

Randi, Eugenio. "Talpe ed extraterrestri: Un inedito di Agostino Trionfo di Ancona sulla plurulita dei mondi," *Rivista di storia della filosofia.* nuova ser., 44, no. 2 (1989): 311–26.

Randle, Kevin, Russ Estes, and William Cone. *The Abduction Enigma: The Truth Behind the Mass Alien Abductions of the Late Twentieth Century.* New York: Thomas Doherty Associates, 1999.

Renard, Jean-Bruno. "The Wild Man and the Extraterrestrial: Two Figures of Evolutionist Fantasy." *Diogenes [International Council for Philosophy and Humanistic Studies]* 127 (1984): 63–81.

Romesberg, Daniel R. "The Scientific Search for Extraterrestrial Intelligence: A Sociological Analysis." Ph.D. dissertation, University of Pittsburgh, 1995.

Ronchi, Lucia Rositani, and Georgio Abetti. "Effetti psico-fisiologici nelle osservazioni astronomiche visuali Il pianeta Marte." *Atti della Fondazione Georgio Ronchi* 19 (1964): 1–33.

Rossi, Paolo. "Nobility of Man and Plurality of Worlds." In *Science, Medicine and Society in the Renaissance,* vol. 2, ed. Allen Debus, 131–62. New York: Science History, 1972.

Rougier, Louis. *Astronomie et religion en Occident.* Paris: Presses universitaires de France, 1980.

Sagan, Carl, and Paul Fox. "The Canals of Mars: An Assessment after Mariner 9." *Icarus* 25 (1975): 602–12.

Schatzberg, Walter. *Scientific Themes in the Popular Literature and the Poetry of the German Enlightenment, 1720–1760.* Bern: Herbert Lang, 1973. Contains useful information on eighteenth-century German authors involved in the debate.

Sheehan, William. "Mars 1909: Lessons Learned." *Sky and Telescope* 76 (1988): 247–49.

Sheehan, William, and Thomas Dobbins. *Epic Moon: A History of Lunar Exploration in the Age of the Telescope.* Richmond, Va.: Willmann-Bell, Inc., 2001. Includes some treatment of ideas of life on the Moon.

Sinnott, Roger W. "Mars Mania of Oppositions Past." *Sky and Telescope* 76 (1988): 244–46.

Storey, Ronald. *The Mammoth Encyclopedia of Extraterrestrial Encounters.* London: Constable and Robinson Ltd., 2002.

Swift, David W. *SETI Pioneers: Scientists Talk about Their Search for Extraterrestrial Intelligence.* Tucson: University of Arizona Press, 1990.

Tatarewicz, Joseph. "Federal Funding and Planetary Astronomy, 1950–75: A Case Study." *Social Studies of Science* 16 (1986): 79–103.

Taylor, Stuart Ross. *Destiny or Chance? Our Solar System and Its Place in the Universe.* Cambridge: Cambridge University Press, 1998.

Tuzet, Hélène. *Le cosmos et l'imagination.* Paris: J. Corti, 1965.

Webb, George E. "The Planet Mars and Science in Victorian America." *Journal of American Culture* 3 (1980): 573–80.

Webb, Stephen. *If the Universe Is Teeming with Aliens . . . , Where Are They? Fifty Solutions to the Fermi Paradox and the Problem of Extraterrestrial Life*. New York: Copernicus Books, 2002. An English physicist presents an engaging study offering fifty answers to the Fermi Paradox: If aliens are out there, why haven't we found evidence of them?

Weston, Anthony. "Radio Astronomy as Epistemology: Some Philosophical Reflections on the Contemporary Search for Extraterrestrial Intelligence." *Monist* 71 (1988): 88–100.

Wheeler, Michael. *Death and the Future Life in Victorian Literature and Theology*. Cambridge: Cambridge University Press, 1990.

Wilford, John Noble. *Mars Beckons: The Mysteries, the Challenges, the Expectations of Our Next Great Adventure in Space*. New York: Knopf, 1990.

Zöckler, Otto. *Geschichte der Beziehungen zwischen Theologie und Naturwissenschaft*, 2 vols. Gütersloh: C. Bertelsmann, 1879.

5. STUDIES OF PARTICULAR INDIVIDUALS

Antoniadi:

McKim, Richard. "The Life and Times of E. M. Antoniadi, 1870–1944: Part 1: An Astronomer in the Making." *Journal of the British Astronomical Association* 203, no. 4 (1993): 164–70.

———. "The Life and Times of E. M. Antoniadi, 1870–1944: Part 2: The Meudon Years." *Journal of the British Astronomical Association* 203, no. 5 (1993): 219–27.

Nomblot, Bernard. "Antoniadi et la planète Mars." *Ciel et espace* 228 (Sept. 1988): 66–69.

Brewster, David:

de Asúa, Miguel. "Sir David Brewster's Changing Ideas on the Plurality of Worlds." *Journal of Astronomical History and Heritage* 9, no. 1 (2006): 83–92.

Bruno:

Bruno, Giordano. *Ash Wednesday Supper*. Translated and introduced by Stanley L. Jaki. The Hague: Mouton, 1975.

Finocchiaro, Maurice. "Philosophy Versus Religion and Science Versus Religion: The Trials of Bruno and Galileo." In *Giordano Bruno: Philosopher of the Renaissance*, ed. Hilary Gatti, 51–95. Aldershot: Ashgate, 2002.

Gatti, Hilary. *Giordano Bruno and Renaissance Science*. Ithaca, N.Y.: Cornell University Press, 1999.

Granada, Miguel A. "L'interpretazione bruniana di Copernico e la 'Narratio prima' di Rheticus." In *Rinascimento: Rivista dell'Instituto nazionale di studi sul Rinascimento*, 2nd ser., 30, 343–65. Firenze: L. S. Olschki, 1990.

———. "Bruno, Digges, Palingenio: Omogeneità ed eterogeneità nella concezione dell'universo infinito." *Rivista di storia della filosofia* 47, no. 1 (1992): 47–73.

Michel, Paul-Henri. *The Cosmology of Giordano Bruno*. Trans. R. E. W. Maddison. Ithaca, N.Y.: Cornell University Press, 1973.

Pantin, Isabelle. "Giordano Bruno, l'impardonable précurseur de Galilée." *Sciences et avenir* 551 (Jan. 1993): 80–84.

Singer, D. W. *Giordano Bruno: His Life and Thought with Annotated Translations of His Work on the Infinite Universe and Worlds.* New York: Henry Schuman, 1950.

Thullier, Pierre. "Martyr de la science or illuminé? Le cas Giordano Bruno." *Recherche* 19 (1988): 510–14.

Yates, Francis A. *Giordano Bruno and the Hermetic Tradition.* New York: Random House, 1975.

Campbell:

DeVorkin, David. "W. W. Campbell's Spectroscopic Study of the Martian Atmosphere." *Quarterly Journal of the Royal Astronomical Society* 18 (1977): 37–53.

Chalmers:

Cairns, David. "Thomas Chalmers's Astronomical Discourses: A Study in Natural Theology." *Scottish Journal of Theology* 9 (1956): 410–21.

Dick:

Astore, William J. *Observing God: Thomas Dick, Evangelicalism, and Popular Science in Victorian Britain and America.* Aldershot: Ashgate, 2001. This is based on Astore's 1995 doctoral dissertation at Oxford University.

Gavine, David. "Thomas Dick, LL.D., 1774–1857." *British Astronomical Association Journal* 84 (1974): 345–50.

Hennessey, Roger. "Thomas Dick's Sublime Science." *Sky and Telescope* 99 (Feb. 2000): 46–49.

Douglass:

Webb, George. *Tree Rings and Telescopes: The Scientific Career of A. E. Douglass.* Tucson: University of Arizona Press, 1983.

Elliot:

Manning, Robert J. "John Elliot and the Inhabited Sun." *Annals of Science* 50 (1993): 349–64.

Flammarion:

Cotardière, Philippe de la, and Patrick Fuentes. *Camille Flammarion.* Paris: Flammarion, 1994.

Fontenelle:

Fontenelle: Actes du colloque tenu à Rouen du 6 au 10 octobre 1987. Publiés par Alain Niderst. Paris: Presses universitaires de France, 1989. Contains among other items a number of articles on Fontenelle's book on the plurality of worlds.

Fourier:

Nathan, Michel. *Le ciel des Fouriéristes: Habitants des étoiles et réincarnations de l'âme.* Lyon: Presses universitaires de Lyon, 1981.

Tuzet, Hélène. "Deux types de cosmogonies ritalistees: 2.— Charles Fourier, hygienisted du cosmos." *Revue des sciences humaines* 101 (1961): 37–53.

Goethe:

Wattenberg, Diedrich. "Goethe und die Sternenwelt." *Goethe: Neue Folge des Jahrbuchs der Goethe-Gesellschaft* 31 (1969): 66–111.

Gruithuisen:

Baum, Richard. "The Man [Gruithuisen] Who Found a City on the Moon." *British Astronomical Association Journal* 102 (June 1992): 157, 159.

Günther, Sigmund. *Kosmo- und Geophysikalische Anschauung eines vergessenen bayerischen Gelehrten (d. i. Gruithuisen).* Munich: Königliche Bayerische Akademie der Wissenschaften, 1914.

Hermann, Dieter B. "Franz von Paula Gruithuisen und seine Analekten für Erd- und Himmeslskunde." *Sterne* 44 (1968): 120–25.

Lecomte, Stéphane. "Gruithuisen, Venus, la Lune et leurs habitants." *L'Astronomie* 112 (Nov./Dec. 1998): 314–15.

Hardy:

Gossin, Pamela. *Thomas Hardy's Novel Universe: Astronomy, Cosmology, and Gender in the Post-Darwinian World.* The Nineteenth Century Series. Aldershot: Ashgate, 2007.

Hegel:

Ferrini, Cinzia. "Feature of Irony and Alleged Errors in Hegel's 'De Orbis Planetarum.'" *Hegel-Jahrbuch 1991*, 459–77. Fernwald (Annerod): Germinal, 1991.

Ross, Joseph. "Kant's and Hegel's Assessment of Analogical Arguments for Extraterrestrial Life." M.A. thesis, University of Notre Dame, 1991.

Herschel, William:

Kawaler, Steven, and J. Veverka. "The Habitable Sun: One of William Herschel's Stranger Ideas." *Journal of the Royal Astronomical Society of Canada* 75 (Jan. 1981): 46–55.

Schaffer, Simon. "'The Great Laboratories of the Universe': William Herschel on Matter Theory and Planetary Life." *Journal for the History of Astronomy* 11 (1980): 81–111.

Huygens:

Bogazzi, Riccardo. "*Kosmotheoros* di Christiaan Huygens." *Physis* 19 (1977): 88–109.

Knight, David. "Celestial Worlds Discover'd." *Durham University Journal* 58 (1965): 23–29.

Nash, J. V. "Some Seventeenth Century Cosmic Speculations." *Open Court* 41 (1927): 476–87.

Seidengart, Jean. "Les théories cosmologiques de Christiaan Huygens." In *Huygens et la France*, 209–22. Paris: J. Vrin, 1982.

Kant:

Ross, Joseph. "Kant's and Hegel's Assessment of Analogical Arguments for Extraterrestrial Life." M.A. thesis, University of Notre Dame, 1991.

Kepler:

Romm, James S. "Lucian and Plutarch as Sources for Kepler's *Somnium.*" *Classical and Modern Literature* 9 (Winter 1989): 97–107.

Lambert:

Jaki, Stanley L. "Introduction." In *Cosmological Letters on the Arrangement of the World Edifice*, Johann Lambert, trans. S. L. Jaki, 1–42. New York: Science History Publications, 1976.

Lewis:

Downing, David C. *Planets in Peril: A Critical Study of C. S. Lewis's Ransom Trilogy.* Amherst: University of Massachusetts Press, 1992.

Lobdell, Jared. *The Scientifiction Novels of C. S. Lewis: Space and Time in the Ransom Stories.* Jefferson, N.C.: McFarland & Co., 2004.

Lowell:

Dobbins, Thomas, and William Sheehan. "The Canals of Mars Revisited." *Sky and Telescope* 107 (March 2004): 114–17.

Hetherington, Norriss. "Lowell's Theory of Life on Mars." *Astronomical Society of the Pacific Leaflet* no. 501, 1971.

———. "Percival Lowell: Scientist or Interloper?" *Journal of the History of Ideas* 42 (1981): 159–61.

Hofling, Charles K. "Percival Lowell and the Canals of Mars." *British Journal of Medical Psychology* 37 (1964): 33–42. Takes a psychoanalytical approach.

Hoyt, William Graves. *Lowell and Mars.* Tucson: University of Arizona Press, 1976. An excellent study of Lowell's work on Mars.

Putnam, William Lowell, et al. *The Explorers of Mars Hill: A Centennial History of Lowell Observatory.* West Kennebunk, Maine: Phoenix Publishing Co., 1994.

Strauss, David. "'Fireflies Flashing in Unison': Percival Lowell, Edward Morse and the Birth of Planetology." *Journal for the History of Astronomy* 24 (1993): 157–69.

———. "Percival Lowell, W. H. Pickering and the Founding of the Lowell Observatory." *Annals of Science* 51 (1994): 37–58.

———. *Percival Lowell: The Culture and Science of a Boston Brahmin.* Cambridge, Mass.: Harvard University Press, 2000. An excellent biography.

Milton:

Tanner, John S. "'And Every Star Perhaps a World of Destined Habitation': Milton and Moonmen." *Extrapolation* 30 (1989): 267–79.

Paine:

Nicolson, Marjorie Hope. "Thomas Paine, Edward Nares, and Mrs. Piozzi's Marginalia." *Huntington Library Bulletin* 10 (1936): 103–33.

Pickering, William:

Plotkin, Howard. "William H. Pickering in Jamaica: The Founding of the Woodlawn and the Studies of Mars." *Journal for the History of Astronomy* 24 (1993): 100–122.

Pope:

Nicolson, Marjorie Hope, and G. S. Rousseau. *"This Long Disease, My Life": Alexander Pope and the Sciences.* Princeton: Princeton University Press, 1976.

Proctor:

Lightman, Bernard. "Astronomy for the People: R. A. Proctor and the Popularization of the Victorian Universe." In *Facets of Faith and Science*, vol. 3, ed. Jitze M. van der Meer, 31–45. Lanham, Md.: Pascal Center and University Press of America, Inc. 1996.

Schiaparelli:

Defrancesco, Sylvia. "Schiaparelli's Determination of the Rotation Period of Mercury: A Reexamination." *Journal of the British Astronomical Association* 98 (1988): 146–50.

Dykstra, Dennis. "Schiaparelli's Martians." *Astronomy* 22 (Oct. 1994): 12. Makes a convincing case that Schiaparelli's word *canali* is used by Italians to refer to human-made structures.

Tempesti, Piero. "Schiaparelli Poeta." *L'Astronomica* 12 (March 1990): 24–31.

Schröter:

Drews, Jörg, and Heinrich Schwier, eds. *Lilienthal oder die Astronomen.* Munich: Edition Text + Kritik, 1984.

Smith, Joseph:

Paul, E. Robert. "Joseph Smith and the Plurality of Worlds Idea." *Dialogue: A Journal of Mormon Thought* 19 (Summer 1986): 12–37.

———. *Science, Religion, and Mormon Cosmology.* Urbana: University of Illinois Press, 1992.

Swedenborg:

Goerwitz, Richard L., III. "Extraterrestrial Life: A Study of the Intellectual Context of Emanuel Swedenborg's *Earths in the Universe.*" *New Philosophy* 88 (1985): 417–46, 477–85.

Twain:

Twain, Mark. *Extract from Captain Stormfield's Visit to Heaven.* New York: Oxford University Press, 1996.

Wallace:

Kevin, James. "Man's Place in the Universe: Alfred Russel Wallace, Teleological Evolution, and the Question of Extraterrestrial Life." M.A. thesis, University of Notre Dame, 1985.

Whewell:

Todhunter, Isaac. *William Whewell.* 2 vols. London: Macmillan, 1876.

Whewell, William. *Of the Plurality of Worlds.* Ed. Michael Ruse. Chicago: University of Chicago Press, 2001.

Wilkins:

Chapman, Allan. "'A World in the Moon': John Wilkins and His Lunar Voyage." *Quarterly Journal of the Royal Astronomical Society* 32 (1991): 121–32.

Mason, Stephen F. "Bishop John Wilkins, F.R.S. (1614–1672): Analogies of Thought-Style in the Protestant Reformation and Early Modern Science." *Notes and Records of the Royal Society of London* 46 (1992): 1–21.

Wordsworth:

Gaull, Marilyn. "Under Romantic Skies: Astronomy and the Poets." *Wordsworth Circle* 21 (Winter 1990): 34–41.

Young:

Odell, D. W. "Young's *Night Thoughts* as an Answer to Pope's *Essay on Man*." *Studies in English Literature, 1500–1900* 12 (1972): 481–501.

Pettit, H. "A Bibliography of Young's *Night Thoughts*." *University of Colorado Studies, Series in Language and Literature* 5 (1954): 1–52.

Young, Edward. *Night Thoughts*. Ed. Stephen Cornford. Cambridge: Cambridge University Press, 1989. Contains a long introduction and a useful bibliography.

6. RELIGIOUS ASPECTS OF THE DEBATE: A SAMPLE

Ashkenazi, Michael. "Not the Sons of Adam: Religious Responses to ETI." *Space Policy* 8 (1992): 341–50.

Balch, Robert W. "Waiting for the Ships: Disillusionment and the Revitalization of Faith in Bo and Peep's UFO Cult." In *The Gods Have Landed: New Religions from Other Worlds*, ed. James R. Lewis, 137–66. Albany, N.Y.: SUNY Press, 1995.

Burgess, Andrew. "Earth Chauvinism." *Christian Century* 93 (Dec. 8, 1976): 1098–1102.

Clarke, Arthur C. "On the Morality of Space." *Saturday Review* 40 (Oct. 5, 1957): 8–10, 35–36.

Crowe, Michael J. "Astronomy and Religion: Some Historical Interactions Regarding Belief in Extraterrestrial Intelligent Life." *Osiris*, 2nd. ser., 16 (2001): 209–26.

Davis, Charles. "The Place of Christ." *Clergy Review*, n.s., 45 (1960): 707–18.

Davis, John Jefferson. "Search for Extraterrestrial Intelligence and the Christian Doctrine of Redemption." *Science and Christian Belief* 9 (1997): 21–34.

Denzler, Brenda. *Lure of the Edge: Scientific Passions, Religious Beliefs, and the Pursuit of UFOs*. Berkeley: University of California Press, 2001.

Dick, Steven J., ed. *Many Worlds: The New Universe, Extraterrestrial Life, and Its Theological Implications*. Philadelphia: Templeton Foundation Press, 2000.

Ellwood, Robert S. "Spiritualism and UFO Religion in New Zealand: The International Transmission of Modern Spiritual Movements." In *The Gods Have Landed: New Religions from Other Worlds*, ed. James R. Lewis, 167–86. Albany, N.Y.: SUNY Press, 1995.

Galloway, Allan D. *The Cosmic Christ*. New York: Harper, 1951.

George, Marie I. "Aquinas on Intelligent Extra-Terrestrial Life." *Thomist* 65 (2001): 239–58.

———. *Christianity and Extraterrestrials? A Catholic Perspective*. New York: iUniverse, Inc., 2005.

Gonzalez, Guillermo, and Jay Wesley Richards. *Privileged Planet: How Our Place in the Cosmos Is Designed for Discovery*. Washington, D.C.: Henry Regnery, 2004.

Heeren, Fred. *Show Me God: What the Message from Space Is Telling Us about God*. Wheeling, Ill.: Day Star, 2000.

Kaiser, Christopher. "Extraterrestrial Life and Extraterrestrial Intelligence." *Reformed Review* 51 (1997–98): 77–91.

Krafft, Fritz. "Von der Einheit der Welt zur Vielfalt der Welten und des Menschen Stellung in Ihnen." In *Naturwissenschaft und Theologie*, ed. H. A. Luyten, 79–117. Düsseldorf: Patmos, 1981.

Lamm, Norman. "The Religious Implications of Extraterrestrial Life." In *Faith and Doubt: Studies in Traditional Jewish Thought*, 107–60. New York: Ktav Pub. House, 1971. A discussion from the point of view of Judaism.

Lewis, C. S. "Onward Christian Spacemen." *Catholic Digest* 27 (August 1963): 90–95.

———. "Religion and Rocketry." In *The World's Last Night and Other Essays*, 83–92. New York: Harcourt, Brace, 1973.

Lovell, Sir Bernard. "Some Reflections on Ethics and the Cosmos." In *The Exploration of Outer Space*, 68–82. New York: Harper and Row, 1962.

Maloney, George A. *The Cosmic Christ: From Paul to Teilhard*. New York: Sheed and Ward, 1968.

McMullin, Ernan. "Persons in the Universe." *Zygon* 15 (1980): 69–89.

Melton, J. Gordon. "The Contactees: A Survey." In *The Gods Have Landed: New Religions from Other Worlds*, edited by James R. Lewis, 1–13. Albany, N.Y.: SUNY Press, 1995.

Melton, J. Gordon, and George M. Eberhart. "The Flying Saucer Contactee Movement, 1950–1994: A Bibliography." In *The Gods Have Landed: New Religions from Other Worlds*, ed. James R. Lewis, 251–332. Albany, N.Y.: SUNY Press, 1995.

New York Times. "Life on Other Planets Is Called Compatible with Jewish Ideas." April 10, 1966, 36.

O'Malley, William J. "Carl Sagan's Gospel of Scientism." *America* 144 (Feb. 7, 1981): 5–8.

O'Meara, Thomas. "Christian Theology and Extraterrestrial Intelligent Life." *Theological Studies* 60 (1999): 3–30.

Palmer, Susan Jean. "Women in the Raelian Movement: New Religious Experiments in Gender and Authority." In *The Gods Have Landed: New Religions from Other Worlds*, ed. James R. Lewis, 104–35. Albany, N.Y.: SUNY Press, 1995.

Peters, Ted. "Exo-Theology: Speculations on Extraterrestrial Life." In *The Gods Have Landed: New Religions from Other Worlds*, ed. James R. Lewis, 187–206. Albany, N.Y.: SUNY Press, 1995.

Puccetti, Roland. *Persons: A Study of Possible Moral Agents in the Universe*. New York: Herder and Herder, 1969. Contains a contemporary restatement of the extraterrestrial intelligent life argument against Christianity.

Ratzsch, Del. "Space Travel and Challenges to Religion." *Monist* 71 (1988): 101–13. This issue of *Monist* contains other relevant essays.

Regis, Edward, Jr., ed. *Extraterrestrials: Science and Alien Intelligence*. Cambridge: Cambridge University Press, 1985. This anthology contains a number of interesting articles, including a reprint of the historical study by L. W. Beck (cited above) as well as some articles on moral issues involving extraterrestrials.

Ross, Hugh, Kenneth Samples, and Mark Clark. *Lights in the Sky and Little Green Men: A Rational Christian Look at UFOs and Extraterrestrials*. Colorado Springs: Navpress, 2002.

Saliba, John A. "Religious Dimensions of UFO Phenomena." In *The Gods Have Landed: New Religions from Other Worlds*, ed. James R. Lewis, 15–64. Albany, N.Y.: SUNY Press, 1995.

———. "UFO Contactee Phenomena from a Sociopsychological Perspective: A Review." In *The Gods Have Landed: New Religions from Other Worlds*, ed. James R. Lewis, 207–50. Albany, N.Y.: SUNY Press, 1995.

Tumminia, Diana, and R. George Kirkpatrick. "Unarius: Emergent Aspects of an American Flying Saucer Group." In *The Gods Have Landed: New Religions from Other Worlds*, ed. James R. Lewis, 85–104. Albany, N.Y.: SUNY Press, 1995.

White, Lynn, Jr. "Christian Myth and Christian History." In *Dynamo and the Virgin Reconsidered*, 33–55. Cambridge, Mass.: MIT Press, 1968.

Whitmore, John. "Religious Dimensions of the UFO Abductee Experience." In *The Gods Have Landed: New Religions from Other Worlds*, ed. James R. Lewis, 65–84. Albany, N.Y.: SUNY Press, 1995.

Wiker, Benjamin D. "Alien Ideas: Christianity and the Search for Extraterrestrial Life." *Crisis* 20, no. 10 (2002): 26–31.

PERMISSIONS

I am indebted to the following publishing companies, institutions, or persons for permission to quote or reprint materials from the sources indicated: Penguin Books Ltd. for passages from R. E. Latham's translation of Titus Lucretius Carus's *On the Nature of the Universe*; Hackett Publishing Company for the excerpt from Thomas Aquinas, *Basic Writings of Saint Thomas Aquinas*, vol. 1, ed. by Anton C. Pegis; University of Wisconsin Press for passages from the translation by Albert Menut of Nicole Oresme, *Le livre du ciel et du monde*, ed. by Albert D. Menut and Alexander J. Denomy, C.S.B.; Green Lion Press, especially to William H. Donahue, for use of his translation of passages from Isaac Newton's *Mathematical Principles of Natural Philosophy*; Stanley L. Jaki for passages from his translation of Johann Lambert's *Cosmological Letters on the Arrangement of the World Edifice*; Anvil Press for the translation by Sally Purcell of Thomas Gray's "Luna habitabilis"; Master and Fellows of Trinity College, Cambridge University for the use of John Herschel's unpublished letter to William Whewell of 1854 and the selections from the letters written to Whewell by James Stephen; Keith Lafortune for his translation of the selection from Auguste Comte's *Cours de philosophie positive*; Robert L. Jones, Jr., for the selection from his unpublished translation of Camille Flammarion's *La pluralité des mondes habités*; and Dover Publications for passages from Michael J. Crowe's *The Extraterrestrial Life Debate, 1750–1900.*

To the following publishers or institutions, I am indebted for permission to include the following illustrations: the Royal Astronomical Society for their picture of a portion of William Herschel's unpublished lunar observation book and their photograph of Eugéne Michael Antoniadi; and the Master and Fellows of Trinity College, Cambridge for the portrait of William Whewell. I also wish to thank Green Lion Press for their assistance in procuring the image of Descartes' vortices, which also appeared in Michael J. Crowe, *Mechanics from Aristotle to Einstein.*

INDEX

MICHAEL J. CROWE
is the Rev. John J. Cavanaugh Professor Emeritus in Humanities in the Program of Liberal Studies and Graduate Program in History and Philosophy of Science at the University of Notre Dame. He has published a number of books, including *The Extraterrestrial Life Debate, 1750–1900: The Idea of a Plurality of Worlds from Kant to Lowell.*